AF361660

Sustainable Agriculture and Soil Conservation

Sustainable Agriculture and Soil Conservation

Editors

Mirko Castellini
Concetta Eliana Gattullo
Anna Maria Stellacci
Mariangela Diacono

MDPI • Basel • Beijing • Wuhan • Barcelona • Belgrade • Manchester • Tokyo • Cluj • Tianjin

Editors

Mirko Castellini
Council for Agricultural
Research and Economics
Research Center for Agriculture
and Environment (CREA-AA)
Bari
Italy

Concetta Eliana Gattullo
Department of Soil, Plant and
Food Sciences
University of Bari "A. Moro"
Bari
Italy

Anna Maria Stellacci
Department of Soil, Plant and
Food Sciences
University of Bari "A. Moro"
Bari
Italy

Mariangela Diacono
Council for Agricultural
Research and Economics
Research Center for Agriculture
and Environment (CREA-AA)
Bari
Italy

Editorial Office
MDPI
St. Alban-Anlage 66
4052 Basel, Switzerland

This is a reprint of articles from the Special Issue published online in the open access journal *Applied Sciences* (ISSN 2076-3417) (available at: www.mdpi.com/journal/applsci/special_issues/ Sustainable_Agriculture_Soil_Conservation).

For citation purposes, cite each article independently as indicated on the article page online and as indicated below:

LastName, A.A.; LastName, B.B.; LastName, C.C. Article Title. *Journal Name* **Year**, *Volume Number*, Page Range.

ISBN 978-3-0365-2477-1 (Hbk)
ISBN 978-3-0365-2476-4 (PDF)

Contents

Editorial

Sustainable Agriculture and Soil Conservation

Mirko Castellini [1,*], Mariangela Diacono [1], Concetta Eliana Gattullo [2] and Anna Maria Stellacci [2]

1 Council for Agricultural Research and Economics—Research Center for Agriculture and Environment (CREA-AA), Via Celso Ulpiani 5, 70125 Bari, Italy; mariangela.diacono@crea.gov.it
2 Department of Soil, Plant and Food Sciences, University of Bari "A. Moro", 70125 Bari, Italy; concettaeliana.gattullo@uniba.it (C.E.G.); annamaria.stellacci@uniba.it (A.M.S.)
* Correspondence: mirko.castellini@crea.gov.it

Abstract: Soil degradation is one of the most topical environmental threats. A number of processes causing soil degradation, specifically erosion, compaction, salinization, pollution, and loss of both organic matter and soil biodiversity, are also strictly connected to agricultural activity and its intensification. The development and adoption of sustainable agronomic practices able to preserve and enhance the physical, chemical, and biological properties of soils and improve agroecosystem functions is a challenge for both scientists and farmers. This Special Issue collects 12 original contributions addressing the state of the art of sustainable agriculture and soil conservation. The papers cover a wide range of topics, including organic agriculture, soil amendment and soil organic carbon (SOC) management, the impact of SOC on soil water repellency, the effects of soil tillage on the quantity of SOC associated with several fractions of soil particles and depth, and SOC prediction, using visible and near-infrared spectra and multivariate modeling. Moreover, the effects of some soil contaminants (e.g., crude oil, tungsten, copper, and polycyclic aromatic hydrocarbons) are discussed or reviewed in light of the recent literature. The collection of the manuscripts presented in this Special Issue provides a relevant knowledge contribution for improving our understanding on sustainable agriculture and soil conservation, thus stimulating new views on this main topic.

Keywords: organic agriculture; soil amendment; soil organic carbon; soil management; soil water repellency; soil contamination; soil remediation; sustainable fruit growing; water conservation practices; multivariate statistical models for SOC prediction

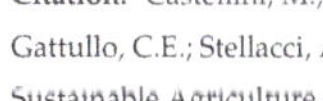

Citation: Castellini, M.; Diacono, M.; Gattullo, C.E.; Stellacci, A.M. Sustainable Agriculture and Soil Conservation. *Appl. Sci.* **2021**, *11*, 4146. https://doi.org/10.3390/app11094146

Received: 25 April 2021
Accepted: 27 April 2021
Published: 1 May 2021

Publisher's Note: MDPI stays neutral with regard to jurisdictional claims in published maps and institutional affiliations.

1. Introduction

The ongoing global climate changes and the ever-increasing world population impose the adoption of agroecological or eco-sustainable production techniques to increase the resilience of the most fragile agrosystems [1]. For this reason, from many directions, a multidisciplinary scientific paradigm is increasingly suggested to better approach the complex problems concerning soil protection and sustainable agriculture [1–3].

The development and adoption of sustainable agronomic practices able to preserve and enhance the physical, chemical and biological properties of soils and improve agroecosystem functions is a challenge for both scientists and farmers [4]. In recent decades, much has been done because, on one hand, scientists have improved and simplified the measurement methods in specific research sectors, and on the other hand, the process of transferring research outputs to farmers has been enhanced. The latter has been considerably extended thanks to research funding granted by the National and Supranational Political Institution, which increasingly involves farmers and stakeholders. From a scientific point of view, much progress has been achieved in the development of sustainable practices based on minimal soil disturbance, the use of cover crops, organic mulching, crop rotations, water and nutrient conservation, recycling crop residues and livestock manure for soil amendment, precision agriculture and so on, and many reviews have summarized the state of the art and traced the future trajectories (among others, refs. [5–9]). However, much remains to be explored, and Special Issues stimulating the scientific community on these main topics are desirable.

The main goal of this Special Issue (SI) is to collect and present recent research on sustainable agriculture and soil conservation. A synthesis of the main results is reported in the following section.

2. Overview of This Special Issue

This SI collects 12 original contributions focused on soil use and management, soil conservation, as well as on the impact of some pollutants on the soil. Specifically, the SI collects nine research papers [10–18], two reviews [19,20] and one methodological work [21]. Eight of the twelve manuscripts considered the effects of soil use and management (topic I) [10–16,21], while the remaining four [17–20] evaluated the impact of some soil contaminants (e.g., crude oil, tungsten, copper and polycyclic aromatic hydrocarbons) (topic II). This grouping was used to detail the contents and for the reader's usefulness.

Regarding the aforementioned nine investigations (i.e., research papers), seven were carried out in the field [10–16], while the others accounted for pot [17] or microcosm [18] experiments. From a geographical point of view, five "field open-air" investigations were carried out in Italy (three in the south [12,13,16], one in the center [15] and one in the north [10] of Italy), while the remaining investigations were in Brazil [14] and China [11].

Topic I comprises eight papers. Pastorelli et al. [10] investigated the short-term effects of digestate on soil properties through a holistic approach, thus investigating the soil's physical, chemical and microbiological properties and their interactions. The digestate effects were evaluated in the two maize growing seasons within a 2-year maize–triticale rotation to feed the biogas plant. The main results showed that the digestate (1) increased the soil's total organic C, total N and K contents, (2) did not affect the soil bulk density while transiently improving the soil aggregate stability, (3) decreased the soil transmission pores (50–500 μm size) proportion as well as increasing the fissures (>500 μm), (4) only transiently affected the soil's microbial community and did not cause soil contamination from Clostridiaceae-related bacteria, (5) significantly impaired seed germination when applied at low dilution ratios and (6) did not improve the crop yield, as it was similar to usual mineral fertilization. In the context of energy crop farming, the authors concluded that the agronomic recycling and usage of digestate from biogas production assured a sustainable crop yield and soil quality. Therefore, digestate was confirmed as a valid resource for sustainable management of the soil's fertility.

Wang et al. [11] investigated the prediction of soil organic matter (SOM) from visible and near-infrared (VNIR) spectra, taking into account the presence of nonlinear relationships and spatial heterogeneity conditions. Specifically, they combined the proposed partial least squares-based multivariate adaptive regression spline (PLS–MARS) method and a regional multivariable associate rule mining and Rank–Kennard–Stone method (MVARC-R-K-S) to construct a nonlinear prediction model to realize local optimality considering spatial heterogeneity. The method, which was applied and validated for a case study in Hubei (China), was able to filter the spatial autocorrelation present in the investigated area. In addition, it showed improvements compared with some available methods in terms of accuracy and robustness, suggesting the reliability of the proposed prediction model.

Losciale et al. [12] compared four soil management strategies at a peach orchard, namely (1) completely tilled (control); (2) mulched with reusable reflective plastic film; (3) mulching with a leguminous cover crop flattened after the peach fruit set; (4) completely tilled and irrigated with the water volumes of the treatment with reflective mulching. The effects on the soil features, water use efficiency (WUE), tree functionality, fruit growth and quality, yield and water productivity were investigated. The main results showed that when using 50% of the regular irrigation, reusable reflective mulching reduced the water loss and soil carbon mineralization while not affecting (and sometimes increasing) the net carbon assimilation, yield and fruit size. Consequently, water productivity was drastically increased. The investigation allowed for the conclusion that a reflective mulching strategy is promising from the point of view of WUE optimization, especially in hot and dry areas with clay soils and low organic matter contents.

De Mastro et al. [13] investigated the effects of different soil management methods (conventional tillage, CT, minimum tillage, MT, and no-tillage, NT), both with fertilization and without fertilization, on the quantity of SOC at different soil depths and soil particle size fractions. Diffuse reflectance infrared Fourier transform spectroscopy (DRIFT) was also performed to obtain information on both the SOC quality and the mineralogical composition of these soil fractions. The results showed that CT provided the highest amount of the finest fraction, while fertilization improved the microbial community with the increase of soil microaggregates. The coarse fraction was highest in the upper soil layer, while the finest fraction was highest in the deepest one. The greatest OC content was observed in the topsoil layer and in the finest soil fraction. The DRIFT analysis could provide information about the quality of the main minerals present in the different soil size fractions.

Lozano-Baez et al. [14] measured both the unsaturated and saturated soil hydraulic conductivity, determined with simple and low-cost field infiltration methods—a minidisk tension infiltrometer, MDTI [22], and Beerkan [23]—for three land covers, namely (1) pastures, (2) 9-year-old restored forest and (3) remnant forest. The paper, which was relevant because it provided the first measurements of soil water repellency (SWR) in the Brazilian Atlantic Forest, proved that MDTI and Beerkan infiltrations gave complementary information, highlighting increasing hydraulic conductivities, especially at the remnant forest plots, when moving from near-saturated to saturated conditions. Moreover, measuring the hydraulic conductivity with MDTI allowed the estimation of the macroscopic capillary length [24], which is a hydrodynamic soil parameter useful for investigating the impact of soil management on water saving [25]. The authors concluded that the applied approach (MDTI + Beerkan) allowed the design of better estimates of the saturated soil hydraulic conductivity under challenging field conditions, such as soil water repellency (SWR).

Amoriello et al. [15] evaluated the impact of soil amendment with biochar from brewers' spent grain (BSG) on hop growth. The field investigation was carried out in Rieti (central Italy). Three different German cultivars of hop plant were considered, and biochar was added at a 20% level. Biochar's effects on root development, shoots, bine length and crop yield were evaluated using multivariate image analysis and general linear models. The results showed that biochar significantly improved root growth. Overall, no variability in shoots, number of leaves or bine length was observed between the two treatments for all cultivars or among the genotypes considered. Moreover, soil amendment significantly improved the yield (number of cones). The authors concluded that BSG-derived biochar could be useful for improving hop plant growth and cone production because it can supply key nutrients for plant growth and improve the soil's properties.

Montemurro et al. [16] evaluated the production capacity of organic horticultural systems and the ex-post sustainability using a new multi-attribute decision model named DEXi-met. Specifically, they compared three horticultural systems in Metaponto (south Italy): (1) ECO (organic system with full implementation of agroecological strategies, agroecological services crops (ASC), strip cultivation and organic amendment), (2) GM (organic system with the introduction of ASC) and (3) no ASC (organic system without agroecological services crops). The qualitative response of the DEXi-met model suggested that the treatments with ASC returned similar total energy outputs, indicating the positive effect of this agroecological practice. Therefore, the authors pointed out that the ECO system could contribute to building up more complex agroecosystems, increasing both the resilience and biodiversity in organic agriculture.

Sofo and Ricciuti [21] presented a standardized method, named Biolog® EcoPlates™ (Biolog Inc., Hayward, CA, USA), to analyze the functional diversity of bacterial communities by means of measuring their ability to oxidize carbon substrates. Specifically, they reported a detailed methodological procedure which was easy to follow and reproduce (i.e., a detailed and step-by-step description of soil preparation, dispersion and dilution, optimal bacterial density of the inoculum and so on), based on previous data and experiments. To this aim, a case study of the soils of a Mediterranean olive orchard was reported for comparing the methods and justifying the methodological protocol proposed. The authors emphasized that the results of this methodological paper could be important for correctly evaluating and comparing the microbiological fertility of the soils managed with sustainable, conservation-based practices or conventional, non-conservation-based ones.

Topic II comprises four papers. Zhang et al. [17] evaluated the impact of differentlevels of crude oil on the soil properties, physiological and chemical parameters of maize leaves and the phenanthrene content in the leaf. The results showed that (1) the soil water content significantly increased as the total petroleum hydrocarbons increased, and the soil electrical conductivity significantly increased compared with the control, and (2) the stomatal length and density, leaf K and Na contents and phenanthrene leaf concentration decreased in the contaminated soil compared with the reference group. The authors concluded that the stomata structure of maize could be influenced by crude oil, thus possibly controlling the accumulation of polycyclic aromatic hydrocarbons in aerial tissues. Practical implications for contaminated soil management were also hypothesized. Indeed, because stomata can play a key role in the aerial uptake of pollutants, regulation of stomatal movement can be usefully applied in phytoremediation of contaminated soil.

Petruzzelli and Pedron [18] investigated the influence of the soil characteristics on the tungsten uptake by maize for three different soils—Histosol, Vertisol and Fluvisol—in the Mediterranean area. Tungsten is largely used in high-tech and military industries. Therefore, soils are increasingly enriched in this element, and the possible transfer in the food chain represents a current issue in agroenvironmental studies. The results showed that tungsten concentrations in the roots and shoots increased with the increasing soil pH and decreased with the increasing organic matter. Further investigations were suggested by the authors to fully understand the mechanisms of tungsten transfer from the soil to the plant and the corresponding environmental impact of this element on human health.

Cesco et al. [19] reviewed the issue of copper (Cu) accumulation in European vineyard soils due to massive application from different sources (among others, metal-contaminated sludges and Cu-based pesticides for crop defense against pathogens). Although Cu is an essential micronutrient for equilibrated crop growth, excessive Cu concentrations in soil may lead to severe symptoms of toxicity, which are beginning to be recognized in crops and, particularly, in acidic soils. After dealing with the state of the art on this topic, they listed and discussed several current challenges, including (1) copper effects on soil agrobiodiversity, (2) rhizosphere management (management of soil–root–microorganism interactions), (3) biotechnologies and breeding for a more resistant plant material and (4) smart viticulture.

Sayara and Sánchez [20] reviewed the topic of bioremediation of contaminated soils due to the accumulation of polycyclic aromatic hydrocarbons (PAHs). Specifically, they focused on the application of compost as a type of biostimulation or bioaugmentation treatment to facilitate the bioremediation of soils contaminated with PAHs. Moreover, the authors reviewed the use of composting as an ex situ bioremediation strategy for PAH-contaminated soils. After dealing with the properties and sources of PAHs, they discussed the bioremediation of PAH-contaminated soils by composting and bioaugmentation (reviewing the effects of PAH characteristics and concentrations, temperatures of treatment, organic co-substrate stability and the mixing ratio), also describing the consequences of PAH bioavailability and biodegradation pathways. The authors concluded that, although the use of compost and composting in several remediation strategies significantly improved the removal of PAHs in contaminated soils, further investigations are needed in this research field.

3. Conclusions

The 12 manuscripts presented in this Special Issue can contribute to improving our understanding of sustainable agriculture and soil conservation through the development and adoption of specific agronomic practices able to preserve and enhance the physical, chemical and biological properties of soils. The derived improvement of agroecosystem functions goes in the direction desired by both scientists and farmers. The investigations presented considered various crops, three different countries and multiple pedoclimatic conditions.

For the reader's convenience, the collected contributions were summarized in two main groups to provide results on the effects of soil use and management (topic I) and on the impact of some soil contaminants (e.g., crude oil, tungsten, copper and polycyclic aromatic hydrocarbons) (topic II).

The common conclusions drawn by the authors of this SI agree on the importance of both improving the available options for soil conservation and developing, implementing and sharing new techniques for the sustainable use of the soil.

Funding: This research received no external funding.

Conflicts of Interest: The authors declare no conflict of interest.

References

1. Leippert, F.; Darmaun, M.; Bernoux, M.; Mpheshea, M. *The Potential of Agroecology to Build Climate-Resilient Livelihoods and Food Systems*; FAO and Biovision: Rome, Italy, 2020. [CrossRef]
2. Wezel, A.; Herren, B.G.; Kerr, R.B.; Barrios, E.; Gonçalves, A.L.R.; Sinclair, F. Agroecological principles and elements and their implications for transitioning to sustainable food systems. A review. *Agron. Sustain. Dev.* **2020**, *40*, 40. [CrossRef]
3. Gargano, G.; Licciardo, F.; Verrascina, M.; Zanetti, B. The Agroecological Approach as a Model for Multifunctional Agriculture and Farming towards the European Green Deal 2030—Some Evidence from the Italian Experience. *Sustainability* **2021**, *13*, 2215. [CrossRef]
4. Diacono, M.; Persiani, A.; Castellini, M.; Giglio, L.; Montemurro, F. Intercropping and rotation with leguminous plants in organic vegetables: Crop performance, soil properties and sustainability assessment. *Biol. Agric. Hortic.* **2021**. [CrossRef]
5. Liu, W.; Shao, X.-F.; Wu, C.-H.; Qiao, P. A systematic literature review on applications of information and communication technologies and blockchain technologies for precision agriculture development. *J. Clean. Prod.* **2021**, *298*, 126763. [CrossRef]
6. Bahinipati, C.S.; Kumar, V.; Viswanathan, P.K. An evidence-based systematic review on farmers' adaptation strategies in India. *Food Sec.* **2021**. [CrossRef]
7. Ke, H.; Zhang, J.; Zeng, Y. Knowledge domain and emerging trends of agricultural waste management in the field of social science: A scientometric review. *Sci. Total Environ.* **2019**, *670*, 236–244.
8. Mak, M.W.; Xiong, X.; Tsang, D.C.W.; Yu, I.K.M.; Poon, C.S. Sustainable food waste management towards circular bioeconomy: Policy review, limitations and opportunities. *Bioresour. Technol.* **2020**, *297*, 122497. [CrossRef]
9. Lozano-Baez, S.E.; Domínguez-Haydar, Y.; Meli, P.; van Meervel, I.; Vásquez, K.V.; Castellini, M. Key Gaps in Soil Monitoring during Forest Restoration in Colombia. *Restor. Ecol.* **2021**. [CrossRef]
10. Pastorelli, R.; Valboa, G.; Lagomarsino, A.; Fabiani, A.; Simoncini, S.; Zaghi, M.; Vignozzi, N. Recycling Biogas Digestate from Energy Crops: Effects on Soil Properties and Crop Productivity. *Appl. Sci.* **2021**, *11*, 750. [CrossRef]
11. Wang, X.; Yang, C.; Zhou, M. Partial Least Squares Improved Multivariate Adaptive Regression Splines for Visible and Near-Infrared-Based Soil Organic Matter Estimation Considering Spatial Heterogeneity. *Appl. Sci.* **2021**, *11*, 566. [CrossRef]
12. Losciale, P.; Gaeta, L.; Manfrini, L.; Tarricone, L.; Campi, P. Orchard Floor Management Affects Tree Functionality, Productivity and Water Consumption of a Late Ripening Peach Orchard under Semi-Arid Conditions. *Appl. Sci.* **2020**, *10*, 8135. [CrossRef]
13. De Mastro, F.; Cocozza, C.; Brunetti, G.; Traversa, A. Chemical and Spectroscopic Investigation of Different Soil Fractions as Affected by Soil Management. *Appl. Sci.* **2020**, *10*, 2571. [CrossRef]
14. Lozano-Baez, S.E.; Cooper, M.; de Barros Ferraz, S.F.; Ribeiro Rodrigues, R.; Lassabatere, L.; Castellini, M.; Di Prima, S. Assessing Water Infiltration and Soil Water Repellency in Brazilian Atlantic Forest Soils. *Appl. Sci.* **2020**, *10*, 1950. [CrossRef]
15. Amoriello, T.; Fiorentino, S.; Vecchiarelli, V.; Pagano, M. Evaluation of Spent Grain Biochar Impact on Hop (*Humulus lupulus* L.) Growth by Multivariate Image Analysis. *Appl. Sci.* **2020**, *10*, 533. [CrossRef]
16. Montemurro, F.; Persiani, A.; Diacono, M. Organic Vegetable Crops Managed with Agro-Ecological Practices: Environmental Sustainability Assessment by *DEXi-met* Decision Support System. *Appl. Sci.* **2019**, *9*, 4148. [CrossRef]
17. Zhang, C.; Huang, H.; Zhou, Y.; Lin, H.; Xie, T.; Liao, C. Stomatal Response of Maize (*Zea mays* L.) to Crude Oil Contamination in Soils. *Appl. Sci.* **2019**, *9*, 4074. [CrossRef]
18. Petruzzelli, G.; Pedron, F. Influence of Increasing Tungsten Concentrations and Soil Characteristics on Plant Uptake: Greenhouse Experiments with *Zea mays*. *Appl. Sci.* **2019**, *9*, 3998. [CrossRef]
19. Cesco, S.; Pii, Y.; Borruso, L.; Orzes, G.; Lugli, P.; Mazzetto, F.; Genova, G.; Signorini, M.; Brunetto, G.; Terzano, R.; et al. A Smart and Sustainable Future for Viticulture Is Rooted in Soil: How to Face Cu Toxicity. *Appl. Sci.* **2021**, *11*, 907. [CrossRef]
20. Sayara, T.; Sánchez, A. Bioremediation of PAH-Contaminated Soils: Process Enhancement through Composting/Compost. *Appl. Sci.* **2020**, *10*, 3684. [CrossRef]

21. Sofo, A.; Ricciuti, P. A Standardized Method for Estimating the Functional Diversity of Soil Bacterial Community by Biolog® EcoPlates™ Assay—The Case Study of a Sustainable Olive Orchard. *Appl. Sci.* **2019**, *9*, 4035. [CrossRef]
22. Alagna, V.; Iovino, M.; Bagarello, V.; Mataix-Solera, J.; Lichner, L. Application of minidisk infiltrometer to estimate water repellency in Mediterranean pine forest soils. *J. Hydrol. Hydromech.* **2017**, *65*, 254–263. [CrossRef]
23. Castellini, M.; Di Prima, S.; Moret-Fernández, D.; Lassabatere, L. Rapid and accurate measurement methods for determining soil hydraulic properties: A review. *J. Hydrol. Hydromech.* **2021**, *69*, 1–19. [CrossRef]
24. Di Prima, S.; Stewart, R.D.; Castellini, M.; Bagarello, V.; Abou Najm, M.R.; Pirastru, M.; Giadrossich, F.; Iovino, M.; Angulo-Jaramillo, R.; Lassabatere, L. Estimating the macroscopic capillary length from Beerkan infiltration experiments and its impact on saturated soil hydraulic conductivity predictions. *J. Hydrol.* **2020**, 125159. [CrossRef]
25. Castellini, M.; Stellacci, A.M.; Mastrangelo, M.; Caputo, F.; Manici, L.M. Estimating the soil hydraulic functions of some olive orchards: Soil management implications for water saving in soils of Salento peninsula (southern Italy). *Agronomy* **2020**, *10*, 177. [CrossRef]

applied sciences

MDPI

Review

A Smart and Sustainable Future for Viticulture Is Rooted in Soil: How to Face Cu Toxicity

Stefano Cesco [1], Youry Pii [1], Luigimaria Borruso [1], Guido Orzes [1,2], Paolo Lugli [1], Fabrizio Mazzetto [1,2], Giulio Genova [1,3], Marco Signorini [1], Gustavo Brunetto [4], Roberto Terzano [5], Gianpiero Vigani [6] and Tanja Mimmo [1,2,*]

[1]   Faculty of Science and Technology, Free University of Bozen-Bolzano, Piazza Università 5, 39100 Bolzano, Italy; stefano.cesco@unibz.it (S.C.); youry.pii@unibz.it (Y.P.); luigimaria.borruso@unibz.it (L.B.); guido.orzes@unibz.it (G.O.); paolo.lugli@unibz.it (P.L.); fabrizio.mazzetto@unibz.it (F.M.); giulio.genova@natec.unibz.it (G.G.); marco.signorini@natec.unibz.it (M.S.)
[2]   Competence Centre for Plant Health, Free University of Bozen-Bolzano, 39100 Bolzano, Italy
[3]   Eurac Research, Institute for Alpine Environment, 39100 Bolzano, Italy
[4]   Department of Soil Science, Universidade Federal de Santa Maria, Av.Roraima, 1000, Camobi, Santa Maria, RS 97105-900, Brazil; brunetto.gustavo@gmail.com
[5]   Department of Soil, Plant and Food Sciences, University of Bari "Aldo Moro", 70126 Bari, Italy; roberto.terzano@uniba.it
[6]   Department of Life Science and System Biology, University of Torino-Turin, 10124 Torino, Italy; gianpiero.vigani@unito.it
*   Correspondence: tanja.mimmo@unibz.it

Citation: Cesco, S.; Pii, Y.; Borruso, L.; Orzes, G.; Lugli, P.; Mazzetto, F.; Genova, G.; Signorini, M.; Brunetto, G.; Terzano, R.; et al. A Smart and Sustainable Future for Viticulture Is Rooted in Soil: How to Face Cu Toxicity. *Appl. Sci.* **2021**, *11*, 907. https://doi.org/10.3390/app11030907

Received: 18 November 2020
Accepted: 13 January 2021
Published: 20 January 2021

Publisher's Note: MDPI stays neutral with regard to jurisdictional claims in published maps and institutional affiliations.

Abstract: In recent decades, agriculture has faced the fundamental challenge of needing to increase food production and quality in order to meet the requirements of a growing global population. Similarly, viticulture has also been undergoing change. Several countries are reducing their vineyard areas, and several others are increasing them. In addition, viticulture is moving towards higher altitudes and latitudes due to climate change. Furthermore, global warming is also exacerbating the incidence of fungal diseases in vineyards, forcing farmers to apply agrochemicals to preserve production yields and quality. The repeated application of copper (Cu)-based fungicides in conventional and organic farming has caused a stepwise accumulation of Cu in vineyard soils, posing environmental and toxicological threats. High Cu concentrations in soils can have multiple impacts on agricultural systems. In fact, it can (i) alter the chemical-physical properties of soils, thus compromising their fertility; (ii) induce toxicity phenomena in plants, producing detrimental effects on growth and productivity; and (iii) affect the microbial biodiversity of soils, thereby influencing some microbial-driven soil processes. However, several indirect (e.g., management of rhizosphere processes through intercropping and/or fertilization strategies) and direct (e.g., exploitation of vine resistant genotypes) strategies have been proposed to restrain Cu accumulation in soils. Furthermore, the application of precision and smart viticulture paradigms and their related technologies could allow a timely, localized and balanced distribution of agrochemicals to achieve the required goals. The present review highlights the necessity of applying multidisciplinary approaches to meet the requisites of sustainability demanded of modern viticulture.

Keywords: copper; rhizosphere; smart agriculture; microbes; vineyard

1. Introduction

In recent decades, the food demand has significantly increased in terms of quality and quantity due to the increase in the global population, which has now reached almost 8 billion people [1]. The growth of the per capita income in developing countries has undoubtedly played a decisive role [2,3]. On the other hand, the arable soil surface has decreased due to soil degradation and the impact of climate change [4]. Indeed, we are experiencing a global annual mean warming of 1 °C above the level of the pre-industrialization period [5]. Moreover, the global non-renewable natural resources used for the production of fertilizers (e.g., rock phosphate) are limited and are expected to be consumed shortly [6]. All these trends highlight the limits and the vulnerability of the current model of economic and social growth based on mass production and consumption, including with respect to soil [7].

For this reason, the United Nations defined 17 Sustainable Development Goals (SDGs) and 169 target actions for the year 2030, to which we are all invited to proactively contribute. These include targets of no poverty, zero hunger, good health and well-being, life below water, life on land, and climate action. In this context, agriculture is called upon to contribute by providing primary production and to reduce its anthropogenic impacts on the environment.

A specific example is represented by copper (Cu) accumulation in agricultural soils due to the massive application of metal-contaminated sludges and/or Cu-based pesticides for crop defense plans against pathogens. While Cu is an essential micronutrient for equilibrated crop growth, excessive Cu concentration, due to its accumulation in soil, leads to severe symptoms of toxicity, which are beginning to be recognized in crops, particularly in acid soils [8,9]. The case of vineyard soils is the most emblematic. Therefore, specific strategies must be adopted to limit and/or mitigate this problem, particularly in the viticulture sector.

This review aims at shedding light on the soil vineyard Cu accumulation and its agricultural and environmental consequences, specifically at the European level. Some approaches to limiting Cu toxicity and the related challenges, considering recent technological innovations (rhizosphere management, biotechnologies and precision/smart viticulture), are discussed. A short description of the role of Cu as an essential nutrient for plants is also included. The review further highlights research gaps that urgently require further study and innovation in order to guarantee sustainable vineyard management, providing soil conditions that will enable quality viticulture in the long term.

1.1. Cu in Agricultural Soils and Crops

Copper is a trace element in soil–plant systems. The average Cu concentration in the Earth's crust is 60 mg kg^{-1} and in soil, it typically ranges between 2 and 50 mg kg^{-1}. Natural soils with a high content of clay minerals (e.g., *Vertisols*) or organic matter (e.g., *Spodosols*, *Histosols*) are usually characterized by a higher Cu content (up to 180 mg kg^{-1}) [10]. Anthropogenic atmospheric depositions (traffic and industry related) and agricultural materials (inorganic fertilizers, agrochemicals, sewage sludge, livestock manure, irrigation water, compost, etc.) can dramatically increase the Cu concentration in soil. Indeed, Cu concentrations higher than 1000 mg kg^{-1} have been recorded in polluted soils, including in agricultural soils [11]. In particular, concentrations exceeding 1000 mg kg^{-1} have been reported for vineyard soils in France and Brazil [9] (see also the following chapter).

The main form of Cu in the soil is Cu^{2+}, typically bound to inorganic and organic ligands, forming both soluble and insoluble compounds. In the solid phase, Cu has a very high affinity for organic matter and manganese oxides, as well as for iron oxides and clay minerals. Because of its high affinity for soil colloids and especially organic matter, Cu shows low mobility in near-neutral soils and is mainly concentrated in the upper soil layers. For instance, in rhizosphere soil (i.e., at the root–soil–microorganism interface), Cu is complexed by low-molecular-weight organic compounds exuded by plants and microbes like carboxylic and phenolic acids, which play an essential role in both external and internal plant tolerance mechanisms [9]. In arable soils, Cu concentration in the soil solution lies in the range 1–300 μg L^{-1} [12].

Concerning the plant acquisition process, roots can use the free ionic forms (Cu^+/Cu^{2+}), although the direct use of Cu complexes with different organic molecules present in the rhizosphere cannot be excluded. Indeed, the identification of an uptake system of Fe-phytosiderophore (i.e., root exudates with a high affinity for Fe) complexes also in dicots [13] further corroborates this hypothesis. Moreover, a reduction step ($Cu^{2+} \rightarrow Cu^+$) is also very likely to occur before the root uptake of the nutrient [9]. Indeed, previous works demonstrated that Fe chelate reductase enzymes and their related gene family (*FRO*) responded to both Cu deficiency and toxicity in *Arabidopsis thaliana* and cucumber plants, respectively [14,15]. The metal is taken up through a specific transmembrane Cu Transporter Protein 1 (COPT1) [16]. However, the involvement of other transporters like Zn/Fe Permeases (ZIPs), Natural Resistance Associated Macrophage Proteins (NRAMP) and Iron Related Transporters (IRT) has also been speculated [17–20]. In this context, the extent of the Cu bioavailable fraction (i.e., the two free ionic Cu forms and the soluble metal complexes with organic ligands) strongly depends on soil properties, especially cation exchange capacity (CEC), pH values, organic matter and clay content [12].

Once taken up by the roots, Cu is loaded into the xylem vessels to be translocated towards the aboveground organs. Despite only a few pieces of information available, the Heavy Metal-transporting P-type ATPase 5 (OsHMA5) has been shown to be involved in the Cu xylem-loading in rice plants [21]. In this respect, the Cu concentration in the xylem sap and in shoots is remarkably limited when the expression of this gene is down-regulated. Moreover, in *A. thaliana*, Yellow Stripe Like 2 (YSL2) transporter was also found to play a pivotal role in the xylem loading of Cu complexed with nicotianamine [22]. Once loaded in the xylem, Cu is bound to an organic ligand, generally amino acids (like histidine and proline in *Brassica carinata*) or nicotianamine (e.g., in tomato and chicory) [23–25]. Once at the leaf level, before being taken up by the cells, the Cu^{2+} complexes need to be reduced to Cu^+ by FRO4 [26] and, only after its transmembrane transport, the metal is delivered to the different organelles (e.g., mitochondria and chloroplasts) through specific transporters [27].

In plant cells, Cu is involved in a plethora of biochemical and physiological processes, including photosynthesis, respiration, oxidative stress responses, cell wall metabolism, and hormone signaling [28,29]. In fact, Cu is an important cofactor in many proteins, such as plastocyanin, cytochrome c-oxidase and amino oxidase [28]. Generally, the Cu concentrations in plant tissues growing in uncontaminated soils range from 5 to 30 mg kg^{-1} dry weight, depending on the type of plant, the growth stage, and the soil characteristics. Deficiency symptoms can appear at Cu concentrations lower than 5 mg kg^{-1} dry weight, while leaf concentration higher than 20 mg kg^{-1} dry weight may result in toxic effects affecting the whole plant development [30–32].

1.2. Current Situation in Vineyards (Including Toxicity in Plants and Effects)

Several studies have reported an accumulation of metals in agricultural soils due to common agricultural practices, such as manure fertilization and spraying for pest control [33–35]. For instance, Cu in apple orchards can have an average annual total increase ranging from 2.5 to 9 mg kg^{-1} year^{-1} deriving from the application of Cu-containing agrochemicals [36]. The accumulation of Cu through time can be influenced not only by the age, but also by other factors, such as soil organic matter content and pH [33]. As shown in Figure 1, vineyard soils are prone to Cu accumulation, since the practice of spraying Cu salts as fungicides dates back at least as far as 1761 [37]. After the introduction in 1885 of the mixture of Cu sulfate and lime, called "Bordeaux mixture" [38], its use became generalized on wine-producing farms. Because of this, the concentration of Cu in vineyards can be many times higher than the background values of forest soils in the same region, reaching up to 3000 mg kg^{-1} [39,40]. Next, high Cu concentration differences can also be found inter-row and intra-row within the same vineyard [41].

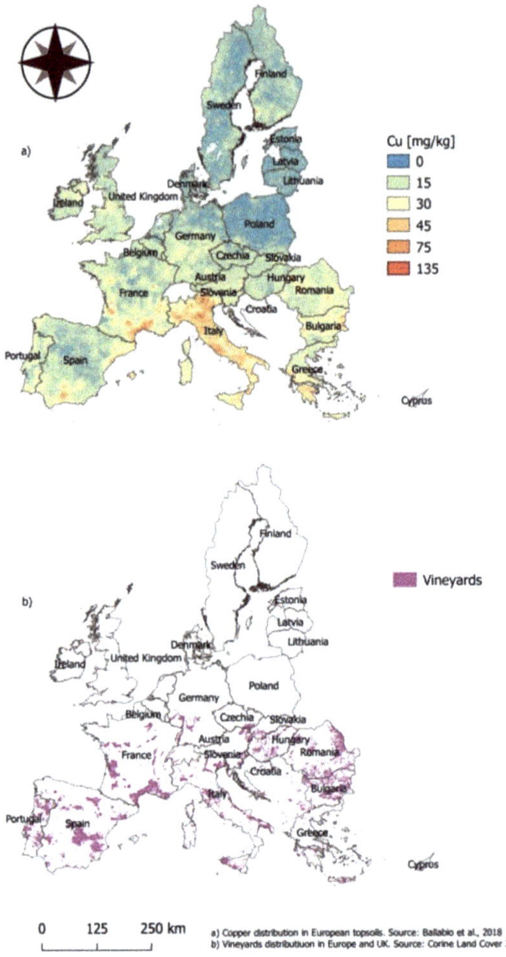

Figure 1. Cu concentration distribution (**a**) compared to the distribution of vineyards (**b**) in Europe [42]. Several high Cu concentration areas perfectly match viticulture areas.

Although much research is currently taking place to find alternatives to Cu, this element is still crucial to fighting fungal diseases in vines, especially in organic farming, limiting the supply of pure Cu to 28 kg ha^{-1} over a period of 7 years [43–45]. For these reasons, the global trends of vine-cultivated areas and the use of Cu should be monitored, particularly in newly planted vineyards. Indeed, the shallow root system of newly planted vine plants might be more affected by high Cu levels than older plants, since Cu mainly accumulates in the upper soil layers (where the concentration of organic matter is higher). In fact, the first effects of Cu phytotoxicity are evident at the root level, with a clear decrease of root elongation, abnormal root branching, thickening and dark coloration [46]. As a consequence of this altered root development, the roots' ability to take up water and other nutrients is clearly impaired [32,47]. In this regard, it is worth mention the negative effect of Cu toxicity on P acquisition by roots [48,49], highlighting the interactions occurring between the different nutrient acquisition mechanisms. Critical Cu concentrations in roots range from 100 to 400 mg kg^{-1} dry weight [10,46]. Like other heavy metals, an excess of Cu may also cause the generation of toxic reactive oxygen species (ROS), which, in turn, can damage several important biomolecules like DNA, proteins and lipids [50]. Usually, plants exhibit an excellent translocation barrier of excessive Cu between roots and shoots. This limits the risk of Cu accumulation in edible plant tissues that exceeds toxic levels to livestock and humans. However, toxic concentration levels, as well as symptoms of Cu toxicity, have also been well described for the aerial part of plants [30–32]. Typical cases of Cu toxicity at the root and shoot levels of vine plants (*Vitis vinifera* L., *cv Glera*) are presented in Figure 2.

Figure 2. Symptoms of Cu toxicity in vine plants grown in a soil with 814 mg kg^{-1} Cu (DTPA extractable fraction: 280 mg kg^{-1} Cu): (**D–F**) canopy, (**C**) leaves and (**H**) roots of grafted plants of *Vitis vinifera* L. (*cv. Glera*); (**G**) canopy of the rootstock SO4 (*V. berlandieri x V. riparia*). As control, canopy (**A**) and leaf (**B**) of healthy vine plants (*Vitis vinifera* L. *cv Glera*).

Considering the soil surface dedicated to viticulture, in 2019, worldwide vine cultivation covered 7.4 million hectares [51], and it has been considered stable in recent years, with a global balance given by heterogeneous evolution in different regions of the world [52]. In Europe (Figure 1), the vine-cultivated area is stable, being balanced between the European Union "grubbing up" program (i.e., the replacement of vine plants with other agricultural crops, [53]) on the one hand, and the possibility of Member States authorizing the planting of up to 1% of the vineyard surface already planted each year [54], on the other hand. Globally, several countries are increasing their vineyard area, namely Russia, New Zealand, Peru, and China, to name a few. Meanwhile, countries like USA, Argentina,

Chile are showing a decline in vineyard area [51]. In addition to wine regulations and the global market [55], Climate Change is playing a fundamental role in directing the spatial distribution of vines, as increasing temperatures are leading to vineyards higher in elevation and at higher latitudes [56,57]. Climate change also affects the incidence of fungal infections in vineyards, requiring increasing use of Cu-based treatments [9]. These changes will then affect the chemical and biological quality of soils of newly planted vines with increasingly higher Cu concentrations. In this regard, it is interesting to note that the annual supply of Cu for pathogen defense plans can range from 1–2 kg ha^{-1} in Europe to 30 kg ha^{-1} in southern Brazil [58] as a function of the infectious pressure of the pathogens. Furthermore, new crops growing where there used to be vineyards might be affected by high Cu concentrations and exhibit symptoms of toxicity.

European countries have different approaches to defining threshold levels, as there is large heterogeneity in the legal systems, the chemical analysis used, the target organisms for toxicity, and background-values and how these are defined. In addition, other soil properties (mainly pH and SOM) and the precipitations/humidity of wine areas can be used as additional data to determine the final threshold level since the need for applying higher quantities of active principle can influence the decision-taking process [59–61]. In Italy, for instance, the management of polluted sites is regulated by the law D. Lgs. 152/2006, and a recent Decree established 200 mg kg^{-1} of Cu as the contamination limit for agricultural soils [62]. A literature review showed a high degree of heterogeneity and a lack of standard approaches and thresholds for heavy metal risk assessment [59]. For example, the Austrian standard ON S 2088-2 defines a two-step evaluation in which, firstly, soil metal content is measured (NH_4NO_3-extract). If certain trigger values are reached, then the possible bioavailability is evaluated in consideration of the metal content and other soil parameters, such as pH and soil organic matter, thus resulting in a final contamination evaluation [59]. The Czech Republic provides indication limits on plant growth inhibition [63] in consideration of the metal content extracted with aqua regia (HCl and HNO_3 at a ratio of 3:1) or with 1 mol L^{-1} of NH_4NO_3. Both extractions must be performed if the limit values for the specific metal are defined. At the global level, the lack of a standardized approach on Cu content thresholds in soils should be addressed in the future both by the scientific community and in public policy. Several soil extraction methods should be evaluated, considering both total and available Cu contents in order to reach a robust decision scheme.

2. Current Challenges

As described in the previous paragraphs, a deeper understanding of the Cu accumulation phenomenon in vineyard soil with an all-encompassing approach, including the main environmental and agricultural risks connected, appears necessary. Concurrently, the identification of approaches and the setup of cultivation practices aimed at mitigating the problem are urgent. Therefore, the following paragraphs describe some examples considering recent technological innovations like rhizosphere management, biotechnologies applied to vine plant breeding and smart viticulture.

2.1. Copper Effects on Soil Agrobiodiversity

Soil contains a vast underground world inhabited by many fungal and bacterial species. It is estimated that 1 g of soil contains a number of bacteria ranging between 100,000 to 1,000,000, and, within this fraction, there may exist different micro-niches in which several ecotypes may live [64,65]. Although only a small fraction of the fungal and bacterial diversity is known, certain fungal and bacterial groups play a crucial role in several agroecosystem services, including plant-growth promotion and cycling of nutrients (e.g., nitrogen, N, and phosphorus, P) [66–68]. However, the accumulation and the excessive availability of heavy metals such as Cu can impact the soil microbial biomass and activity, thus affecting some microbial-driven soil processes, including N fixation and mineralization [69]. Moreover, Cu has been demonstrated to impact microbial communities' composition and functionality, with higher effects on bacterial than fungal communities [70–74]. Fungi and bacteria have evolved different strategies against high Cu concentrations in soils (Figure 3). For instance, some filamentous fungal and yeasts species act against the overload soil Cu, directly at the cytosol level [75]. Herein, the sequestration of the free Cu ions occurs due to metallothioneins, a family of cysteine-rich proteins (e.g., Cup1 and Crs5) that chelate Cu [76–78]. Furthermore, when the metal in the cytosol is in excess, fungal cells act through organellar compartmentation of the metal, mainly directing it to the vacuole. Concerning this latter strategy, no specific Cu transporter catalyzing the transmembrane transport into the vacuole has yet been identified [79]. In addition, cytoplasmic chaperones' activity is essential for mitigating Cu stress in the cytosol. Indeed, while the chaperone Atx1 delivers Cu to the trans-Golgi network, Ccs1 delivers Cu to the Cu-Zn Superoxide dismutase Sod1 [80,81].

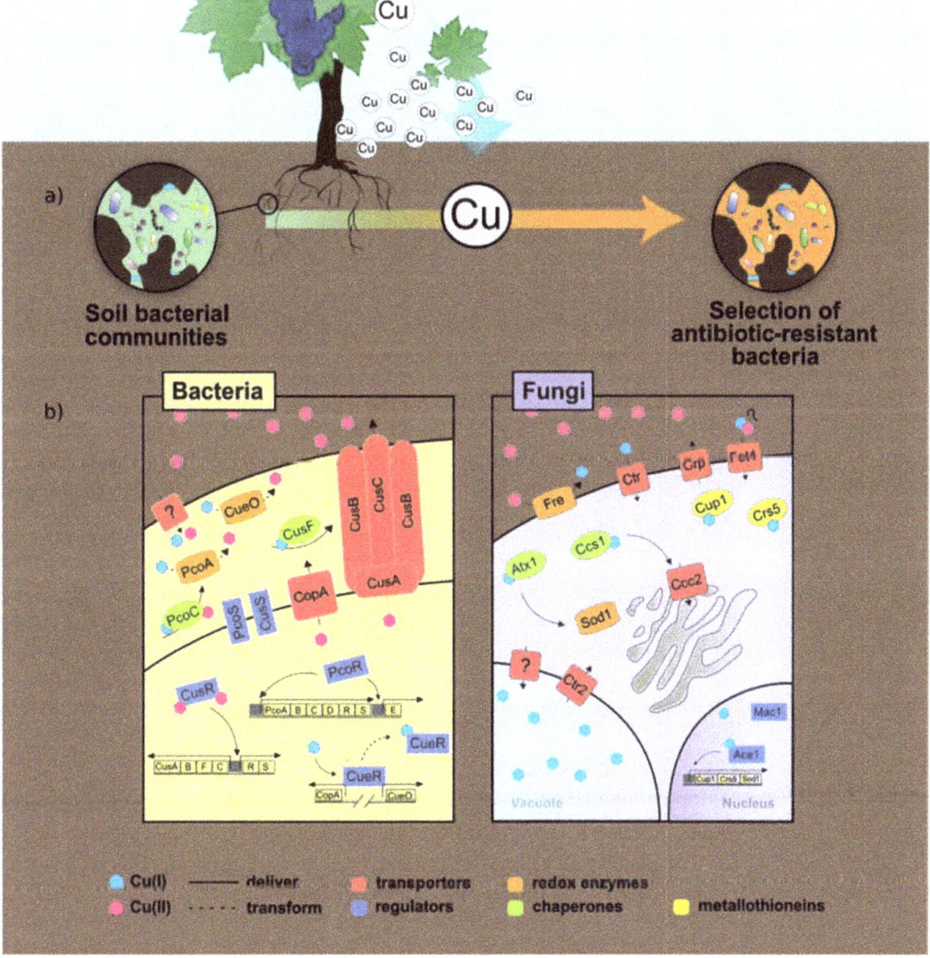

Figure 3. Exposure of soil microorganisms to Cu in vineyards. (**a**) Selection of Cu-resistant and antibiotic-resistant genes due to the Cu application (**b**) Schematic representation of the bacterial and fungal strategies for coping with excessive Cu inputs;

In contrast, in bacteria, the detoxification strategy relies mainly on a Cu export system based on the functionality of two different mechanisms for Gram-positive and Gram-negative bacteria [82]. In Gram-positive bacteria, proteins belonging to the Cop family are at the basis of cytosolic Cu disposition. The system includes a Cu chaperone (CopZ) that delivers Cu first to CopY (copper responsive repressor), activating the transcription of *copA* and *copB* (genes encoding for ATPase pumps), which extrude Cu^+ and Cu^{2+} from the cell, respectively. In Gram-negative bacteria, the presence of periplasmic space determines the evolution of specific mechanisms for extruding Cu in the extracellular space. Specifically, the Cop-protein system is located on the inner membrane and in the inner cell, and its activation is mediated by the MerR-type Cu-responsive transcriptional activator CueR [83]. Additionally, while Gram-positive bacteria mainly use the Cop system, Gram-negative bacteria exploit a trans-envelope extrusion system (CusABC ATPase pumps system) [82]. Additionally, the periplasmic multicopper oxidases (PcoA, CueO), oxidizing Cu^+ to Cu^{2+} and activating the production of siderophores to sequester the cytoplasmic Cu, play a crucial role in the Cu detoxification process [84,85].

Moreover, the Cu accumulation in agricultural soils other than bacterial Cu resistance can also promote the occurrence of antibiotic resistance, with these traits being linked together [86–90] (Figure 3). Therefore, at the molecular level, the Cu-resistant and antibiotic-resistant genes can be reasonably co-selected. Thus, they can commonly occur on the same mobile genetic element (i.e., plasmid, integron or transposon) [91,92]. Additionally, the bacteria cells can have a cross-resistance molecular mechanism in which the same gene confers resistance to different antimicrobial agents, such as antibiotics and Cu [91]. Indeed, the repeated application of Cu to agricultural soils over the years, given that the metal is not degradable, can dramatically enhance the frequency and dissemination of Cu-resistant and antibiotic-resistant genes [92]. In this context, Cu-contaminated vineyard soils could even be considered a reservoir of antibiotic-resistance traits that can be transferred among the different bacterial species via vertical and horizontal gene transfer (e.g., plasmid-mediated conjugation, integrons) [36]. However, more detailed studies should be carried out to elucidate the possible long-term impact of Cu on the bacterial communities inhabiting agricultural soils linked to the antibiotic resistance. Indeed, there is a general lack of knowledge about the potential route for the transmission of antibiotic-resistant bacteria from soil to crops, animals and humans. Specifically, it will be crucial to investigate the potential health risk assessment to prevent further development of pathogenic-resistant bacteria in agricultural ecosystems.

2.2. Soil Management Considering the Root–Microorganism Interactions (Rhizosphere Management)

Considering the ever-increasing incidence of soils contaminated with Cu, it is evident that rhizosphere dynamics must be taken into account when setting up and applying soil management practices to mitigate the Cu stress in these agricultural soils. In fact, the biogeochemical cycles of the nutrients completely differ at the soil–root interface compared to non-rooted soil (i.e., bulk soil). Indeed, the highly dynamic interactions between roots, soil and microorganisms occurring in the rhizosphere control not only the speciation and the consequent availability of the nutrients (and, in turn, the extent of their root acquisition), but also determine the level of availability/toxicity of heavy metals like Cu. For instance, in the hotspot that is the rhizosphere, pH value modifications and the release of low- and high-molecular-weight organic compounds by both roots and microorganisms are the main drivers of Cu availability.

Concerning soil pH values, root-induced changes of this chemical parameter are mainly triggered by plant nutrient uptake. This process is often coupled with an active proton extrusion catalyzed by the plasma membrane H^+-ATPase [28]. Furthermore, an unbalanced cation/anion uptake of nutrients might lead to either acidification (i.e., due to an excessive uptake of cations) or alkalization (i.e., due to an excessive uptake of anions) of the rhizosphere. These imbalances might occur naturally due to different plant requirements, depending on the physiological status of the plants, microbial interactions and/or competition with neighboring plants, as well as potentially being caused by the unbalanced availability of some nutrients in the growth medium [93,94]. Indeed, in vineyards, when cover crops are co-cultivated with other plants like vine plants, nutrient competition might induce an unbalanced uptake of elements with consequent root-induced changes of the pH. As is to be expected, this phenomenon impacts Cu mobility/availability and, consequently, its toxicity. On the other hand, excessive concentrations of metals like Cu might also be themselves responsible for an unbalanced nutrient uptake. Recent studies have highlighted that Cu stress can induce synergism and/or antagonisms with many other nutrients like P, Fe, Zn and Mn, causing further root-induced pH changes both in annual grass species and in perennial plants like grapevine [15,95]. The complexity is further exacerbated by the fact that rhizosphere effects (like acidification and alkalization processes) are not always constant over time [96] and are strongly plant species and genotype/rootstock dependent [15,95,96].

As mentioned earlier, besides pH values, rhizosphere organic compounds are the main drivers shaping Cu speciation in agricultural soils. Indeed, specific compound classes such as phenolic compounds and carboxylic acids play a fundamental role in both internal and external Cu tolerance strategies of plants [97]. Recent studies have shown that perennial plants like grapevines also trigger their root exudation when exposed to excessive Cu concentrations, yet the phenomenon strongly depends on the type of the rootstock and the growing conditions [15]. Indeed, when grapevines are intercropped with annual grass species like oat, the exudation pattern is completely modified. Furthermore, the enhanced exudation reduces the Cu accumulation in grapevine plants, making this a valuable agronomic strategy for mitigating Cu stress [15]. It is important to highlight that the alleviating effect is highly plant species dependent, and thus the intercropping approach should be evaluated for each specific rootstock. In addition, the intermingling of roots in intercropping systems might induce a competition between plant species leading on one side to a modified exudation profile and on the other side to an unbalanced nutrient uptake ultimately affecting the effect of Cu stress.

In addition, soil management also comprises fertilization strategies, which also strongly affect rhizosphere dynamics. For instance, the source of N supply (NO_3^- or NH_4^+) affects the pH at the soil–root interface both in annual and perennial plant species [28]. Nitrate-based fertilizers induce an enhanced anion root uptake with proton consumption and a consequent alkalization of the rhizosphere. On the other hand, ammonium-based fertilizers trigger enhanced proton release by roots, resulting in significant rhizosphere acidification. Even though nitrate usually predominates over ammonium in agricultural soils due to the very rapid microbial nitrification processes, preferential ammonium uptake might still occur in acid soils, at low soil temperatures, or shortly after the application of ammonium fertilizers, organic fertilizers and/or nitrification inhibitors [28]. Soil management should thus comprise fertilization-induced root activities, since Cu availability in soil is strongly pH dependent and might undergo mobilization processes, exacerbating its toxicity. Moreover, an adequate nutritional balance for a crop seems to be decisive in mitigating the effects of Cu toxicity when the metal is already present in relevant quantities. Vineyards in calcareous soils represent a clear example. In fact, in this case, the edaphic conditions (essentially soil pH values) should theoretically restrain the availability of Cu to the plants. However, if the Fe availability for crops becomes limited as a consequence of the chemical–mineralogical nature of the soil, the activation of the plant's adaptive responses to this deficiency (including the acidification of the rhizosphere [98]) could transform the Cu problem from potential to real.

The complexity of rhizosphere dynamics highlights that soil management, and particularly the management of nutrients, in agricultural agroecosystems such as vineyards needs a comprehensive overview and an appropriate knowledge of the soil–plant system and its dynamics. Indeed, such knowledge could be crucial in counteracting the negative effects of toxic concentrations of heavy metals like Cu and/or exploiting beneficial synergism between elements to benefit the agricultural production. Future studies should therefore focus on a holistic approach that also includes the third main actor involved in rhizosphere processes, i.e., microorganisms. Even though microorganisms are known to increase plant growth and stress resilience, little is known about the effect of Cu stress on their composition and functionality in the rhizosphere in real agroecosystems such as vineyards. To date, most studies have been lab-based and recreate artificial metal stress conditions; thus, they still provide an insufficient understanding of the microbiome functioning and the mechanisms of plant–microbiome–soil interactions. Large-scale experiments at the field level are thus essential in order to give a complete overview of the effect of Cu on the complex interplay between soil, plants and microorganisms involved in the biogeochemical cycles of nutrients at the agroecosystem level.

2.3. Biotechnologies and Breeding for a More Resistant Plant Material

Grapevine cultivation and wine production worldwide face challenges posed by the high pressure of fungal and fungal-like diseases, causing production losses [99]. The most common and severe problems in viticulture include those presented by downy mildew and powdery mildew infections, caused by *Plasmopara viticola* and *Erisyphe necator*, respectively [100]. Both pathogens are commonly controlled by the application of fungicides in vineyards, mainly based on Cu [101]. However, the implementation of innovative strategies, such as the use of pathogen-resistant vine genotypes, is highly desirable with the aim of achieving a more sustainable viticulture. For this reason, breeding programs have been implemented in order to transfer the so-called resistance (*R*) genes from resistant *Vitis* species to sensitive *V. vinifera* varieties [102].

The production of resistant grapevine hybrids began in the 19th Century, yet the first varieties, obtained from traditional breeding carried a high percentage of non-*V. vinifera* genetic material and presented undesirable "foxy" flavors in the resulting wine, a distinctive feature of *V. labrusca* [101]. In the post-genomic era, new techniques, such as marker-assisted backcrossing, marker-assisted background selection and marker-assisted pyramidization, contributed to the development of breeding strategies allowing the precise introgression of wild alleles in susceptible genomes, while reducing the undesired residual genetic material [103]. At present, up to 27 Quantitative Trait Loci (QTL), known as *Rpv* genes, are associated with the resistance to *P. viticola*, while up to 13 QTL (*Ren* and *Run* genes) are related to resistance against *E. necator* [102]. Following infections, *Rpv* genes can induce different defense responses (e.g., hypersensitive response, the accumulation of phenolic compounds in the infected tissues, the accumulation of callose and lignin, the synthesis of phytoalexins, the induction of either cell necrosis or peroxidase activity; [104] and the references therein), while *Ren* and *Run* loci have been shown to encode small gene families of receptor-like proteins that are thought to directly or indirectly interact with pathogen-specific effectors. This interaction elicits a signaling cascade that leads to the transcriptional reprogramming of the host plant and the expression of plant defense genes [105]. Despite the potential economic and environmental benefits achievable, the diffusion of resistant hybrids is being prevented by the wine-making industry, which preferentially chooses traditional genotypes, despite their being susceptible, over resistant ones, mainly because of current regulations [102,106]. A remarkable opportunity to overcome the issues related to traditional breeding might be offered by the genome editing approach, directed towards the susceptibility (*S*) genes encoded in the *V. vinifera* genome [107]. The functionality of *S* genes is required for successful infections by both *E. necator* [108,109] and *P. viticola* [106,110]. Indeed, the targeted inactivation of *S* genes in *V. vinifera* plants might potentially lead to the generation of resistant clones, while still preserving the genetic identity of the parental genotypes, which would be highly appreciated by the wine producing industry [106].

Interestingly, these new varieties can be defined as being tolerant to the pathogens; however, in cases of very high infection pressure, the endogenous plant responses might not be sufficient for counteracting disease, making the application of agrochemicals like Cu necessary for controlling and/or contrasting pathogen diffusion within the canopy

(Figure 4). Indeed, despite not completely avoiding the application of fungicides, resistant varieties might contribute to strongly decreasing the use of agrochemicals in open fields.

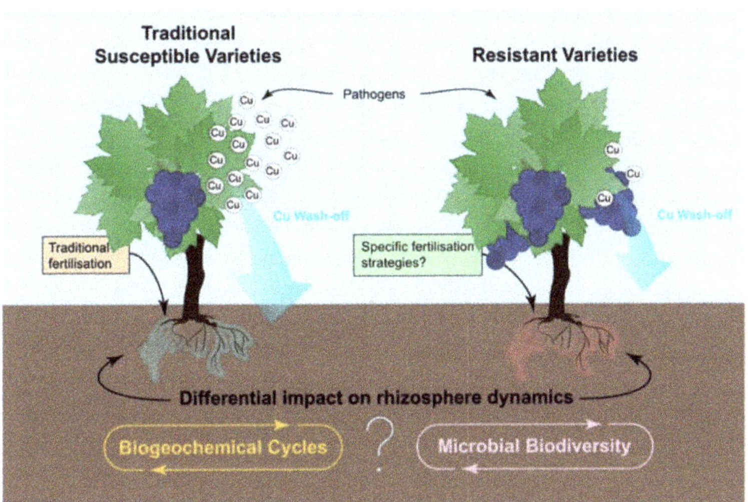

Figure 4. Traditional susceptible varieties vs. resistant varieties in viticulture. The protection of traditional susceptible varieties from pathogen infections requires the repeated application of Cu-based fungicides, which, through drift phenomena and rain wash-offs, can cause a consistent accumulation of Cu in vineyard soils. The increased Cu concentration in soil can, on one hand, induce vine plants to modify their exudation to cope with Cu toxicity, and, on the other hand, promote an alteration of soil microbial biodiversity. The combination of modified exudation patterns with an altered rhizosphere microbial community might impact the biogeochemical cycles of mineral nutrients, thus modifying their availability to plants. In resistant varieties, the protection against pathogens requires lower application rates of agrochemicals, thus also reducing the input of Cu. Nonetheless, the expression of the resistance gene(s) might alter the physiology of vine plants at transcriptomic, proteomic and metabolomic levels, in this case also causing a modification in the exudation pattern. In addition, evidence shows that resistant varieties might require specific fertilization strategies in order to fully present the resistance traits. Again, an altered exudation pattern combined with the input, albeit minimal of Cu and fertilizers may impact the rhizosphere dynamics, affecting both soil microbial biodiversity and the biogeochemical cycles of mineral elements.

Nevertheless, both the introgression of foreign genetic material (such as for traditional breeding) and the silencing of genes (such as for the genome editing approach) could potentially alter the physiology of plants. The modification of vine plants' genetic material might produce, for instance, organisms that could require particular edaphic conditions (e.g., specific fertilization practices and/or specific provision of micro/macronutrients) in order to express disease resistance at optimum levels (Figure 4). In fact, specific nutrient fluxes within the leaf blade seem to be the basis for the expression of the pathogen response in these resistant varieties [111]. In addition, transcriptional reprogramming, as in the case of either *Rpv* or *Ren/Run* loci, can alter the metabolome [112,113], which might mirror modifications in qualitative and quantitative root exudation patterns. Root rhizodeposition plays a crucial role in the transformations and fluxes of nutrients from soil to plant [114], considerably affecting the extent of their acquisition by plants, and therefore, the crop yield. Moreover, root exudates represent, at least in part, the mechanism by which plants can recruit soil microbiota and shape the rhizosphere microbial community [115]. As mentioned earlier, this could either directly or indirectly impact the structure, the diversity and the functionality of soil and rhizosphere microbial communities, which play a pivotal role in the processes occurring below ground (Figure 3) [68,115–117]. In this regard, the accidental selection of pathogenic microbes, which could represent a future challenge for viticulture in terms of new plant diseases, cannot be excluded. Therefore, the suitability assessment of these new plant materials should not be limited to those traits that are strictly connected with the expression of the pathogen resistance genes for the purpose of overcoming old (i.e., the classical and well-recognized vine diseases) challenges. Indeed, it is necessary to evaluate resistant varieties also with respect to putative new challenges (i.e., new pathogens and the related diseases) in order to establish the sustainability of these viticultural practices in both the short and the long term. Furthermore, the availability of rootstocks with traits of tolerance to Cu toxicity could be advantageous in viticultural areas. Finally, the increase in phenotype and genotype data of the scions and the rootstocks could significantly benefit from the application of machine learning to accelerate the crop breeding process [118] in press).

2.4. Smart Viticulture

Cu accumulation in soil can be restrained through direct (e.g., using more resistant plant species, as described in the previous paragraph) and indirect actions (e.g., decreasing the availability in the rhizosphere and/or limiting the quantity of Cu that reaches the soil via new applications). Since the resistant varieties of vine plants currently available are essentially tolerant, and therefore a certain level of plant defense using pesticides is essential, indirect actions are still important not only for the traditional varieties, but also for the new ones. In this respect, the main goal is to avoid drift phenomena during the application of pesticides and to limit Cu accumulation in the soil. The Cu applied to the vine canopy will indeed reach the soil due to rain, foliar irrigation and/or autumn leaf fall.

In the last two decades, a wide range of technologies have been increasingly applied in viticulture, such as monitoring technologies, decision support systems and operating technologies. **Monitoring technologies** include geolocation, remote sensing (using satellites, aircrafts, and unmanned aerial vehicle/drone images), and proximal sensing during production (using wireless sensor networks to measure soil moisture, leaf wetness, grape temperature, sap flow, etc.) and harvesting phases (using volumetric grape sensors and optical sensors for grape "quality"). **Decision support systems** (DSS) make it possible to consider the spatial variability (i.e., the variability of soil properties, landscape features, crop stresses, yield and quality in the different areas of a field [119]) in process optimization (irrigation, fertilization, or chemical treatments) and harvesting. Finally, operating technologies include **variable-rate technologies (VRT)** and **agricultural robots**, which make it possible to implement the activities defined with the help of the DSS, while also limiting drift phenomena [120].

These technologies were proposed in the 1990s, initially for their ability to manage crops according to site-specific logic, which makes it possible to deal with the spatial variability of soil, nutrient and phytosanitary conditions. They also include geo-referencing, which allows producers to micro-manage soil and plant processes within small areas of a single field. Despite positive expectations, these techniques' diffusion is still relatively limited, as several authors have reported [121–125]. In several European countries, the use of ITC in agriculture remains at less than 10% of farms, and, of these, only 50% of farms that complete site-specific applications. The current circumstances are considered to be very favorable for the growth of the sector, not only for the new generation, with habits more rooted in the use of digital technologies, but also for the "cultural dragging" caused by the so-called "Industry 4.0" revolution [126–128]. In fact, this revolution has brought significant attention to the new technological frontiers connected to the "Internet of Things" (IoT), Big Data, Cloud Computing, hyper-connectivity, cybernetics and augmented reality. As a consequence of this technological trend and its effects, the terms *agriculture 4.0* and *smart agriculture* are coming to be preferred to the term *precision agriculture* [129].

To the best of our knowledge, there are no studies measuring the exact impact of precision or smart viticulture techniques on the limitation of Cu accumulation/toxicity at the soil level. However, various papers have highlighted that the goal of these techniques is to improve not only the yield and quality of grape production, but also its environmental sustainability (i.e., reducing the chemical treatments) [120,130]. Indeed, these techniques have shown that it is possible to reduce the use of Cu-based fungicides by 25–50% [131–133]. Therefore, it can be expected that the contribution to the limitation of Cu accumulation/toxicity at the soil level will be remarkable. Thus, a set of examples of *precision* and *smart viticulture* technologies (which might be used by practitioners and scholars as an initial reference point for implementation projects and precise impact analyses) is discussed in the following paragraphs. Other examples can be found in the detailed reviews of literature and field applications of precision viticulture [120] or precision agriculture [134–136].

While soil protection has mostly been studied using a mono-disciplinary approach (e.g., soil chemistry, plant defense), the use of precision/smart viticulture technologies in combination with an interdisciplinary approach involving technology experts (from the fields of sensors, electronics, computer science, and agricultural machinery), domain-experts (e.g., from the fields of soil chemistry, plant pathology, genetics, and agronomy) and sustainability experts (or Lifecycle assessment experts) is increasingly needed.

In particular, *precision* and *smart viticulture* technologies could contribute positively to the following aspects: (1) crop monitoring, i.e., using optical sensors on vehicles (e.g., tractors, autonomous vehicles, or drones) to perform a sort of "early digital scouting" to be carried out promptly to keep the phytosanitary status of wide-crop areas under control [137,138]; (2) operational monitoring, i.e., using the so-called FORK systems (Field Operation Register Keeping) [129,139,140] which allow the detection and automatic recording of how a field operation is carried out; (3) implementation of site-specific techniques with retrofit components, i.e., adaptable to existing farm machinery, thus avoiding the need to make large investments in new equipment (see Figure 5). The first two aspects concern information management actions, necessary for medium- and long-term quality decision-making processes at the farm, based on targeted information. The third aspect allows a direct intervention with immediate control effects, especially if the retrofit device is also equipped with sensors capable of locally estimating the volumes of canopy to be treated and then enabling the adjustment of the doses to be distributed accordingly.

Sensors installed both on agricultural machines and within the vineyards (both at the plant and soil level) allow the collection and monitoring of data (e.g., the color, shape and dimensions of leaves and grapes, temperature and weather conditions, soil pH, and the presence of pathogens [120,130,131,141]). These data can then be processed (cleaned, filtered, aggregated, represented and archived) and analyzed to extract the relevant information. These analyses might be carried out by a DSS at the farm or at external servers accessible over the Internet (i.e., in the Cloud) [133,141]. Moreover, they might be automated through Artificial Intelligence and/or neural networks algorithms, and they might also be used to simulate/prevent future behaviors/events (e.g., a digital twin of the vineyards can be created where the impact of different interventions/strategies can be simulated) [142]. The data analyses might also combine data from multiple farms/vineyards, as well as other big (information assets characterized by such a high volume, velocity and variety that it requires specific technology and analytical methods for its transformation into value [143]) data from several sources [144,145]. Finally, the results obtained can lead to an action, such as process optimization (modification of irrigation, fertilization, or chemical treatments), or to a report for internal or external users [129,131,133]. Such actions can also be implemented in a more precise and/or automatic way through straddle tractors/machines, autonomous vehicles/robots, and drones [120,146]. Finally, the application of these technologies might help the winegrower to spray, fertilize or irrigate (1) when it is needed; (2) exactly where it is needed (also limiting the drift); and (3) at the quantity needed in the different areas of the field and times (i.e., considering the horizontal, vertical, and spatial variability), thus potentially minimizing the Cu accumulation process in the soil. The technologies presented above are (mostly) already available on the market and have been adopted by some vineyards. An interesting example of this is PlantCT^{TM} (formerly known as Smartvineyard), a system that continuously monitors the plants and the environment using a wide range of sensors (i.e., spore and pest detectors, as well as leaf surface, light, humidity, and temperature, wind, solar radiation, precipitation and soil sensors) and suggests to the winegrower the interventions/actions that should be implemented (https://plantct.com/). Another example—more focused on the topic of this paper—is Coptimizer, a model-driven DSS that might help to optimize and track the use of Cu-based fungicides in viticulture. Other similar examples include VineSens [141], GRover [147], and FeelGrid^{TM} (https://www.feelgrid.com/). For a detailed review of smartphone applications for smart agriculture, see [148]. There are then several companies producing variable-rate (or smart) straddle tractors/machines and spraying drone systems (also) for vineyard applications (e.g., Pellenc, New Holland, Durand-Wayland, Tecnovit, Fly Dragon Drone Tech., Chouette). Some of these systems have been presented in detail by [120].

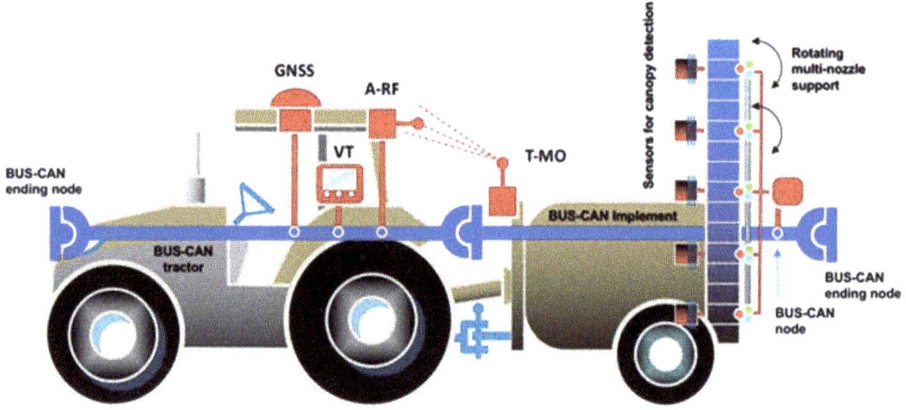

Figure 5. Design of a possible retrofit device to be applied to sprayers already present at the vineyard to enable them to operate according to site-specific logics. The retrofit application concerns a vertical spraying tower equipped with a set of rotating supports for independent nozzle selection and activation. The control is supported by a set of volumetric ultrasonic sensors able to detect the size of the canopy to be processed. The solution even includes a Field Operation Register Keeping (FORK) system based on an identification device enabling the tractor to automatically detect the identity of the coupled machine through a simple Radio Frequency (RF) system (T-MO: implement transmitting code; A-RF: tractor receiving antenna; identification is enabled only on distances <20 m). On board data transmission is performed via Controller Area Network (BUS-CAN), supposing this is already available on-board. In case of older tractors (no BUS-CAN), a direct wire connection can be easily established.

The necessary investments and/or operating costs required by the abovementioned systems—as well as more generally for the implementation smart farming solutions—are, however, rather high, particularly for small vineyards [149,150]. Furthermore, specific know-how is often needed to properly use them [149]. The Industry 3.0 phase (indicatively occurring between 1970 and 2010) was characterized by digital culture and technologies in the management of industrial production processes. On the other hand, the Agriculture 3.0 phase (occurring between 1980 and 2000) introduced technological innovations mainly focused on the improvement of the quality of mechanization in the fields of electronics, ergonomics and safety, but with limited results in the field of information technology. The "digital gap" in agriculture compared with the industrial sector is particularly relevant. Therefore, the key challenge is to adopt these systems and technologies for economic sustainability (the retrofit idea mentioned above could, for instance, be an interesting solution) and to explain their potential benefits to winegrowers. Another possible idea could be the creation of some consortia or cooperative forms to cover the high investments/costs. In any case, with the application of these strategies, it is possible to significantly limit the supply of Cu to the vineyard, thus limiting the metal accumulation process in the wine-growing context.

However, future research is needed on the smart viticulture topic at different levels to contribute to the general challenge. First, while well-developed and robust solutions for specific purposes exist (e.g., grape yield monitors, soil and leaf monitors and precision sprayers), they still tend to work as separate silos. Future research should therefore try to develop visions as well as practical architectures/applications for an integrated smart (or digital) management of vineyards. Second, there is a strong need to identify the competencies needed by farmers (and by wine-growing consultants) in order to correctly and effectively use the technologies available and to consequently update the study programs. Finally, smart viticulture's impact on the three sustainability dimensions (i.e., environmental, social and economic) should be further investigated. [151] showed that both positive and negative impacts of Industry 4.0 technologies can be observed in the industrial field. This might also be applied in viticulture (and, more generally, in agriculture) where, to date, research has mostly focused on the environmental dimension [130,131].

3. Conclusions

The repeated application of copper (Cu)-based fungicides in conventional and organic farming has caused a significant increase in the total Cu contents in vineyard soils, posing agricultural and environmental threats. The goal of our review was to shed light on this issue and to discuss some approaches that might be adopted to limit Cu toxicity.

We presented the current Cu accumulation situation, with particular reference to the European context (see Figure 1), and discussed its potential risks in viticulture (see Figure 2), considering the chemical-physical-microbiological properties of soils, as well as the toxicity phenomena in plants. Furthermore, some approaches and setups of cultivation practices aimed at mitigating the Cu accumulation problem (namely, rhizosphere management, biotechnologies and breeding for more resistant plant material, as well as precision/smart viticulture techniques for a timely, localized and balanced distribution of agrochemicals) were discussed. We summarized the current literature for all these topics and highlighted a set of research gaps that urgently require future studies and innovation. In particular, research is needed to shed light on the possible long-term impacts of Cu on selecting antibiotic-resistant bacterial species of agricultural soils and on their potential threats to animal and human health. Future studies should then focus on vineyard rhizosphere dynamics more holistically, i.e., considering the microorganisms and the effect of Cu stress on their composition and functionality, as well as relying on large-scale experiments at the field level. Third, the assessment of "new" resistant plant material (scions) should not be limited to the traits strictly connected with the expression of the pathogen resistance genes for overcoming current vine diseases, but should also consider putative new pathogens and related diseases. From this perspective, particular attention should also be paid to rootstocks with tolerance traits for high Cu concentrations in soils. Finally, future research is needed (1) to develop visions as well as practical architectures/applications for an integrated smart (or digital) management of the vineyard; (2) to identify the competencies needed by farmers and wine-growing consultants to use the technologies available correctly and effectively; and (3) to shed empirical light on the impact of smart viticulture on the three sustainability pillars (i.e., environmental, social and economic).

More generally, our review showed that the setting up of a more sustainable soil management in viticulture requires a multidisciplinary approach involving all the players throughout the whole production chain. In this context, the availability of plant material more resistant to pathogens and physiologically more suitable for the edaphic environment is essential. Moreover, Cu-induced antibiotic resistance in soil microorganisms highlights the relevance of the *One Health* concept, where the protection of plant, animal and human health is considered to be one. For instance, the frequent contamination of the water sources used for crop irrigation with synthetic compounds for human care (such as drugs like ibuprofen, [152] clearly indicates that the defense plan against pathogens of agricultural crops like vines using agrochemicals should not be considered independently of animal and human health, and vice versa. Actually, thanks to an approach like this, it is possible to effectively contribute to both the continuation of the viticulture (and more generally that of agricultural production) and the preservation of the environment as a whole, in particular the soil ecosystem.

Author Contributions: S.C., T.M. and G.O. conceived and designed the review; S.C., Y.P., L.B., G.O., P.L., F.M., G.G., M.S., G.B., R.T., G.V. and T.M. wrote the manuscript. All authors have read and agreed to the published version of the manuscript.

Funding: This research was supported by grants from Free University of Bolzano (TN5056, TN2612, TN2081). We acknowledge support by the unibz Open Access publishing fund, Free University of Bolzano.

Conflicts of Interest: The authors declare that the research was conducted in the absence of any commercial or financial relationships that could be construed as a potential conflict of interest.

References

1. ONU. *World Population Prospects 2019*; Department of Economic and Social Affairs: New York, NY, USA, 2019.
2. Meyers, W.H.; Kalaitzandonakes, N. World Population, Food Growth, and Food Security Challenges. *Front. Econ. Glob.* **2015**, *15*, 161–177. [CrossRef]
3. Conijn, J.G.; Bindraban, P.S.; Schröder, J.J.; Jongschaap, R.E.E. Can our global food system meet food demand within planetary boundaries? *Agric. Ecosyst. Environ.* **2018**, *251*, 244–256. [CrossRef]
4. Indoria, A.K.; Sharma, K.L.; Reddy, K.S. Hydraulic properties of soil under warming climate. *Clim. Chang. Soil Interact.* **2020**, 473–508.
5. González-Alemán, J.J.; Pascale, S.; Gutiérrez-Fernández, J.; Murakami, H.; Gaertner, M.Á.; Vecchi, G.A. Potential Increase in Hazard From Mediterranean Hurricane Activity With Global Warming. *Geophys. Res. Lett.* **2019**, *46*, 1754–1764. [CrossRef]
6. Koppelaar, R.H.; Weikard, H. Assessing phosphate rock depletion and phosphorus recycling options. *Glob. Environ. Chang.* **2013**, *23*, 1454–1466. [CrossRef]
7. Xie, H.; Huang, Y.; Chen, Q.; Zhang, Y.; Wu, Q. Prospects for Agricultural Sustainable Intensification: A Review of Research. *Land* **2019**, *8*, 157. [CrossRef]
8. Miotto, A.; Ceretta, C.A.; Brunetto, G.; Nicoloso, F.T.; Girotto, E.; Farias, J.G.; Tiecher, T.L.; De Conti, L.; Trentin, G. Copper uptake, accumulation and physiological changes in adult grapevines in response to excess copper in soil. *Plant Soil* **2013**, *374*, 593–610. [CrossRef]
9. Brunetto, G.; De Melo, G.W.B.; Terzano, R.; Del Buono, D.; Astolfi, S.; Tomasi, N.; Pii, Y.; Mimmo, T.; Cesco, S. Copper accumulation in vineyard soils: Rhizosphere processes and agronomic practices to limit its toxicity. *Chemosphere* **2016**, *162*, 293–307. [CrossRef]
10. Oorts, K. Copper. In *Heavy Metals in Soils: Trace Metals and Metalloids in Soils and Their Bioavailability*; Springer Science + Business Media: Dordrecht, The Netherlands, 2013; pp. 367–394.
11. Shabbir, Z.; Sardar, A.; Shabbir, A.; Abbas, G.; Shamshad, S.; Khalid, S.; Natasha, N.; Murtaza, G.; Dumat, C.; Shahid, M. Copper uptake, essentiality, toxicity, detoxification and risk assessment in soil-plant environment. *Chemosphere* **2020**, *259*, 127436. [CrossRef]
12. Blume, H.-P.; Brümmer, G.W.; Fleige, H.; Horn, R.; Kandeler, E.; Kögel-Knabner, I.; Kretzschmar, R.; Stahr, K., Wilke, B.-M. Chemical Properties and Processes. In *Scheffer/Schachtschabel Soil Science*; Springer Science and Business Media LLC: Berlin/Heidelberg, Germany, 2016; pp. 123–174.
13. Astolfi, S.; Pii, Y.; Mimmo, T.; Lucini, L.; Miras-Moreno, B.; Coppa, E.; Violino, S.; Celletti, S.; Cesco, S. Single and Combined Fe and S Deficiency Differentially Modulate Root Exudate Composition in Tomato: A Double Strategy for Fe Acquisition? *Int. J. Mol. Sci.* **2020**, *21*, 4038. [CrossRef]
14. Bernal, M.; Casero, D.; Singh, V.; Wilson, G.T.; Grande, A.; Yang, H.; Dodani, S.C.; Pellegrini, M.; Huijser, P.; Connolly, E.L.; et al. Transcriptome Sequencing Identifies SPL7-Regulated Copper Acquisition Genes FRO4/FRO5 and the Copper Dependence of Iron Homeostasis in Arabidopsis. *Plant Cell* **2012**, *24*, 738–761. [CrossRef] [PubMed]
15. Marastoni, L.; Sandri, M.; Pii, Y.; Valentinuzzi, F.; Brunetto, G.; Cesco, S.; Mimmo, T. Synergism and antagonisms between nutrients induced by copper toxicity in grapevine rootstocks: Monocropping vs. intercropping. *Chemosphere* **2019**, *214*, 563–578. [CrossRef]
16. Yuan, M.; Li, X.; Xiao, J.; Wang, S. Molecular and functional analyses of COPT/Ctr-type copper transporter-like gene family in rice. *BMC Plant Biol.* **2011**, *11*, 69. [CrossRef]
17. Korshunova, Y.O.; Eide, D.; Clark, W.G.; Guerinot, M.L.; Pakrasi, H.B. The IRT1 protein from Arabidopsis thaliana is a metal transporter with a broad substrate range. *Plant Mol. Biol.* **1999**, *40*, 37–44. [CrossRef]

18. Wintz, H.; Fox, T.; Wu, Y.-Y.; Feng, V.; Chen, W.; Chang, H.-S.; Zhu, T.; Vulpe, C. Expression Profiles of Arabidopsis thalianain Mineral Deficiencies Reveal Novel Transporters Involved in Metal Homeostasis. *J. Biol.* **2003**, *278*, 47644–47653.
19. Yruela, I. Copper in plants. *Braz. J. Plant Physiol.* **2005**, *17*, 145–156. [CrossRef]
20. Tsai, H.-H.; Schmidt, W. One way. Or another? Iron uptake in plants. *New Phytol.* **2017**, *214*, 500–505. [CrossRef]
21. Deng, F.; Yamaji, N.; Xia, J.; Ma, J.F. A Member of the Heavy Metal P-Type ATPase OsHMA5 Is Involved in Xylem Loading of Copper in Rice. *Plant Physiol.* **2013**, *163*, 1353–1362. [CrossRef]
22. DiDonato, R.J.; Roberts, L.A.; Sanderson, T.; Eisley, R.B.; Walker, E. Arabidopsis Yellow Stripe-Like2 (YSL2): A metal-regulated gene encoding a plasma membrane transporter of nicotianamine-metal complexes. *Plant J.* **2004**, *39*, 403–414. [CrossRef]
23. Irtelli, B.; Petrucci, W.A.; Navari-Izzo, F. Nicotianamine and histidine/proline are, respectively, the most important copper chelators in xylem sap of Brassica carinata under conditions of copper deficiency and excess. *J. Exp. Bot.* **2008**, *60*, 269–277. [CrossRef]
24. Liao, M.T.; Hedley, M.J.; Woolley, D.J.; Brooks, R.R.; Nichols, M.A. Copper uptake and translocation in chicory (Cichorium intybus L. cv. Grasslands Puna) and tomato (Lycopersicon esculentum Mill. cv. Rondy) plants grown in NFT system. I. Copper uptake and distribution in plants. *Plant Soil* **2000**, *221*, 135–142. [CrossRef]
25. Cao, Y.; Ma, C.; Chen, H.; Zhang, J.; White, J.C.; Chen, G.; Xing, B. Xylem-based long-distance transport and phloem remobilization of copper in Salix integra Thunb. *J. Hazard. Mater.* **2020**, *392*, 122428. [CrossRef] [PubMed]
26. Ryan, B.M.; Kirby, J.K.; Degryse, F.; Harris, H.; McLaughlin, M.J.; Scheiderich, K. Copper speciation and isotopic fractionation in plants: Uptake and translocation mechanisms. *New Phytol.* **2013**, *199*, 367–378. [CrossRef] [PubMed]
27. Printz, B.; Lutts, S.; Hausman, J.-F.; Sergeant, K. Copper Trafficking in Plants and Its Implication on Cell Wall Dynamics. *Front. Plant Sci.* **2016**, *7*, 601. [CrossRef]
28. Marschner, H.; Nutrition, M.; Plants, H.; Africa, W.; Marschner, P. *Marschner's Mineral Nutrition of Higher Plants*; Elsevier: Amsterdam, The Netherlands, 2012.
29. Taiz, L.; Zeiger, E. *Plant Physiology and Development*; Sinauer Associates, Inc.: Sunderland, MA, USA, 2015.
30. Kabata-Pendias, A. *Trace Elements in Soils and Plants*, 4th ed.; CRC Press: Boca Raton, FL, USA, 2010.
31. Tiecher, T.L.; Tiecher, T.; Ceretta, C.A.; Ferreira, P.A.; Nicoloso, F.T.; Soriani, H.H.; Tassinari, A.; Paranhos, J.T.; De Conti, L.; Brunetto, G. Physiological and nutritional status of black oat (Avena strigosa Schreb.) grown in soil with interaction of high doses of copper and zinc. *Plant Physiol. Biochem.* **2016**, *106*, 253–263. [CrossRef]
32. Ambrosini, V.G.; Rosa, D.; De Melo, G.W.B.; Zalamena, J.; Cella, C.; Simão, D.G.; Da Silva, L.S.; Dos Santos, H.P.; Toselli, M.; Tiecher, T.L.; et al. High copper content in vineyard soils promotes modifications in photosynthetic parameters and morphological changes in the root system of 'Red Niagara' plantlets. *Plant Physiol. Biochem.* **2018**, *128*, 89–98. [CrossRef]
33. Morgan, R.K.; Bowden, R. Copper accumulation in soils from two different-aged apricot orchards in Central Otago, New Zealand. *Int. J. Environ. Stud.* **1993**, *43*, 161–167. [CrossRef]
34. Novak, J.M.; Watts, D.; Stone, K.C. Copper and zinc accumulation, profile distribution, and crop removal in coastal plain soils receiving long-term, intensive applications of swine manure. *Trans. ASAE* **2004**, *47*, 1513–1522. [CrossRef]
35. Huang, S.; Jin, J.-Y. Status of heavy metals in agricultural soils as affected by different patterns of land use. *Environ. Monit. Assess.* **2007**, *139*, 317–327. [CrossRef]
36. Li, B.; Qiu, Y.; Song, Y.; Lin, H.; Yin, H. Dissecting horizontal and vertical gene transfer of antibiotic resistance plasmid in bacterial community using microfluidics. *Environ. Int.* **2019**, *131*, 105007. [CrossRef]
37. Horsfall, J.G. Fungicides and Their Action. *J. AOAC Int.* **1946**, *29*, 116–117. [CrossRef]
38. Millardet, A.; Gayon, U. *The Discovery of Bordeaux Mixture: Three Papers: I. Treatment of Mildew and Rot. II. Treatment of Mildew with Copper Sulphate and Lime Mixture. III. Concerning the History of the Treatment of Mildew with Copper Sulphate*; American Phytopathological Society: Saint Paul, MN, USA, 1933.
39. Mirlean, N.; Roisenberg, A.; Chies, J.O. Metal contamination of vineyard soils in wet subtropics (southern Brazil). *Environ. Pollut.* **2007**, *149*, 10–17. [CrossRef] [PubMed]
40. Fernández-Calviño, D.; Nóvoa-Muñoz, J.C.; Díaz-Raviña, M.; Arias-Estévez, M. Copper accumulation and fractionation in vineyard soils from temperate humid zone (NW Iberian Peninsula). *Geoderma* **2009**, *153*, 119–129. [CrossRef]
41. Mackie, K.; Müller, T.; Zikeli, S.; Kandeler, E. Long-term copper application in an organic vineyard modifies spatial distribution of soil micro-organisms. *Soil Biol. Biochem.* **2013**, *65*, 245–253. [CrossRef]
42. Ballabio, C.; Panagos, P.; Lugato, E.; Huang, J.-H.; Orgiazzi, A.; Jones, A.; Fernández-Ugalde, O.; Borrelli, P.; Montanarella, L. Copper distribution in European topsoils: An assessment based on LUCAS soil survey. *Sci. Total Environ.* **2018**, *636*, 282–298. [CrossRef]
43. Dagostin, S.; Schärer, H.-J.; Pertot, I.; Tamm, L. Are there alternatives to copper for controlling grapevine downy mildew in organic viticulture? *Crop. Prot.* **2011**, *30*, 776–788. [CrossRef]
44. Regulation 1981/2018. Renewing the Approval of the Active Substances Copper Compounds, as Candidates for Substitution, in Accordance with Regulation (EC) No 1107/2009 of the European Parliament and of the Council Concerning the Placing of Plant Protection Products on the Market, and Amending the Annex to Commission Implementing Regulation (EU) No 540/2011. European Commission, Council of the European Union. Available online: https://eur-lex.europa.eu/legal-content/EN/TXT/?uri=CELEX%3A32018R1981 (accessed on 9 January 2021).

45. La Torre, A.; Righi, L.; Iovino, V.; Battaglia, V. Control of late blight in organic farming with low copper dosages or natural products as alternatives to copper. *Eur. J. Plant Pathol.* **2019**, *155*, 769–778. [CrossRef]

46. Kopittke, P.M.; Menzies, N.W. Effect of Cu Toxicity on Growth of Cowpea (Vigna unguiculata). *Plant Soil* **2006**, *279*, 287–296. [CrossRef]

47. Guimarães, P.R.; Ambrosini, V.G.; Miotto, A.; Ceretta, C.A.; Simão, D.G.; Brunetto, G. Black Oat (Avena strigosa Schreb.) Growth and Root Anatomical Changes in Sandy Soil with Different Copper and Phosphorus Concentrations. *Water Air Soil Pollut.* **2016**, *227*, 1–10. [CrossRef]

48. Baldi, E.; Miotto, A.; Ceretta, C.A.; Quartieri, M.; Sorrenti, G.; Brunetto, G.; Toselli, M. Soil-applied phosphorous is an effective tool to mitigate the toxicity of copper excess on grapevine grown in rhizobox. *Sci. Hortic.* **2018**, *227*, 102–111. [CrossRef]

49. Feil, S.B.; Pii, Y.; Valentinuzzi, F.; Tiziani, R.; Mimmo, T.; Cesco, S. Copper toxicity affects phosphorus uptake mechanisms at molecular and physiological levels in Cucumis sativus plants. *Plant Physiol. Biochem.* **2020**, *157*, 138–147. [CrossRef] [PubMed]

50. Del Buono, D.; Terzano, R.; Panfili, I.; Bartucca, M.L. Phytoremediation and detoxification of xenobiotics in plants: Herbicide-safeners as a tool to improve plant efficiency in the remediation of polluted environments. A mini-review. *Int. J. Phytoremediation* **2020**, *22*, 789–803. [CrossRef] [PubMed]

51. International Organization of Vine and Wine, 2020. Current Situation Of The Vitivinicultural Sector At A Global Level. Available online: http://www.oiv.int/js/lib/pdfjs/web/viewer.html?file=/public/medias/7260/en-oiv-press-conference-april-2020-press-release.pdf (accessed on 18 November 2020).

52. Benoît, L.; William, A.; Lindsey, H.; Adrienne, L.F.; Marianne, M.W. *Wine Sector: Definitions and Nuances from Global to Country Analysis—A Comparison between Old World, New World, and Emerging Wine Countries from 2005 to Current*; Elsevier BV: Amsterdam, The Netherlands, 2019; pp. 7–32.

53. European Commission. Council Regulation (EC) No 479/2008. *Off. J. Eur. Union.* **2008**, *148*, 1–61.

54. European Parliament and of the Council. EU regulation No 1308/2013. *Off. J. Eur. Communities* **2008**, *347*, 1–22.

55. Meloni, G.; Anderson, K.; Deconinck, K.; Swinnen, J. Wine Regulations. *Appl. Econ. Perspect. Policy* **2019**, *41*, 620–649. [CrossRef]

56. Jones, G.V.; Webb, L.B. Climate Change, Viticulture, and Wine: Challenges and Opportunities. *J. Wine Res.* **2010**, *21*, 103–106. [CrossRef]

57. Vigl, L.E.; Schmid, A.; Moser, F.; Balotti, A.; Gartner, E.; Katz, H.; Quendler, S.; Ventura, S.; Raifer, B. Upward shifts in elevation—a winning strategy for mountain viticulture in the context of climate change? *E3S Web Conf.* **2018**, *50*, 02006. [CrossRef]

58. Couto, R.D.R.; Benedet, L.; Comin, J.J.; Filho, P.B.; Martins, S.R.; Gatiboni, L.C.; Radetski, M.; De Valois, C.M.; Ambrosini, V.G.; Brunetto, G. Accumulation of copper and zinc fractions in vineyard soil in the mid-western region of Santa Catarina, Brazil. *Environ. Earth Sci.* **2014**, *73*, 6379–6386. [CrossRef]

59. Carlon, C. Derivation methods of soil screening values in Europe: A review and evaluation of national procedures towards harmonization. In *JRC Scientific and Technical Reports*; Office for Official Publications of the European Communities: Luxembourg, 2007.

60. Komárek, M.; Čadková, E.; Chrastný, V.; Bordas, F.; Bollinger, J.-C. Contamination of vineyard soils with fungicides: A review of environmental and toxicological aspects. *Environ. Int.* **2010**, *36*, 138–151. [CrossRef]

61. Mackie, K.; Müller, T.; Kandeler, E. Remediation of copper in vineyards—A mini review. *Environ. Pollut.* **2012**, *167*, 16–26. [CrossRef]

62. Italian Ministry for Environment Land and Sea Protection. *DECRETO 1 Marzo 2019, n. 46*; Italian Ministry for Environment Land and Sea Protection: Rome, Italy, 2019.

63. Vácha, R.; Sanka, M.; Hauptman, I.; Zimova, M.; Čechmánková, J. Assessment of limit values of risk elements and persistent organic pollutants in soil for Czech legislation. *Plant Soil Environ.* **2014**, *60*, 191–197. [CrossRef]

64. Dykhuizen, D. Species Numbers in Bacteria. *Proc. Calif. Acad. Sci.* **2005**, *56*, 62–71.

65. Schmidt, T.; Waldron, C. Microbial Diversity in Soils of Agricultural Landscapes and Its Relation to Ecosystem Function. In *The Ecology of Agricultural Landscapes: Long-term Research on the Path to Sustainability*; Oxford University Press: New York, NY, USA, 2015; pp. 135–157.

66. Bennett, A.E.; Classen, A.T. Climate change influences mycorrhizal fungal–plant interactions, but conclusions are limited by geographical study bias. *Ecology* **2020**, *101*, e02978. [CrossRef]

67. Pii, Y.; Mimmo, T.; Tomasi, N.; Terzano, R.; Cesco, S.; Crecchio, C. Microbial interactions in the rhizosphere: Beneficial influences of plant growth-promoting rhizobacteria on nutrient acquisition process. A review. *Biol. Fertil. Soils* **2015**, *51*, 403–415. [CrossRef]

68. Rolli, E.; Marasco, R.; Saderi, S.; Corretto, E.; Mapelli, F.; Cherif, A.; Borin, S.; Valenti, L.; Sorlini, C.; Daffonchio, D. Root-associated bacteria promote grapevine growth: From the laboratory to the field. *Plant Soil* **2016**, *410*, 369–382. [CrossRef]

69. Abdu, N.; Abdullahi, A.A.; Abdulkadir, A. Heavy metals and soil microbes. *Environ. Chem. Lett.* **2017**, *15*, 65–84. [CrossRef]

70. Dell'Amico, E.; Mazzocchi, M.; Cavalca, L.; Allievi, L.; Andreoni, V. Assessment of bacterial community structure in a long-term copper-polluted ex-vineyard soil. *Microbiol. Res.* **2008**, *163*, 671–683. [CrossRef]

71. Giller, K.E.; Witter, E.; McGrath, S.P. Heavy metals and soil microbes. *Soil Biol. Biochem.* **2009**, *41*, 2031–2037. [CrossRef]

72. Lejon, D.P.; Pascault, N.; Ranjard, L. Differential copper impact on density, diversity and resistance of adapted culturable bacterial populations according to soil organic status. *Eur. J. Soil Biol.* **2010**, *46*, 168–174. [CrossRef]

73. Borruso, L.; Zerbe, S.; Brusetti, L. Bacterial community structures as a diagnostic tool for watershed quality assessment. *Res. Microbiol.* **2015**, *166*, 38–44. [CrossRef]

74. Cavani, L.; Manici, L.M.; Caputo, F.; Peruzzi, E.; Ciavatta, C. Ecological restoration of a copper polluted vineyard: Long-term impact of farmland abandonment on soil bio-chemical properties and microbial communities. *J. Environ. Manag.* **2016**, *182*, 37–47. [CrossRef]

75. Vest, K.E.; Zhu, X.; Cobine, P.A. Copper Disposition in Yeast. In *Clinical and Translational Perspectives on WILSON DISEASE*; Elsevier BV: Amsterdam, The Netherlands, 2019; pp. 115–126.

76. Winge, D.R.; Nielson, K.B.; Gray, W.R.; Hamer, D.H. Yeast metallothionein. *J. Biol. Chem.* **1985**, *260*, 14464–14470. [CrossRef]

77. Culotta, V.C.; Howard, W.R.; Liu, X.F. CRS5 encodes a metallothionein-like protein in Saccharomyces cerevisiae. *J. Biol. Chem.* **1994**, *269*, 25295–25302. [CrossRef]

78. Jensen, L.T.; Howard, W.R.; Strain, J.J.; Winge, D.R.; Culotta, V.C. Enhanced Effectiveness of Copper Ion Buffering by CUP1 Metallothionein Compared with CRS5 Metallothionein inSaccharomyces cerevisiae. *J. Biol. Chem.* **1996**, *271*, 18514–18519. [CrossRef]

79. Gerwien, F.; Skrahina, V.; Kasper, L.; Hube, B.; Brunke, S. Metals in fungal virulence. *FEMS Microbiol. Rev.* **2018**, *42*, 1–21. [CrossRef]

80. Carroll, M.C.; Girouard, J.B.; Ulloa, J.L.; Subramaniam, J.R.; Wong, P.C.; Valentine, J.S.; Culotta, V.C. Mechanisms for activating Cu- and Zn-containing superoxide dismutase in the absence of the CCS Cu chaperone. *Proc. Natl. Acad. Sci. USA* **2004**, *101*, 5964–5969. [CrossRef]

81. Raffa, N.; Osherov, N.; Keller, N.P. Copper Utilization, Regulation, and Acquisition by Aspergillus fumigatus. *Int. J. Mol. Sci.* **2019**, *20*, 1980. [CrossRef]

82. Solioz, M. *Copper Disposition in Bacteria*; Elsevier BV: Amsterdam, The Netherlands, 2019; pp. 101–113.

83. Outten, F.W.; Outten, C.E.; Hale, J.; O'Halloran, T.V. Transcriptional Activation of anEscherichia coliCopper Efflux Regulon by the Chromosomal MerR Homologue, CueR. *J. Biol. Chem.* **2000**, *275*, 31024–31029. [CrossRef]

84. Djoko, K.Y.; Xiao, Z.; Wedd, A.G. Copper Resistance in *E. coli*: The Multicopper Oxidase PcoA Catalyzes Oxidation of Copper(I) in CuICuII-PcoC. *ChemBioChem* **2008**, *9*, 1579–1582. [CrossRef]

85. Rademacher, C.; Masepohl, B. Copper-responsive gene regulation in bacteria. *Microbiology* **2012**, *158*, 2451–2464. [CrossRef]

86. Berg, J.; Tom-Petersen, A.; Nybroe, O. Copper amendment of agricultural soil selects for bacterial antibiotic resistance in the field. *Lett. Appl. Microbiol.* **2005**, *40*, 146–151. [CrossRef]

87. Borruso, L.; Harms, K.; Johnsen, P.J.; Nielsen, K.M.; Brusetti, L. Distribution of class 1 integrons in a highly impacted catchment. *Sci. Total Environ.* **2016**, *566–567*, 1588–1594. [CrossRef]

88. Hao, X.; Luthje, F.L.; Qin, Y.; McDevitt, S.F.; Lutay, N.; Hobman, J.L.; Asiani, K.; Soncini, F.C.; German, N.; Zhang, S.; et al. Survival in amoeba—a major selection pressure on the presence of bacterial copper and zinc resistance determinants? Identification of a "copper pathogenicity island". *Appl. Microbiol. Biotechnol.* **2015**, *99*, 5817–5824.

89. Shan, Y.; Tysklind, M.; Hao, F.; Ouyang, W.; Chen, S.; Lin, C. Identification of sources of heavy metals in agricultural soils using multivariate analysis and GIS. *J. Soils Sediments* **2013**, *13*, 720–729. [CrossRef]

90. Zhang, S.; Abbas, M.; Rehman, M.U.; Huang, Y.; Zhou, R.; Gong, S.; Yang, H.; Chen, S.; Wang, M.; Cheng, A. Dissemination of antibiotic resistance genes (ARGs) via integrons in Escherichia coli: A risk to human health. *Environ. Pollut.* **2020**, *266*, 115260. [CrossRef]

91. Baker-Austin, C.; Wright, M.S.; Stepanauskas, R.; McArthur, J. Co-selection of antibiotic and metal resistance. *Trends Microbiol.* **2006**, *14*, 176–182. [CrossRef]

92. Hu, H.H.; Wang, J.; Li, J.; Li, J.-J.; Ma, Y.-B.; Chen, D.; He, J. Field-based evidence for copper contamination induced changes of antibiotic resistance in agricultural soils. *Environ. Microbiol.* **2016**, *18*, 3896–3909. [CrossRef]

93. Tomasi, N.; Kretzschmar, T.; Espen, L.; Weisskopf, L.; Fuglsang, A.T.; Palmgren, M.G.; Neumann, G.; Varanini, Z.; Pinton, R.; Martinoia, E.; et al. Plasma membrane H+-ATPase-dependent citrate exudation from cluster roots of phosphate-deficient white lupin. *Plant Cell Environ.* **2009**, *32*, 465–475. [CrossRef]

94. Tomasi, N.; Mimmo, T.; Terzano, R.; Alfeld, M.; Janssens, K.; Zanin, L.; Pinton, R.; Varanini, Z.; Cesco, S. Nutrient accumulation in leaves of Fe-deficient cucumber plants treated with natural Fe complexes. *Biol. Fertil. Soils* **2014**, *50*, 973–982. [CrossRef]

95. Bravin, M.N.; Martí, A.L.; Clairotte, M.; Hinsinger, P. Rhizosphere alkalisation—a major driver of copper bioavailability over a broad pH range in an acidic, copper-contaminated soil. *Plant Soil* **2009**, *318*, 257–268. [CrossRef]

96. Faget, M.; Blossfeld, S.; Von Gillhaussen, P.; Schurr, U.; Temperton, V.M. Disentangling who is who during rhizosphere acidification in root interactions: Combining fluorescence with optode techniques. *Front. Plant Sci.* **2013**, *4*, 1–9. [CrossRef]

97. Leitenmaier, B.; Kupper, H. Compartmentation and complexation of metals in hyperaccumulator plants. *Front. Plant Sci.* **2013**, *4*, 374. [CrossRef]

98. Dell'Orto, M.; Santi, S.; Cesco, S.; Varanini, Z.; Zocchi, G.; De Nisi, P.; Pinton, R. Development of Fe?deficiency responses in cucumber (Cucumis sativus L.) roots: Involvement of plasma membrane H + ?ATPase activity. *J. Exp. Bot.* **2000**, *51*, 695–701. [CrossRef]

99. Earmijo, G.; Eschlechter, R.; Eagurto, M.; Emuñoz, D.; Enúñez, C.; Arce-Johnson, P. Grapevine Pathogenic Microorganisms: Understanding Infection Strategies and Host Response Scenarios. *Front. Plant Sci.* **2016**, *7*, 382. [CrossRef]

100. Zendler, D.; Schneider, P.; Töpfer, R.; Zyprian, E. Fine mapping of Ren3 reveals two loci mediating hypersensitive response against Erysiphe necator in grapevine. *Euphytica* **2017**, *213*, 1–23. [CrossRef]

101. Gessler, C.; Pertot, I.; Perazzolli, M. Plasmopara viticola: A review of knowledge on downy mildew of grapevine and effective disease management. *Phytopathol. Mediterr.* **2011**, *50*, 3–44. [CrossRef]

102. Zini, E.; Dolzani, C.; Stefanini, M.; Gratl, V.; Bettinelli, P.; Nicolini, D.; Betta, G.; Dorigatti, C.; Velasco, R.; Letschka, T.; et al. R-Loci Arrangement Versus Downy and Powdery Mildew Resistance Level: A Vitis Hybrid Survey. *Int. J. Mol. Sci.* **2019**, *20*, 3526. [CrossRef]

103. Di Gaspero, G.; Foria, S. *Molecular Grapevine Breeding Techniques*; Elsevier BV: Amsterdam, The Netherlands, 2015; pp. 23–37.

104. Bove, F.; Rossi, V. Components of partial resistance to Plasmopara viticola enable complete phenotypic characterization of grapevine varieties. *Sci. Rep.* **2020**, *10*, 1–12. [CrossRef]

105. Qiu, W.; Feechan, A.; Dry, I.B. Current understanding of grapevine defense mechanisms against the biotrophic fungus (Erysiphe necator), the causal agent of powdery mildew disease. *Hortic. Res.* **2015**, *2*, 15020. [CrossRef]

106. Giacomelli, L.; Zeilmaker, T.; Malnoy, M.; Van Der Voort, J.R.; Moser, C. Generation of mildew-resistant grapevine clones via genome editing. *Acta Hortic.* **2019**, *1248*, 195–200. [CrossRef]

107. Dalla Costa, L.; Malnoy, M.; Lecourieux, D.; Deluc, L.; Lecourieux, F.O.-; Thomas, M.R.; Torregrosa, L. The state-of-the-art of grapevine biotechnology and new breeding technologies (NBTS). *OENO One* **2019**, *53*, 189–212. [CrossRef]

108. Feechan, A.; Jermakow, A.M.; Torregrosa, L.; Panstruga, R.; Dry, I.B. Identification of grapevine MLO gene candidates involved in susceptibility to powdery mildew. *Funct. Plant Biol.* **2008**, *35*, 1255–1266. [CrossRef]

109. Winterhagen, P.; Howard, S.F.; Qiu, W.; Kovács, L.G. Transcriptional Up-Regulation of Grapevine MLO Genes in Response to Powdery Mildew Infection. *Am. J. Enol. Vitic.* **2008**, *59*, 159–168.

110. Pirrello, C.; Zeilmaker, T.; Bianco, L.; Giacomelli, L.; Moser, C.; Vezzulli, S. Mining downy mildew susceptibility genes: A diversity study in grapevine. *bioRxiv* **2020**. [CrossRef]

111. Cesco, S.; Tolotti, A.; Nadalini, S.; Rizzi, S.; Valentinuzzi, F.; Mimmo, T.; Porfido, C.; Allegretta, I.; Giovannini, O.; Perazzolli, M.; et al. Plasmopara viticola infection affects mineral elements allocation and distribution in Vitis vinifera leaves. *Sci. Rep.* **2020**, *10*, 1–18. [CrossRef]

112. Chitarrini, G.; Riccadonna, S.; Zulini, L.; Vecchione, A.; Stefanini, M.; Larger, S.; Pindo, M.; Cestaro, A.; Franceschi, P.; Magris, G.; et al. Two-omics data revealed commonalities and differences between Rpv12- and Rpv3-mediated resistance in grapevine. *Sci. Rep.* **2020**, *10*, 1–15. [CrossRef]

113. Malacarne, G.; Vrhovsek, U.; Zulini, L.; Cestaro, A.; Stefanini, M.; Mattivi, F.; Delledonne, M.; Velasco, R.; Moser, C. Resistance to Plasmopara viticola in a grapevine segregating population is associated with stilbenoid accumulation and with specific host transcriptional responses. *BMC Plant Biol.* **2011**, *11*, 114. [CrossRef]

114. Badalucco, L.; Nannipieri, P. Nutrient transformations in the rhizosphere. In *The Rhizosphere Biochemistry and Organic Substances at the Soil-plant Interface*; CRC Press: Boca Raton, FL, USA, 2007; Volume 11, pp. 111–133.

115. Terrazas, R.A.; Giles, C.; Paterson, E.; Robertson-Albertyn, S.; Cesco, S.; Mimmo, T.; Pii, Y.; Bulgarelli, D. Plant–Microbiota Interactions as a Driver of the Mineral Turnover in the Rhizosphere. *Adv. Clin. Chem.* **2016**, *95*, 1–67. [CrossRef]

116. Pii, Y.; Borruso, L.; Brusetti, L.; Crecchio, C.; Cesco, S.; Mimmo, T. The interaction between iron nutrition, plant species and soil type shapes the rhizosphere microbiome. *Plant Physiol. Biochem.* **2016**, *99*, 39–48. [CrossRef]

117. Mimmo, T.; Pii, Y.; Valentinuzzi, F.; Astolfi, S.; Lehto, N.; Robinson, B.; Brunetto, G.; Terzano, R.; Cesco, S. Nutrient availability in the rhizosphere: A review. *Acta Hortic.* **2018**, *1217*, 13–28. [CrossRef]

118. Bayer, P.E.; Edwards, D. Machine learning in agriculture: From silos to marketplaces. *Plant Biotechnol. J.* **2020**. [CrossRef]

119. Mulla, D.J. Spatial Variability in Precision Agriculture. In *Encyclopedia of GIS*; Springer Science and Business Media LLC: Berlin/Heidelberg, Germany, 2016; pp. 1–8.

120. Matese, A.; Di Gennaro, S.F. Technology in precision viticulture: A state of the art review. *Int. J. Wine Res.* **2015**, *7*, 69–81. [CrossRef]

121. Landonio, S.; (ARVAtec srl, Rescaldina, Italy). Personal Communication, 2020.

122. Maheswari, R.; Ashok, K.R.; Prahadeeswaran, M. Precision farming technology, adoption decisions and produc-tivity of vegetables in resource-poor environments. *Agric. Econ. Res. Rev.* **2008**, *21*, 415–424.

123. McBride, W.D.; Daberkow, S.G. Information and the adoption of precision farming technologies. *J. Agri-Bus.* **2003**, *21*, 21–38.

124. Kutter, T.; Tiemann, S.; Siebert, R.; Fountas, S. The role of communication and co-operation in the adoption of precision farming. *Precis. Agric.* **2011**, *12*, 2–17. [CrossRef]

125. Vecchio, Y.; Agnusdei, G.P.; Miglietta, P.P.; Capitanio, F. Adoption of Precision Farming Tools: The Case of Italian Farmers. *Int. J. Environ. Res. Public Health* **2020**, *17*, 869. [CrossRef]

126. Rodríguez-Molano, J.I.; Triana-Casallas, J.A.; Contreras-Bravo, L.E. Modeling and Simulation of Integration of Internet of Things and Manufacturing Industry 4.0. *Programmieren für Ingenieure und Naturwissenschaftler* **2018**, *916*, 231–241. [CrossRef]

127. Shamim, S.; Cang, S.; Yu, H.; Li, Y. Management approaches for Industry 4.0: A human resource management perspective. In Proceedings of the 2016 IEEE Congress on Evolutionary Computation (CEC), Vancouver, BC, Canada, 24–29 July 2016; pp. 5309–5316.

128. Wan, J.; Cai, H.; Zhou, K. Industrie 4.0: Enabling technologies. In Proceedings of the 2015 International Conference on Intelligent Computing and Internet of Things, Harbin, China, 17–18 January 2015; pp. 135–140.

129. Mazzetto, F.; Gallo, R.; Riedl, M.; Sacco, P. Proposal of an ontological approach to design and analyse farm information systems to support Precision Agriculture techniques. *IOP Conf. Ser. Earth Environ. Sci.* **2019**, *275*, 012008. [CrossRef]

130. Balafoutis, A.T.; Koundouras, S.; Anastasiou, E.; Fountas, S.; Arvanitis, K. Life Cycle Assessment of Two Vineyards after the Application of Precision Viticulture Techniques: A Case Study. *Sustainability* **2017**, *9*, 1997. [CrossRef]
131. Brunori, E.; Maesano, M.; Moresi, F.V.; Scarascia Mugnozza, G.; Biasi, R. Towards sustainable viticulture: Key role of vineyard's precision monitoring. In Proceedings of the International Symposium on Technologies for Smart City, Malaga, Spain, 11–12 November 2019; p. 6.
132. Kuflik, T.; Prodorutti, D.; Frizzi, A.; Gafni, Y.; Simon, S.; Pertot, I. Optimization of copper treatments in organic viticulture by using a web-based decision support system. *Comput. Electron. Agric.* **2009**, *68*, 36–43. [CrossRef]
133. Román, C.; Llorens, J.; Uribeetxebarria, A.; Sanz, R.; Planas, S.; Arnó, J. Spatially variable pesticide application in vineyards: Part II, field comparison of uniform and map-based variable dose treatments. *Biosyst. Eng.* **2020**, *195*, 42–53. [CrossRef]
134. Lindblom, J.; Lundström, C.; Ljung, M.; Jonsson, A. Promoting sustainable intensification in precision agriculture: Review of decision support systems development and strategies. *Precis. Agric.* **2017**, *18*, 309–331. [CrossRef]
135. Santesteban, L. Precision viticulture and advanced analytics. A short review. *Food Chem.* **2019**, *279*, 58–62. [CrossRef]
136. Zarco-Tejada, P.J.; Hubbard, N.; Loudjani, P. *Precision Agriculture: An Opportunity for eu Farmers—Potential Support with the Cap 2014–2020*; Joint Research Centre (JRC) of the European Commission: Ispra, Italy, 2014.
137. Mazzetto, F.; Calcante, A.; Mena, A.; Vercesi, A. Integration of optical and analogue sensors for monitoring canopy health and vigour in precision viticulture. *Precis. Agric.* **2010**, *11*, 636–649. [CrossRef]
138. Vidoni, R.; Gallo, R.; Ristorto, G.; Carabin, G.; Mazzetto, F.; Scalera, L.; Gasparetto, A. ByeLab: An agricultural mobile robot prototype for proximal sensing and precision farming. In Proceedings of the ASME International Mechanical Engineering Congress and Exposition, Tampa, FL, USA, 3–9 November 2017.
139. Mazzetto, F.; Gallo, R.; Importuni, P.; Petrera, S.; Sacco, P. Automatic filling of field activities register, from challenge into reality. *Chem. Eng. Trans.* **2017**, *58*, 667–672. [CrossRef]
140. Lee, S.-G.; Yang, A.; Jeon, B.-H.; Park, H.-D. A structure of scalable and configurable interface for sensor and actuator devices in smart farming system. *Int. J. Inn. Tech. Exp. Eng.* **2019**, *8*, 2779–2786.
141. Pérez-Expósito, J.P.; Fernández-Caramés, T.M.; Fraga-Lamas, P.; Castedo, L. VineSens: An Eco-Smart Decision-Support Viticulture System. *Sensors* **2017**, *17*, 465. [CrossRef]
142. Fuentes, S.; Tongson, E. Advances and requirements for machine learning and artificial intelligence applications in viticulture. *Wine Vitic. J.* **2018**, *33*, 3–47.
143. De Mauro, A.; Greco, M.; Grimaldi, M. A formal definition of Big Data based on its essential features. *Libr. Rev.* **2016**, *65*, 122–135. [CrossRef]
144. Delgado, J.A.; Short, N.M.; Roberts, D.P.; Vandenberg, B. Big Data Analysis for Sustainable Agriculture on a Geospatial Cloud Framework. *Front. Sustain. Food Syst.* **2019**, *3*. [CrossRef]
145. Lioutas, E.D.; Charatsari, C.; La Rocca, G.; De Rosa, M. Key questions on the use of big data in farming: An activity theory approach. *NJAS Wagenng. J. Life Sci.* **2019**, *90*, 100297. [CrossRef]
146. Penn, C. Twelve innovations from Vinitech 2018. *Aust. N. Z. Grapegrow. Winemak.* **2019**, *660*, 56.
147. Siebers, M.H.; Edwards, E.J.; Jimenezberni, J.A.; Thomas, M.R.; Salim, M.; Walker, R. Fast Phenomics in Vineyards: Development of GRover, the Grapevine Rover, and LiDAR for Assessing Grapevine Traits in the Field. *Sensors* **2018**, *18*, 2924. [CrossRef]
148. Mendes, J.; Pinho, T.M.; Dos Santos, F.N.; Sousa, J.J.; Peres, E.; Cunha, J.B.; Cunha, M.; Morais, R. Smartphone Applications Targeting Precision Agriculture Practices—A Systematic Review. *Agronomy* **2020**, *10*, 855. [CrossRef]
149. Caffaro, F.; Cremasco, M.M.; Roccato, M.; Cavallo, E. Drivers of farmers' intention to adopt technological innovations in Italy: The role of information sources, perceived usefulness, and perceived ease of use. *J. Rural. Stud.* **2020**, *76*, 264–271. [CrossRef]
150. Tey, Y.S.; Brindal, M. Factors influencing the adoption of precision agricultural technologies: A review for policy implications. *Precis. Agric.* **2012**, *13*, 713–730. [CrossRef]
151. Bai, C.; Ciano, M.P.; Orzes, G.; Sarkis, J. Industry 4.0 technologies assessment: A sustainability perspective. *Int. J. Prod. Econ.* **2020**, *229*, 107776. [CrossRef]
152. Luo, Y.; Guo, W.; Ngo, H.H.; Nghiem, L.D.; Hai, F.I.; Zhang, J.; Liang, S.; Wang, X.C. A review on the occurrence of micropollutants in the aquatic environment and their fate and removal during wastewater treatment. *Sci. Total Environ.* **2014**, *473–474*, 619–641. [CrossRef]

applied sciences

Article

Recycling Biogas Digestate from Energy Crops: Effects on Soil Properties and Crop Productivity

Roberta Pastorelli [1], Giuseppe Valboa [1], Alessandra Lagomarsino [1], Arturo Fabiani [1], Stefania Simoncini [1,†], Massimo Zaghi [2] and Nadia Vignozzi [1,*]

1   Research Centre Agriculture and Environment, Consiglio per la Ricerca in Agricoltura e L'analisi Dell'economia Agraria (CREA-AA), Via di Lanciola, 12/A, 50125 Firenze, Italy; roberta.pastorelli@crea.gov.it (R.P.); giuseppe.valboa@crea.gov.it (G.V.); alessandra.lagomarsino@crea.gov.it (A.L.); arturo.fabiani@crea.gov.it (A.F.); stefania.simoncini@crea.gov.it (S.S.)
2   Cooperativa Agroenergetica Territoriale (CAT), Via Fossa Faiella 6/A, 42015 Correggio (Reggio Emilia), Italy; massimo.zaghi@libero.it
*   Correspondence: nadia.vignozzi@crea.gov.it; Tel.: +39-055-2492254
†   Current address: Research Centre Plant Protection and Certification, Consiglio per la Ricerca in Agricoltura e L'analisi Dell'economia Agraria (CREA-DC), Via di Lanciola, 12/A, 50125 Firenze, Italy.

Abstract: Digestate from biogas production can be recycled to the soil as conditioner/fertilizer improving the environmental sustainability of the energy supply chain. In a three-year maize-triticale rotation, we investigated the short-term effects of digestate on soil physical, chemical, and microbiological properties and evaluated its effectiveness in complementing the mineral fertilizers. Digestate soil treatments consisted of combined applications of the whole digestate and its mechanically separated solid fraction. Digestate increased soil total organic C, total N and K contents. Soil bulk density was not affected by treatments, while aggregate stability showed a transient improvement due to digestate treatments. A decrement of the transmission pores proportion and an increment of fissures was observed in digestate treated soils. Soil microbial community was only transiently affected by digestate treatments and no soil contamination from Clostridiaceae-related bacteria were observed. Digestate can significantly impair seed germination when applied at low dilution ratios. Crop yield under digestate treatment was similar to ordinary mineral-based fertilization. Overall, our experiment proved that the agronomic recycling of digestate from biogas production maintained a fair crop yield and soil quality. Digestate was confirmed as a valid resource for sustainable management of soil fertility under energy-crop farming, by combining a good attitude as a fertilizer with the ability to compensate for soil organic C loss.

Keywords: anaerobic digestion residues; soil amendment; soil fertilization; soil organic C; soil porosity; soil microbial community

Citation: Pastorelli, R.; Valboa, G.; Lagomarsino, A.; Fabiani, A.; Simoncini, S.; Zaghi, M.; Vignozzi, N. Recycling Biogas Digestate from Energy Crops: Effects on Soil Properties and Crop Productivity. *Appl. Sci.* **2021**, *11*, 750. https://doi.org/10.3390/app11020750

Received: 21 December 2020
Accepted: 11 January 2021
Published: 14 January 2021

Publisher's Note: MDPI stays neutral with regard to jurisdictional claims in published maps and institutional affiliations.

1. Introduction

Interest in biogas production has grown significantly in the past two decades, following the need to reduce fossil fuel consumption in favour of renewable energy sources. To encourage biogas market penetration, EU policy issued financial incentives [1] which have led to a significant increase in the number of biogas plants. More than 18,000 biogas plants were registered by October 2020 [2] with an overall installed electric capacity (IEC) of 13,520 MW estimated at the end of 2019 [3]. Currently, in Europe, Italy and Germany rank first in terms of the number of active biogas plants, with most Italian plants (1900 units with an IEC of around 1000 MW) located in the Po Valley and other northern regions [4,5].

Biogas production from anaerobic digestion mainly relies on four types of biomass sources: (i) biomass wastes from farms (animal slurries and crop residues) and households (municipal solid waste and food waste); (ii) agro-industrial by-products; (iii) sewage sludges; (iv) biomass from dedicated energy crops [6].

The energy derived from anaerobic digestion is considered to be almost "carbon-neutral" and to bring environmental and social benefits, contributing to a reduction of greenhouse gas emissions (allowed by replacement of fossil fuels) and organic wastes [7], and supporting rural development and new employment opportunities [8]. Against these benefits, biogas production from energy crops generates several issues and conflicts that are under the political attention on a world scale, since the shift of farmland to non-food systems creates doubts concerning the security of food supply and the environmental impact of energy crops cultivation [9]. One main concern is the environmental sustainability of energy crop cultivation as large amounts of organic matter and plant nutrients are removed with the crop biomass from the field. Depletion of organic matter and plant nutrients from the agricultural system can lead to soil degradation if not balanced by appropriate replenishments. Secondly, since the number of biogas plants in many European countries has increased significantly in recent years, the disposal of residues from anaerobic digestion has become of growing concern [10]. From a sustainability perspective of the biogas supply chain, since a wide range of undecomposed organic compounds and plant nutrients removed from the field (mainly ammonia and phosphate) are retained in the digestate [11–14], the direct land application of digestate is an economical option for residues disposal and soil amendment/fertilization [15–17]. The risk of a potential transfer of organic pollutants, such as herbicides and fungicides, from digestate to rotational crops and feedstuffs is considered very low by the European Food Standards Agency [16].

A third concern is that energy crops require resources (land, water and energy) which inevitably become no longer available for food production [8,9,18,19]. For cereal crop-based productions, the "food vs. fuel" conflict would be overcome if the grain was excluded from the biogas feedstock and used for livestock feed, while in general, the conflict would not exist at all if energy-crop cultivation was carried out on soils unsuitable for food production (marginal land) [20]. In this context, energy crop farming is an effective and profitable strategy to prevent the land from abandonment and degradation while promoting rural investments and new job opportunities [8].

Digestate can be applied to the soil without further processing (whole digestate, WD) [17] or after mechanical separation to obtain a solid fibrous material (solid digestate, SD) which can be directly spread to the field, composted, or dried for intermediate storage and transport [17,21]. Both WD and SD are sources of organic carbon and plant nutrients but since they exhibit quantitative and qualitative differences, they are expected to contribute differently to soil organic matter turnover [22], plant nutrients availability, and soil physical properties [23]. Typically, SD exhibits a great percentage (38–75%) of highly stable organic matter and a low NH_4-N to total-N (TN) ratio [23], which make it suitable for use as a soil conditioner rather than as a source of readily available N. The use of digestate as a soil amendment can contribute to soil carbon sequestration, especially in intensively cultivated soils where crop residues are removed [24]. Organic matter addition is beneficial to soil fertility, since it may improve soil structure, increase plant nutrient retention, and water holding capacity and stimulate microbial activity [25]. A higher microbial activity, in turn, may enhance the release of plant nutrients from added residues and soil organic matter itself [26]. Conversely, the low organic matter concentration and the high NH_4-N/TN ratio in WD makes it more suitable for use as an N-fertilizer [22,23]. The efficiency of digestate as N fertilizer changes with the features of digestate itself, soil type, crop and time of spreading [4]. Like any other fertilizer, WD should be applied at appropriate rates and times during the crop growing season, to ensure optimum plant nutrient uptake and to avoid phytotoxic effect and pollution of groundwater [16].

Research on digestate suitability for land application is relatively recent and is focused, on the one hand, on agricultural benefits of digestate as soil fertilizer and/or improver, and on the other hand, on the environmental risks associated with digestate use. Overall, many studies have investigated the potential of digestate as N fertilizer and the fate of N in the soil after land application [27,28], as well as the effect of digestate on soil organic matter and chemical properties [28–30], while there is still little knowledge about the impact of digestate on soil physical [27,28,31] and biological properties (bacterial and fungal communities) [32,33], which are key factors of soil functioning. Knowledge gaps about appropriate rates and soil-digestate interactions still exist, and the research field is very broad and complex, involving different kinds of feedstocks, crops, soils, environments, and agricultural management.

The main goal of our research was to understand the short-term effects of digestate on soil properties through a holistic approach, investigating soil physical, chemical, and microbiological properties and their interactions. Furthermore, we evaluated the effectiveness of digestate in replacing mineral fertilizers and as a resource to compensate for carbon depletion due to biomass removal in a three-year energy crop rotation. The study included both the whole digestate and its mechanically-separated solid fraction.

2. Materials and Methods

2.1. Digestate

The whole digestate was obtained from the biogas plant of the Cooperativa Agroenergetica Territoriale (CAT) in Correggio (Reggio Emilia, Italy). The digester was fed with the above-ground biomass from energy crops, including maize, triticale and sorghum silages, combined with by-products from the agricultural industry (i.e., stalks of grapes and sugar beet pulps), and cattle slurry from Parmesan cheese farms [34]. The solid fraction (SD) was retrieved from the whole digestate (WD) through a mechanical solid/liquid separation system following the digestion. SD was rich in organic C (44.4% of air-dry digestate) but relatively poor in N, P and K, whereas WD had a very low organic C content (1.1% of air-dry digestate) and a low C/N ratio (3.1), with a large proportion of NH_4-N in the total amount of N (about 60% of air-dry digestate) (Table S1).

For a more in-depth characterization of digestate, we performed molecular-level analyses of microbial communities (see the paragraph on soil sampling and analysis) on WD, SD and two additional fractions, one collected directly from the fermentation silos, the other one from the mechanically-separated liquor (LD).

2.2. Study Site and Experimental Design

The experimental field was a 35 × 130 m area belonging to the R.G.R. Farm (CAT cooperative partner) located in the lower Po Valley (Correggio, Reggio Emilia, Italy; 44°49′ N–10°45′ E). The land use of the area had been converted from sugar beet cultivation to a 2-year maize-triticale rotation to feed the biogas plant, according to the set-aside scheme introduced by the Common Agriculture Policy. The trial was carried out from January 2011 to October 2013, maize cultivated from spring to summer 2011 and 2013, and triticale from autumn 2011 to summer 2012. The effects of digestate application on soil properties were investigated in the two maize growing seasons, using the whole digestate (WD) as a partial or total replacement of mineral fertilizer, and the digestate solid fraction (SD) as a soil amendment. Nitrogen fertilization was performed during maize post-emergency stage as follows: D0 plots, with mineral fertilizer only (control); D50 plots, based on WD + mineral fertilizer; D100 plots, with WD only. The SD fraction was applied to the WD-fertilized plots (D50 and D100) between one crop cycle and the next.

The treatments were assigned to 4 × 10 m field plots according to a randomized block design with three blocks (replicates). 1 m between plots and 5 m between blocks were kept free to avoid disturbance during soil tillage and to allow machinery operations. During the trial period, the mean annual air temperature was 14.2 °C and precipitation 681 mm (Figure S1). The experimental soil was a Hypocalcic Hypovertic Calcisols [35], with a silty-clay texture (Table S2). The main soil physical and chemical characteristics at the start of the trial (time t_0) are given in Table S2.

In September 2010, the field was ploughed and harrowed for seedbed preparation. On 2 April 2011, maize (*Zea mais* L., cv. Kalumet) was sown at a density of 7 plants/m^2 and all plots fertilized with urea (125 kg N ha^{-1}). At the plant emergence (20 May 2011), an additional N fertilization was applied as follows: D0 plots, urea (125 kg N ha^{-1}); D50 plots, urea (62.5 kg N ha^{-1}) plus WD (17,400 L ha^{-1} = 62.5 kg N ha^{-1}); D100 plots, WD only (34,700 L ha^{-1} = 125 kg N ha^{-1}). WD was spread on the soil surface along the maize rows using mobile equipment (Figure S2) specifically developed by CAT and Cavazzuti Franco (Carpi, Modena, Italy), consisting of a 1 m^3 tanker mounted on a tractor and connected to a boom with 4 trailing hoses, with a 2.80 m working width. The tanker was equipped with a pump-loading apparatus for filling. Maize was irrigated on 26 May, 13 June and 6 July 2011, and harvested at the wax ripeness stage (17 August 2011). On 16 September 2011, the D50 and D100 plots received 40 m^3 ha^{-1} SD (equivalent to 10 t ha^{-1}), applied by a solid manure spreader (Vaschieri, Solignano di Castelvetro, Modena, Italy) and incorporated into the soil by ploughing and harrowing. Triticale (x *Triticosecale* Wittm., cultivar Agrano) was sown in November 2011 at a density of 240 kg seeds ha^{-1} and fertilized in a single operation in April 2012 by urea only (30 kg ha^{-1}). Due to the high plant density and the lack of suitable equipment for WD application in the standing crop, no WD top-dressing treatment was possible for triticale. The option of a pre-sowing WD treatment was discarded because of the low N use efficiency in the autumn-winter period and the related risk of N leaching [4]. Triticale biomass was harvested on 24 June 2012. In October 2012, the D50 and D100 plots were amended with the SD fraction (40 m^3 ha^{-1}) and the whole field was prepared for maize sowing as previously described (19 April 2013). The trial continued with a maize cycle according to the same practices as for the first experimental year. Due to unfavourable weather conditions (Figure S1), sowing, fertilization and harvesting operations needed to be delayed for about one month, respectively. Maize was harvested at the wax ripeness stage on 3 September 2013.

The whole above-ground biomass yielded at the end of the crop cycles was harvested and used as feedstock for biogas production.

The combination of both SD and WD with the agricultural management (fertilization factor) and sampling data (time factor) were the factors considered for the evaluation of differences in soil physical-chemical and biological characteristics.

2.3. Seed Germination Bioassay

Extracts of the two digestate fractions (WD and SD) collected from biogas plant at the beginning of experimentation were prepared by adding 25 g digestate to 100 mL of sterile deionized water. The suspensions were shaken for two hours and then centrifuged at 5000 g for 30 min. The supernatants from each digestate were used to prepare test solutions with digestate concentrations of 100% (pure), 50%, 25%, 12.5% and 0% (distilled water as control). Petri dishes of 9 cm diameter were prepared, each containing twenty maize seeds placed upon two sheets of Whatman N. 1 filter paper pre-treated with 10 mL of the test solution. The dishes were transferred to a germination chamber under controlled temperature (20 °C) in the dark. There were five replicates for each treatment.

The number of seeds germinated in each Petri dish was counted after three days and after one week of incubation, and the germination index (GI) was calculated as a percentage relative to the control [26]. Seedling root elongation was measured after 1 week.

2.4. Crop Yield

Crop yield just before harvest (in August for maize and in June for triticale) was estimated by collecting biomass at ground level from three randomly selected point in each plot spaced 30 cm from the edges to avoid border effects. In each sampling point, maize was harvested from 1 m in length row sections (including 6–7 plants), while triticale was harvested from 0.5 m^2 areas. After weighing, the biomass was oven-dried at 70 °C until constant weight (about 56 h for maize and 48 h for triticale) to determine the dry weight.

2.5. Soil

2.5.1. Sampling

Soil samples were collected before maize sowing (25 March 2011 = t_0; at the beginning of the trial); after maize harvesting (17 November 2011 = t_1); before sowing in the second maize cycle (14 April 2013 = t_2); at the end of the trial (3 October 2013 = t_3).

For soil chemical, biochemical, microbiological and particle size analyses, each plot was sampled by auger to a depth of 20 cm in three selected points, collecting soil cores of 5 cm in diameter. The three cores were then mixed thoroughly providing a single composite sample per plot (3 replicates for each treatment, as a whole). Before chemical and biochemical analyses, the soil was air-dried, ground and sieved through a 2 mm mesh size. The samples for microbiological analyses were stored untreated at $-80\,^{\circ}C$ until analysis.

For soil bulk density (BD) and macro-porosity measurements, three undisturbed soil samples were collected from the central part of each plot, at depth increments of 0–10 and 10–20 cm, using a hammer-driven linear sampler. Samples for BD were collected at each sampling time whereas those for macro-porosity analysis were taken only at t_0 and t_2.

Soil aggregate stability was determined at t_0, t_1 and t_2 on a single composite sample per plot, obtained from three spatially separated sub-samples of soil aggregates collected down to 10 cm depth.

2.5.2. Chemical Analyses

Soil pH was measured potentiometrically in a 1:2.5 soil-water suspension. Soil cation exchange capacity (CEC) and exchangeable base concentrations (Ca, Mg, K and Na) were determined on $BaCl_2$ triethanolamine (pH 8.2) extracts by flame atomic absorption spectrometry [36]. Soil available Cu, Zn, Fe and Mn were extracted and quantified according to Lindsay and Norvell [37]. Soil total organic carbon (TOC) and total nitrogen (TN) in the bulk soil were measured by dry combustion using a Thermo Flash 2000 CN elemental analyser (Thermo Fisher Scientific, Walthman, MA, USA). The analysis was performed on 20 to 40 mg of soil weighed into Ag-foil capsules and pre-treated with 10% HCl until complete removal of carbonates.

2.5.3. Biochemical Analyses

Microbial biomass carbon and nitrogen (MBC and MBN, respectively) were determined following the fumigation extraction method [38]. Two aliquots from each soil sample were brought to 60% of water holding capacity (WHC), 24 h before the analysis; a first aliquot was immediately extracted with K_2SO_4 (0.5 M) and then filtered with Whatman n. 42 filter paper; the second aliquot was fumigated for 24 h at $25\,^{\circ}C$ with $CHCl_3$ and extracted as the first one. The organic C and N concentration in the extracts was then determined by Thermo Flash 2000 CN elemental analyser (Thermo Fisher Scientific). MBC and MBN were calculated as the difference between the C and N extracted from the fumigated samples and those extracted from non-fumigated samples, respectively.

Soil microbial respiration was determined according to Badalucco et al. [39]. Each sample was incubated at $28\,^{\circ}C$ in a flask sealed with a stopper. The CO_2 developed during incubation was trapped in NaOH solution after 1, 3, 7, 10, 14, 21 and 28 days and then measured by titration with HCl (0.1 M). The cumulative amount of CO_2 produced over 28 days of incubation (MRcum) was regarded as the potentially mineralizable C.

2.5.4. Microbiological Analyses

Soil RNA was extracted using the RNA PowerSoil^{TM} Total Isolation Kit (MoBio, Solano Beach, CA, USA), following the manufacturer's instructions with the minor modification of adding Na-EDTA (0.5 M) to the lyses solution to improve the DNA desorption from clay particles [40]. RNA was eluted in nuclease-free water (Promega, Madison, WI, USA) and then DNA was co-extracted by the RNA PowerSoil^{TM} DNA Elution Accessory Kit (MoBio). The extracted RNA was subsequently subjected to DNase digestion using the

RQ1 RNase-free DNase (Promega) and complementary cDNA was generated by reverse transcription (RT) using the ImProm-IITM Reverse Transcriptase System (Promega).

For Denaturing Gradient Gel Electrophoresis (DGGE) analysis of microbial communities, the extracted DNA and the generated cDNA were amplified using specific primers for bacterial and archaeal 16S rDNA, and for fungal 18S rDNA (Table S3). Amplification and DGGE procedures were carried out following Pastorelli et al. [41] and Lazzaro et al. [42].

Representative bands from archaea and Clostridiaceae-related DGGEs were excised, eluted from gels and screened according to Pastorelli et al. [43]. Selected bands were subjected to direct sequencing by Macrogen Service (Macrogen Ltd., Amsterdam, The Netherlands). The nucleotide sequences collected in this study were deposited in the GenBank database under the accession numbers MF415444-MF415489.

2.5.5. Physical Analyses

Soil texture was determined by the pipette method [44]. Soil bulk density (BD) was measured by the core method according to Blake and Hartge [45].

Soil macro-porosity was determined by the micro-morphometric method [46]. This method allows the characterization and quantification of soil macro-porosity according to pore shape, size distribution, irregularity, orientation and continuity from vertically oriented thin sections of 5.5×8.5 cm size, obtained from undisturbed soil samples. A 2.82×3.54 cm area of each thin section was captured with a video camera avoiding the edges where disruption could have occurred. The images collected were then analysed by Image-Pro Plus software (Media Cybernetics, Silver Spring, MD, USA), set up specifically to measure pores larger than 50 μm. The total porosity and pore distribution were calculated from the measurement of pore shape and size [46]. From a functional point of view, the elongated pores of 50–500 μm were described as transmission pores and the pores with >500 μm size as fissures [47]. The thin sections were also examined by a Zeiss "R POL" microscope at a $25\times$ magnification to characterize soil structure.

Soil aggregate stability was determined by the wet sieving method and the calculation of the mean weight diameter of water-stable aggregates (MWD) [48]. Soil aggregates from each composite sample were air-dried, weighed and separated into different size fractions (10–20, 4.75–10, 2–4.75, 1–2, <1 mm) using a vibrating sieve shaker (Retsch, Germany). The most representative aggregate size fraction was used to perform wet sieving. Twenty grams of aggregates from the most abundant size class (4.75–10 mm) were directly soaked for 5 min on the top of nests of 4.75, 2, 0.25- and 0.05-mm diameter sieves immersed in water (fast wetting). The nest of sieves with its content was then vertically shaken in water by an electronically controlled machine with a stroke of 40 mm per 10 min, at a rate of 30 complete oscillations per minute. For each sample, 3 repetitions were performed.

2.6. Statistical Analyses

The results of soil physical, chemical and microbiological (richness and α-diversity indices) analyses were processed by analysis of variance (ANOVA) followed by Fisher's least significant difference (LSD) test at the significance level $p \leq 0.05$, using the Statistica software (Palo Alto, CA, USA). Pearson correlation analysis was performed among physico-chemical properties of soil by Statistica software.

Band migration distance and intensity for each DGGE profile were obtained using the Gel Compare II software v 4.6 (Applied Maths, Saint-Martens-Latem, Belgium). The number of bands (species richness) and their relative abundance (Shannon index, H' and Simpson index, D) were used as proxies of richness and α-diversity of soil microbial communities, as described by Pastorelli et al. [43]. The banding patterns of bacterial and fungal DGGE profiles were converted into presence/absence band matching tables and imported into PAST3 software [49]. Non-metric multidimensional scaling (MDS) based on the Dice coefficient was performed to represent the distance between the DGGE profiles in the two-dimensional space. Analysis of similarity (ANOSIM) based on Dice similarity coefficient and 9999 permutational tests were run to assess the statistical significance in microbial community structure due to fertilizer/amendment treatments.

Nucleotide sequence chromatograms were edited using Chromas Lite software v2.1.1 (Technelysium Pty Ltd., Tewantin, Old, AU) to verify the absence of ambiguous peaks and to convert them to FASTA format. The DECIPHER's Find Chimeras web tool [50] was used to uncover chimaeras hidden in nucleotide sequences. The Web-based BLAST tools was used to identify closely related nucleotide sequences within those stored in the GenBank database.in order Microbial taxonomic identification was achieved by means of different sequence similarity thresholds as described by Webster et al. [51].

3. Results

3.1. Germination Index Bioassay

GI was lowest when maize seeds were treated with undiluted SD and WD soluble extracts (57% and 34.9%, respectively; Table 1) but increased with increasing dilution ratio. According to McLachlan et al. [52], GI values above 70%, as those observed under 50%, 25% and 12.5% digestate concentrations, indicate the absence of toxicity.

Table 1. Relative seed germination index (GI) of maize under different digestate concentrations, expressed as the percentage of germinated seeds relative to the control GI (distilled water).

Digestate Concentration (%)		GI (%)	
		SD	WD
100	(undiluted)	57.0	34.9
50	(1:2)	75.6	75.6
25	(1:4)	84.9	84.9
12.5	(1:8)	104.7	81.4

SD = solid digestate; WD = whole digestate.

After one-week incubation there were significant differences in rootlet lengthening between SD and WD treated maize seeds, as well as between the different digestate used concentrations. The rootlet length was lowest in the undiluted extracts and greatest under 12.5% (SD) and 25% (WD) concentrations (Table 2). With a 12.5% SD concentration, the root elongated more than in the control, although not significantly.

Table 2. Root length (mm) of maize seedlings after 1 week of incubation under different digestate concentrations (means followed by standard error in brackets). Different Latin letters within a column indicate statistically significant differences between digestate concentrations at $p \leq 0.05$ (Fisher LSD test); different Greek letters within a row indicate statistically significant differences between digestate types at $p \leq 0.05$ (Fisher LSD test).

Digestate Concentration (%)		Root Length (mm)					
		SD			WD		
0	(water)	25.4 (1.8)	ab		25.4 (1.8)	a	
100	(undiluted)	11.2 (1.5)	d	α	3.6 (0.7)	d	β
50	(1:2)	19.8 (2.1)	c	α	12.3 (1.4)	c	β
25	(1:4)	22.9 (2.0)	bc	α	16.4 (1.5)	b	β
12.5	(1:8)	30.5 (2.0)	a	α	15.9 (1.8)	bc	β

SD = solid digestate; WD = whole digestate.

3.2. Crop Biomass Yield

Neither maize nor triticale biomass showed significant differences between treatments (Table 3). In 2013, due to abundant rainfall (Figure S1), the growth of maize suffered a marked delay compared to 2011, along with a reduction in the biomass yield irrespective of treatment (8.8–12.8 t ha^{-1} against 19.6–22.1 t ha^{-1}, respectively).

Table 3. Maize and triticale above-ground biomass (t dry weight ha^{-1}) under different experimental treatments (means followed by standard error in brackets).

	Maize 2010–2011		Triticale 2011–2012		Maize 2012–2013	
D0	22.1	(1.0)	12.7	(0.4)	11.9	(2.3)
D50	20.4	(2.7)	11.6	(0.7)	12.8	(3.7)
D100	19.6	(1.2)	11.9	(0.4)	8.8	(2.2)

D0 = 100% N as urea; D50 = 50% N as urea + 50% N as WD; D100 = 100% N as WD.

3.3. Soil Chemical Properties

As reported in Table 4, the average TOC content in the plots under digestate treatment generally showed slight increases compared to that of the control plots, with no difference between the application rates.

Table 4. Soil total organic carbon (TOC), total nitrogen (TN) and C/N ratio under the different experimental treatments at different sampling times (means followed by standard error in brackets). Different letters indicate statistically significant differences at $p \leq 0.05$ (Fisher LSD test).

Plots	Time	TOC g kg^{-1}			TN g kg^{-1}			C/N		
D0	t_0	10.6	(0.2)	abcd	1.15	(0.03)	bcde	9.2	(0.1)	abc
	t_1	11.2	(0.4)	abcd	1.25	(0.02)	abcd	9.0	(0.2)	abcd
	t_2	11.4	(0.2)	abc	1.17	(0.08)	bcde	9.8	(0.6)	a
	t_3	9.7	(0.2)	d	1.10	(0.01)	de	8.8	(0.2)	abcd
D50	t_0	9.8	(0.8)	cd	1.07	(0.08)	e	9.2	(0.1)	abc
	t_1	12.2	(0.6)	a	1.28	(0.08)	ab	9.6	(0.2)	ab
	t_2	11.0	(0.5)	abcd	1.30	(0.04)	ab	8.5	(0.7)	cd
	t_3	11.1	(0.2)	abcd	1.26	(0.03)	abcd	8.8	(0.1)	abcd
D100	t_0	10.3	(0.9)	bcd	1.12	(0.07)	cde	9.2	(0.3)	abc
	t_1	12.2	(0.6)	a	1.27	(0.03)	abc	9.5	(0.2)	ab
	t_2	11.0	(0.3)	abcd	1.34	(0.05)	a	8.2	(0.2)	d
	t_3	11.7	(0.9)	ab	1.28	(0.07)	abc	9.2	(0.4)	abc

D0 = 100% N as urea; D50 = 50% N as urea + 50% N as WD; D100 = 100% N as WD.

TN followed a different trend but, overall, it was well correlated with TOC (0.723 ***), confirming a positive digestate effect in the third experimental year (t_2 and t_3 sampling). The TOC to TN ratio did not change with treatment, except for t_1 sampling which showed lower C/N values under digestate application (Table 4).

Soil CEC values, exchangeable Ca and exchangeable Mg concentrations did not differ among treatments for the entire duration of the trial (Table 5). In contrast, at the end of the first year (t_1) the exchangeable K concentration was increased by D50 and D100 regardless of the application rate. At t_2, K showed a significant decrease in all plots as compared to t_0 and t_1 contents. The available Cu, Zn, Fe and Mn contents were not affected by treatments (Table 5).

Table 5. Soil cation exchange capacity (CEC), exchangeable bases (K, Na, Mg, Ca) and available metal content (Cu, Zn, Fe, Mn) in the soil (depth = 0–20 cm) under different fertilization treatments (D0, D50, D100) and at different sampling times (t_0, t_1, t_2) (means followed by standard error in brackets). Different letters indicate significant differences between soil samples at $p \leq 0.05$ (Fisher LSD test).

Plots		CEC cmol(+) kg^{-1}		K mg kg^{-1}			Na mg kg^{-1}			Mg mg kg^{-1}		Ca mg kg^{-1}		Cu mg kg^{-1}		Zn mg kg^{-1}		Fe mg kg^{-1}			Mn mg kg^{-1}		
D0	t_0	21.3	(0.7)	274.5	(3.1)	b	20.6	(2.1)	bc	196.6	(8.9)	3781.0	(157.9)	28.4	(7.4)	1.8	(0.3)	15	(0.8)	a	16.6	(1.4)	a
	t_1	21.2	(0.6)	269.6	(9.8)	b	24.1	(1.5)	abc	207.8	(7.4)	3752.0	(118.5)	27.7	(7.5)	1.9	(0.2)	14.6	(0.2)	abc	15.1	(0.5)	ab
	t_2	20.7	(0.4)	225.7	(20.2)	c	15.9	(1.6)	d	201.1	(8.8)	3691.9	(74.4)	27.5	(7.7)	1.7	(0.3)	14.0	(0.7)	bcd	13.2	(4.8)	bc
D50	t_0	20.1	(1.1)	275.7	(6.8)	b	25.9	(4.1)	abc	212.3	(46.2)	3583.3	(292.3)	26.9	(3.0)	1.9	(0.3)	15.3	(0.3)	ab	16.1	(0.3)	ab
	t_1	21.4	(0.3)	330.0	(20.4)	a	29.9	(2.2)	a	236.3	(34.3)	3705.2	(73.1)	25.1	(1.6)	1.9	(0.3)	14.2	(0.6)	abcd	16.7	(1.6)	a
	t_2	21.0	(0.4)	220.0	(12.2)	c	17.3	(2.7)	d	200.3	(27.2)	3747.1	(92.3)	24.9	(1.5)	1.6	(0.2)	13.3	(0.5)	cd	11.5	(2.3)	c
D100	t_0	21.3	(0.2)	285.7	(12.7)	b	19.6	(1.5)	cd	196.1	(17.9)	3788.5	(20.2)	28.9	(6.0)	1.6	(0.1)	15.0	(0.3)	ab	16.1	(1.0)	ab
	t_1	21.1	(0.7)	328.8	(11.1)	a	26.5	(1.5)	ab	210.9	(10.8)	3683.6	(131.0)	27.8	(6.2)	2.2	(0.3)	14.4	(0.7)	abcd	17.3	(1.0)	a
	t_2	20.9	(0.1)	216.9	(12.2)	c	16.3	(1.3)	d	192.8	(15.2)	3750.1	(36.5)	28.7	(6.1)	1.8	(0.0)	12.8	(0.1)	d	10.2	(0.4)	c

D0 = 100% N as urea; D50 = 50% N as urea + 50% N as WD; D100 = 100% N as WD.

3.4. Soil Physical Properties

Soil BD did not change significantly with treatments and was stable over time (Figure 1).

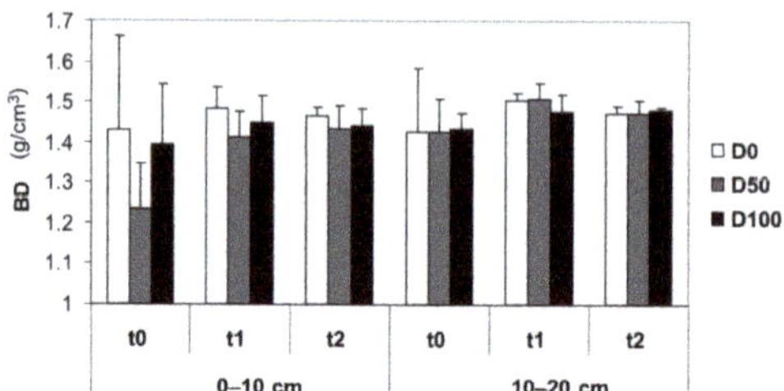

Figure 1. Soil bulk density (BD; g cm^{-3}) at 0–10 and 10–20 cm depth, under the different treatments and at the different sampling times.

Macro-porosity ranged within moderate levels in the surface layer (10–25%) while it averaged less than 5% in the deeper layer, indicating a very compact soil, as defined according to the micro-morphometric method [53] (Figure 2). Differences between treatments were significant in the surface layer only. The t_0 sampling showed a certain degree of field variability for soil macro-porosity, with D0 plots showing a higher macro-porosity than D50 and D100 plots (related to a larger number of fissures) and D50 plots featuring a higher proportion of irregular pores compared to D0 and D100 plots. In the t_{0-2} time frame, soil total macro-porosity increased under D100 with an increase in the percentage of >500 μm elongated pores (fissures) and a reduction in that of 50–500 μm elongated pores. Over the same period, macro-porosity remained quantitatively unchanged under D50, showing a decrease in the 50–500 μm elongated pores.

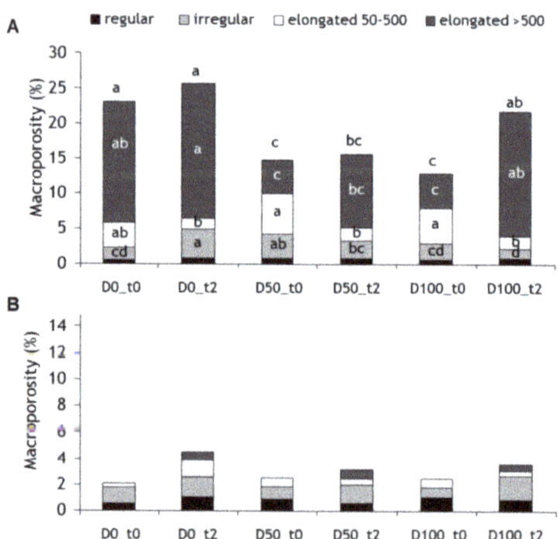

Figure 2. Soil macroporosity (pores size >50 μm) expressed as a percentage of area occupied by pores of different shape (regular, irregular and elongated pores) at 0–10 cm (**A**) and 10–20 cm (**B**) depth and at two different sampling times (t_0 and t_2). Different letters above bars indicate statistically significant differences between the % of fissures (elongated pores, size > 500 μm) in relation to the total macro-porosity; different letters inside the bars indicate significant differences within each shape or size class of pores at $p \leq 0.05$ (Fisher LSD test).

Soil aggregate stability was very low at the beginning of the trial (MWD < 2.5 mm, against a theoretical MWD maximum of 7.375 mm for the 4.75–10 mm size class aggregate) but increased over time regardless of treatment (Figure 3). The effect of digestate treatment was significant only at t_1, soon after WD distribution. After two years (t_2), the differences in soil aggregate stability between treatments were not significant.

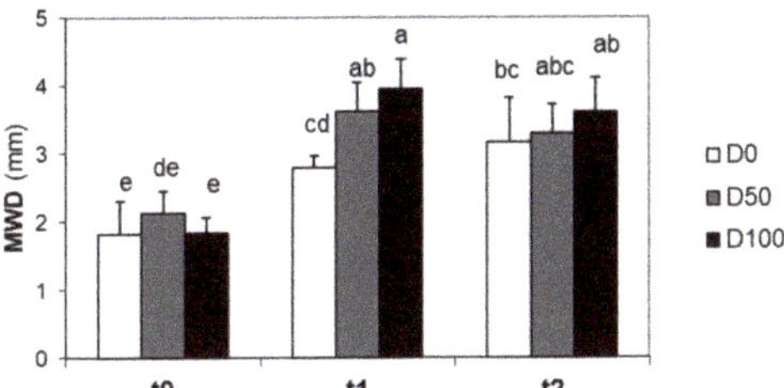

Figure 3. Soil aggregate stability as expressed by the mean weight diameter index (MWD; mm), under the different treatments and at the different sampling times. Different letters indicate statistically significant differences between treatments and sampling times at $p \leq 0.05$ (Fisher LSD test).

3.5. Soil Microbial Biomass and Respiration

Soil MBC was relatively stable over time with a small but significant increase ($p \leq 0.05$) only in D50 plots. The MBN decreased from t_1 to t_2 regardless of treatments (Table 6).

Table 6. Soil microbial biomass C (MCB), microbial N (MBN) and cumulative microbial respiration (MRcum) under different fertilization treatments (D0, D50, D100) and at different sampling times (t_0, t_1, t_2) (means followed by standard error in brackets). Different letters indicate significant differences between soil samples at $p \leq 0.05$ (Fisher LSD test).

Plots	Time	MBC $\mu g\ g^{-1}$			MBN $\mu g\ g^{-1}$			MRcum $\mu g\ C\text{-}CO_2\ g^{-1}$		
D0	t_0	159.8	(27.4)	ab	20.1	(8.1)	abc	426.5	(28.1)	b
	t_1	135.5	(23.7)	ab	32.4	(6.6)	ab	424.4	(15.9)	b
	t_2	157.5	(26.6)	ab	7.0	(2.7)	c	482.8	(17.8)	ab
D50	t_0	137.1	(10.1)	ab	19.8	(1.7)	abc	443.0	(26.8)	ab
	t_1	143.6	(17.5)	ab	33.0	(7.5)	ab	507.8	(47.8)	a
	t_2	203.5	(31.1)	a	16.2	(3.7)	c	469.1	(7.9)	ab
D100	t_0	133.3	(32.5)	b	35.4	(5.9)	a	462.2	(21.3)	ab
	t_1	153.4	(10.7)	ab	18.4	(5.7)	bc	470.5	(25.4)	ab
	t_2	174.4	(14.4)	ab	11.8	(2.4)	c	508.3	(20.3)	a

D0 = 100% N as urea; D50 = 50% N as urea + 50% N as WD; D100 = 100% N as WD.

The C mineralization potential (after 28 days of incubation) did not change significantly either in relation to treatment or time, except for D50 plots where it was higher than in the control plots at t_1 (Table 6).

3.6. DGGE Analysis of Total Bacterial and Fungal Communities

The abundance (richness) and α-diversity (Shannon–Weiner and Simpson indices) calculated from DGGE profiles showed that the soil bacterial community was overall richer and more diverse than the fungal community (Table S4). When considering all groups independently (12 groups: 4 sampling time combined with three digestate treatments), there were significant differences between soil samples for both bacterial and fungal communities (Table S4). Multifactorial ANOVA (Table S4) showed that the species richness and α-diversity indices of the bacterial community were significantly influenced by the interaction between sampling time and digestate treatment. Differently, only the sampling time had a significant effect on the species richness and α-diversity indices of the fungal community (Table S4).

At t_0, MDS ordination showed a low inter-specific variation between the bacterial communities from the differently treated plots (Figure 4A). At t_1, the D50 and D100 bacterial communities were clearly separated from the D0 ones, which grouped with t_0 communities. Bacterial communities at t_2 and t_3 grouped together regardless of treatment and were well separated from the t_0 and t_1 ones (Figure 4A). Conversely, at t_0 the fungal community showed a higher inter-specific variation than bacterial community. In the following sampling, the fungal community showed a progressive change of its structure, which seems to be independent of the treatments (Figure 4B).

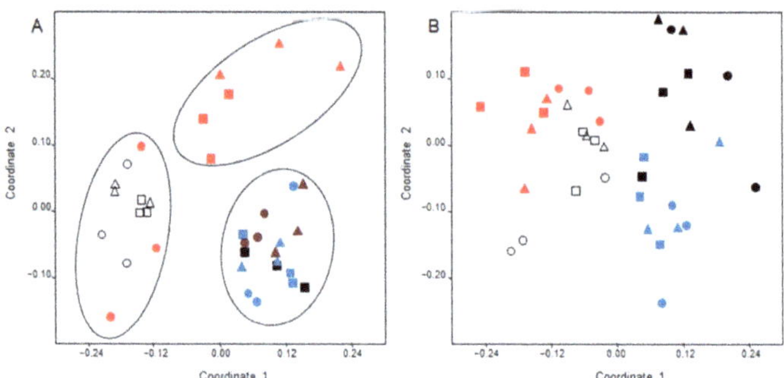

Figure 4. MDS ordination plots of bacterial 16S rDNA (**A**; stress = 0.218) and fungal 18S rDNA (**B**; stress = 0.282). Symbols: circle = D0 treatment; triangle = D50 treatment; square = D100 treatment. Colours: white = time t_0; red = time t_1; blue = time t_2; black = time t_3.

Due to the poor reliability of MDS ordination results, especially for 18S-DGGE (stress = 0.282), DGGE profiles were further analysed by multivariate analysis. When testing all groups independently (sampling time × fertilizer treatment), the one-way ANOSIM global test revealed significant differences in both bacterial and fungal DGGE profiles (Table 7), although R values were not sufficiently reliable. According to the outcomes of the two-way crossed ANOSIM test, the differences in both bacterial and fungal community composition were greater in relation to the sampling time (R = 0.822 and 0.808 for bacteria and fungi, respectively) than in relation to the digestate treatment (R = 0.448 and 0.275 for bacteria and fungi, respectively) (Table 7).

Table 7. Summary of ANOSIM analysis based on 16S- and 18S-rDNA Dice similarity matrices. In the one-way ANOSIM groups were analysed independently (three digestate treatments vs. four sampling time), whereas the two factors (sampling time and digestate treatments) were analysed by a two-way analysis.

| | One-Way Global Test | | Two-Way Crossed | | | |
| | | | Sampling Time | | Digestate Treatment | |
DGGE	R	P	R	P	R	P
16S-rDNA	0.575	0.0001	0.822	0.0001	0.448	0.0001
18S-rDNA	0.637	0.0001	0.808	0.0001	0.275	0.0009

3.7. DGGE Analysis of Total Active Bacterial Community

The active bacterial community was analysed by matching the t_0 DGGE profiles with those obtained at t_1 (different seasons within the same year: March vs. November) and t_2 (different years under the same field conditions: before maize sowing, 2011 vs. 2013).

The abundance (richness), α-diversity (Shannon–Weiner and Simpson indices), and composition of the active bacterial community were more influenced by sampling time than by digestate treatment or sampling time × digestate treatment interaction (Data not shown). The separation between active bacterial communities was stronger when they were compared according to the different season (t_0 vs. t_1) than to the different year (t_0 vs. t_2) (Table 8).

Table 8. Summary of ANOSIM analysis based on the 16S-cDNA Dice similarity matrices. In the one-way ANOSIM groups were analysed independently (three fertilizer treatments vs. four sampling time), whereas the two factors (sampling time and fertilizer treatments) were analysed by a two-way analysis.

	One-Way Global Test		Two-Way Crossed			
			Sampling Time		Digestate Treatment	
DGGE	**R**	**P**	**R**	**P**	**R**	**P**
t_0 vs. t_1	0.686	0.0001	0.901	0.0016	0.467	0.0005
t_0 vs. t_2	0.404	0.0001	0.494	0.0032	0.267	0.0050

3.8. DGGE Analysis of Archaea and Clostridiaceae-Related Communities

The DGGE profiles from the different digestate fractions were very similar to each other and quite different from those of the soil (Figure S3). Digestate-based treatments had no substantial effect on soil archaeal (Figure S3a) and Clostridiaceae-related bacterial (Figure S3b–e) communities. Some additional dominant bands were found at t_1 in D50 and D100 DGGE profiles obtained with the primer sets specific for Clostridiaceae-cluster I and -cluster IV. In particular, a group of γ-Proteobacteria-related bands appeared in the Clostridiaceae-cluster I DGGE profiles (Figure S3b), while a group of bands phylogenetically related to *Caproiciproducens galactitolivorans* (similarity ranged from 93% to 94%) appeared in the Clostridiaceae-cluster IV DGGE profiles (Figure S3d). These bands were almost undetectable in the t_2 DGGE profiles.

Overall, the digestate DGGE profiles were characterized by one or more all-time dominant bands related to Clostridiacea (Figure S3b–e; Table S5). However, none of the primer sets was specific enough to detect only *Clostridium*-related species, since several DGGE bands revealed to be related to β-, δ- and γ-Proteobacteria divisions, Acidobacteria group or Actinobacteria phylum (Table S5).

4. Discussion

4.1. Effects of Digestate on Soil Chemical, Physical and Microbiological Properties

In this trial, digestate treatments provided consistent results in the two years of maize cultivation. The soil TOC tended to be slightly higher in plots treated with digestate than in plots under mineral fertilization, in agreement with results obtained using digestate or other different bioenergy by-products as a soil amendment or fertilizer [28,29]. It is possible that, to some extent, soil organic C enrichment was limited by tillage practices, due the dilution of the organic matter across the ploughed layer and the exposure of physically protected organic compounds to enhanced mineralization [54].

Functional properties of organic residues as amendments are related to the organic matter stability, i.e., the ratio of recalcitrant to labile organic fractions [55] and how these interact with soil features, climate and crop management. There is consistent laboratory evidence of lower carbon mineralization of digestate compared to undigested feedstock, due to an increase of the recalcitrance of organic matter during digestion [28]. However, results from previous short-term experiments on the effects of digestate on soil carbon and nitrogen and crop yield are contrasting, probably due to the various chemical characteristics of digestate and different type of soil used in the experimentations [13,33,56,57].

Overall, the role of soil organic matter in soil fertility and plant nutrition may be summed up in its ability to supply and store plant nutrients [58]. This role is expressed through the release of organically-bound plant nutrients by microbial mineralization and the contribution of organo-mineral complexes to the retention of plant nutrients as available cations [58]. As indicated by the close similarity between soil TN and TOC distribution patterns, soil organic matter contributed to the overall N pool. In addition, the determination coefficient of the relationship between soil TN and TOC ($R^2 = 0.723$ ***) suggests that additional factors may account for TN variations, namely the dynamics of mineral N supplied by fertilizers (WD, urea) and soil organic matter mineralization.

According to crop yield performances, digestate treatments were at least as effective as mineral fertilization in supporting crop requirements. There was no evidence of a significant contribution of the organic fraction of the digestate to the cation exchange capacity of the soil, which is explained by the modest TOC variations found after digestate treatments and the fine-textured composition of the soil mineral fraction [59]. This agrees with previous studies showing that the effects of organic amendment on soil CEC were generally more pronounced in coarse-textured soils than in clayey soils [59].

The whole digestate (WD) also proved to be a valuable source of K, by increasing the available K pool of the soil by 22% during the first year of trial. To further support the high potential of digestate as a substitute for mineral K-fertilizers, numerous experimental evidences demonstrate very high rates of K recovery during anaerobic digestion (above 94%) from a wide range of feedstock materials [60].

The unexpected decrease in soil K and Na content across the experimental field before sowing in the second maize cycle (compared to their average content in the previous sampling times) can be ascribed to a leaching effect (Figure S1), which conversely left Ca and Mg concentrations unchanged due to their lower water-solubility and the buffering effect of soil carbonates [61].

Soil BD was not affected by digestate treatments, which disagrees with results by other authors who found a reduction of BD under organic amendments in both compacted and uncompacted soils [62]. BD and organic matter are linked by a close relationship involving physical and chemical interactions between organic substances and soil mineral particles [63]. Usually, due to a lower density of the organic matter compared to that of the mineral fraction, the average BD of a mixture mineral fraction/organic matter decreases as the organic matter content increases. In the present experiment, several factors may have interfered with these relationships, i.e., an experimental period too short compared to the time required for soil structure formation and a contrasting effect of soil tillage on aggregates formation and stabilization. This was reflected in the pattern of soil pore size distribution, with a decrease in the amount of transmission pores, which are of primary importance for optimal soil–water–plant relationships, and an increase in the proportion of fissures mainly involved in water and air flows but related to poor structure and physical degradation when they are (as in D100) over 70% of total porosity [64].

The stability of soil aggregates is a key indicator of soil physical quality, affecting the ability of the soil to retain its structure and the related physical and hydraulic functions against degradative forces [65]. Soil aggregate stability relies on a complex range of factors involving soil texture and mineralogy, the chemistry of soil cations and soil organic matter content and quality [64]. At the beginning of the trial, aggregate stability (expressed as MWD) was quite low possibly due to the high silt proportion and the low organic C content [66,67]. However, aggregate stability was improved by digestate treatment during the first experimental year, consistently with findings of other authors [27,31]. In addition, it correlated positively with TOC (Figure 5), in line with the role of soil organic C as a driver of soil aggregate formation [66–68].

With respect to soil biological parameters, biochemical analysis revealed just a slight (statistically not significant) increase over time in the soil MBC under digestate treatment. This increase was consistent with the trend of TOC, suggesting that part of the organic C supplied by digestate could have been converted into MBC [69]. The small extent of MBC increase was expected from a short term digestate treatment, due to the relatively high recalcitrance of the organic matter in SD and the low organic C concentration in WD [26,70]. This evidence confirms SD application as a valuable tool to improve soil C sequestration [71] and to compensate for C depletion associated to crop biomass removal.

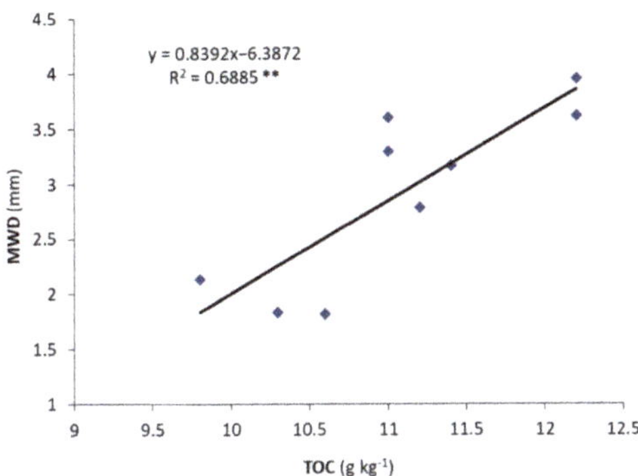

Figure 5. Correlation between soil aggregate stability based on the mean weight diameter (MWD) and total organic carbon (TOC) (** $p \leq 0.01$).

Microorganisms are crucial for soil fertility [25]. They drive the turnover of organic substrates and their abundance and diversity can be affected by soil management as well as by the quality of soil amendments/fertilizers. A reduction in microorganism number and diversity may impair their ability to perform specific functions as well as to withstand soil perturbations from a long-term perspective [72]. In this trial, "time" was the main factor affecting the α-diversity of soil microbial communities. Differences in the microbial community structure in response to digestate addition were showed when t_0 and t_1 were compared. Conversely, the microbial community structure remained quite similar when t_2 and t_3 were compared. In addition, active bacterial communities resulted more affected by season than by digestate treatments, contrasting many reports which indicated an enhanced soil microbial activity after field applications of digestates [32,33]. Calbrix et al. [73], in a study dealing with the impact of organic amendments on soil bacterial communities over a 12-month period, observed that changes in soil bacterial community structure were only temporary and that seasonal variations had the greatest effect on microbial community composition. Accordingly, in our study, digestate showed to have only a transient effect on the microbial community structure. Successively, the soil microbial communities developed new stability and equilibrium over time in both digestate treatments, thus strengthening the hypothesis of a resilience of microbial communities to anthropogenic changes [74–76]. Likewise, the Archaea and Clostridiaceae-related bacteria revealed remarkably stable soil resident communities, with negligible and temporary changes after the introduction of allochthones species (Figure S3).

4.2. Effects of Digestate on Seed Germination and Crop Yield

The GI bioassays revealed that highly concentrated SD and WD extracts impaired seed germination, whereas <50% digestate concentrations had no phytotoxicity. This suggests that the use of digestate should follow appropriate rates and timings of application to avoid the direct contact with seeds, as also described by Alburquerque et al. [26]. According to our experimental plan, we can exclude any phytotoxic interference of digestate with maize seed germination under field condition, as the SD fraction was applied several months before sowing and WD in the post-emergence stage.

Interestingly, the 12.5% SD concentration increased the germination index and the relative root elongation as compared to the control, which can be explained by assuming a stimulating effect of plant nutrients, growth enhancers or even phytormone-like compounds contained in SD as suggested by other authors [26,77].

In the first two years of the trial, both maize and triticale biomass yields were consistent with the average yields in the area [78], which is promising in the perspective of agricultural use of digestate, alone or combined with mineral fertilizers. The implementation of conservation tillage management [79] may further improve the efficiency of digestate as an amendment and/or fertilizer.

With regard to the pronounced decrease in maize yield across the whole experimental field in the third experimental year (second maize growth season), it was most likely due to a combination of adverse climate and soil physical conditions arising from the abundant rainfall between January and April (Figure S1), which caused a shift of the entire growing season. At the same time, the fine texture of the soil combined with the low organic matter content might have led to a stronger impact of the heavy machinery on soil structure and hydrological behaviour, resulting in insufficient drainage, extended water stagnation and overall poor soil physical conditions for seed germination and shoot development [80].

5. Conclusions

With a focus on the environmental sustainability of the bioenergy supply chain, the application of digestate to the soil can meet the need to safely dispose and recycle the residues coming from anaerobic digestion and, at the same time, to compensate for soil C and plant nutrient depletion due to crop biomass removal.

From our results, digestate application in a three-year maize-triticale rotation cycle proved to be as effective as 100% mineral fertilization in maintaining crop productivity level. Moreover, the increase in soil TOC following digestate treatments confirmed digestate effectiveness to compensate for carbon depletion.

Further research is needed to increase the knowledge about the optimum dose of digestate to be applied in relation to soil/crop specificities and the best application method to minimize potential negative effects of digestate to the soil and environment quality. The pattern and extent with which the effects of digestate treatments were expressed and their temporal fluctuations underline complex dynamics of chemical, physical and biological processes affecting the material brought to the soil. This suggests that a more or less long period of time is needed during which the achievement of a new stable equilibrium in the soil functions is regulated by the interaction between the amount and quality of biomass supplied, the impact of mechanical operations associated with crop management and climate trend.

Further expected benefits from digestate as amendment, such as improvement of soil bulk density and porosity, were not observed, possibly due to a counteracting interference of soil tillage operations. The effectiveness of a soil amendment and the sustainability of the use of digestate can be strongly conditioned by the crop management system as a whole, and in particular, by those cultivation practices that have a direct impact on the soil and the dynamics of the organic matter and nutrients supplied with the treatment. For this reason, to fully exploit digestate potential, its use should be integrated within an overall more conservative soil management system, involving reduced soil mechanical disturbance. This would be essential to prevent soil physical degradation and excessive organic matter mineralization, thus allowing the organic compounds of digestate to perform their chemical, physical and biological functions and minimize the risk of N loss by leaching and/or gas emissions.

Additional considerations regard the cultivation of energy crops in marginal lands or set-aside areas; this could be a solution to the "food vs. fuel conflict" and, at the same time, would promote rural investments and new job opportunities.

Supplementary Materials: The following are available online at https://www.mdpi.com/2076-341 7/11/2/750/s1. References [81–94] are cited in the supplementary materials.

Author Contributions: Conceptualization, R.P., G.V., A.L., A.F. and N.V.; formal analysis, R.P., G.V., A.L., S.S. and N.V.; investigation, R.P., G.V., A.L. and N.V.; resources, M.Z.; data curation, R.P., G.V., A.L. and N.V.; writing—original draft preparation, R.P., G.V., A.L. and N.V.; writing—review and editing, R.P., G.V., A.L. and N.V. All authors have read and agreed to the published version of the manuscript.

Funding: This study was supported by the Italian Ministry of Agriculture (BIOMASSVAL project).

Institutional Review Board Statement: Not applicable.

Informed Consent Statement: Not applicable.

Data Availability Statement: Data is contained within the article or supplementary material.

Acknowledgments: We acknowledge the technical support of Azienda Agraria R.G.R. that hosts the field experiment and the CAT cooperative that supplies the digestate. In particular, we greatly acknowledge Gabriele Santi, president of CAT, that coordinated the agronomic management of the field trial. The technical work during sampling of Raimondo Piccolo and Ilaria Secciani is highly appreciated.

Conflicts of Interest: The authors declare no conflict of interest.

References

1. Beurskens, L.W.M.; Hekkenberg, M. *Renewable Energy Projections as Published in the National Renewable Energy Action Plans of the European Member States*; ECN/European Environment Agency: Petten, NL, USA, 2011.
2. European Biogas Association (EBA). *Biogas Success Stories 2020*; Renewable Energy House: Brussels, Belgium, 202. Available online: https://www.europeanbiogas.eu/biogas-success-stories-2020 (accessed on 1 December 2020).
3. International Renewable Energy Agency (IRENA). Renewable Energy Statistics. 2020. Available online: https://www.irena.org/publications/2020/Jul/Renewable-energy-statistics-2020 (accessed on 1 December 2020).
4. Italian Biogas Consortium (CIB). ISAAC-PROJECT. *Deliverable D2.1. State of the Art and Best Practices Collection*. 2017. Available online: https://andrianeparla.it/wp-content/uploads/2017/05/D2.2-State-of-the-art-and-best-practices-collection.pdf (accessed on 1 December 2020).
5. Barbera, E.; Menegon, S.; Banzato, D.; D'Alpaos, C.; Bertucco, A. From biogas to biomethane: A process simulation-based techno-economic comparison of different upgrading technologies in the Italian context. *Renew. Energy* **2019**, *135*, 663–673. [CrossRef]
6. Schievano, A.; D'Imporzano, G.; Orzi, V.; Colombo, G.; Maggiore, T.; Adani, F. Biogas from dedicated energy crops in Northern Italy, Electric energy generation costs. *GCB Bioenergy* **2015**, *7*, 899–908. [CrossRef]
7. Braun, R.; Weiland, P.; Wellinger, A. Biogas from Energy Crop Digestion; International Energy Agency (IEA) Bioenergy; Task 37, Energy from Biogas. 2010. Available online: https://www.iea-biogas.net (accessed on 1 December 2020).
8. Murphy, J.D.; Braun, R.; Weiland, P.; Wellinger, A. Update Biogas from Crop Digestion; International Energy Agency (IEA) Bioenergy; Task 37, Energy from Biogas. 2011. Available online: https://www.iea-biogas.net (accessed on 1 December 2020).
9. Smeets, E.M.W.; Faaij, A.P.C.; Lewandowski, I.M.; Turkenburg, W.C. A bottom-up assessment and review of global bioenergy potentials to 2050. *Prog. Energy Combust. Sci.* **2007**, *33*, 56–106. [CrossRef]
10. Fuchs, W.; Drosg, B. Assessment of the state of the art of technologies for the processing of digestate residue from anaerobic digesters. *Water Sci. Technol.* **2013**, *67*, 1984–1993. [CrossRef]
11. Gijzen, H.J. Anaerobic digestion for sustainable development, a natural approach. *Water Sci. Technol.* **2002**, *45*, 321–328. [CrossRef]
12. Teglia, C.; Tremier, A.; Martel, J.L. Characterization of solid digestates, Part 1; review of existing indicators to assess solid digestates agricultural use. *Waste Biomass Valori.* **2011**, *2*, 43–58. [CrossRef]
13. Möller, K.; Müller, T. Effects of anaerobic digestion on digestate nutrient availability and crop growth, a review. *Eng. Life Sci.* **2012**, *12*, 242–257. [CrossRef]
14. Diacono, M.; Persiani, A.; Testani, F.; Ciaccia, C. Recycling agricultural wastes and by-products in organic farming: Biofertilizer production, yield performance and carbon footprint analysis. *Sustainability* **2020**, *11*, 3824. [CrossRef]
15. Svensson, K.; Odlare, M.; Pell, M. The fertilizing effect of compost and biogas residues from source separated household waste. *J. Agric. Sci.* **2004**, *142*, 461–467. [CrossRef]
16. Al Seadi, T.; Lukehurst, C. Quality Management of Digestate from Biogas Plants Used as Fertiliser; International Energy Agency (IEA) Bioenergy; Task 37: Energy from Biogas. 2012. Available online: http://www.iea-biogas.net (accessed on 1 December 2020).
17. Drosg, B.; Fuchs, W.; Al Seadi, T.; Madsen, M.; Linke, B. Nutrient Recovery by Biogas Digestate Processing; International Energy Agency (IEA) Bioenergy; Task 37, Energy from Biogas. 2015. Available online: http://www.iea-biogas.net (accessed on 1 December 2020).
18. Karp, A.; Halford, N.G. Energy crops, introduction. In *Energy Crops*; Karp, A., Halford, N.G., Eds.; The Royal Society of Chemistry: Cambridge, UK, 2010; pp. 1–12.
19. González-García, S.; Bacenetti, J.; Negri, M.; Fiala, M.; Arroja, L. Comparative environmental performance of three different annual energy crops for biogas production in Northern Italy. *J. Clean. Prod.* **2013**, *43*, 71–83. [CrossRef]
20. Sims, R.E.H.; Hastings, A.; Schlamadinger, B.; Taylors, G.; Smith, P. Energy crops, current status and future prospects. *Glob. Chang. Biol.* **2006**, *12*, 2054–2076. [CrossRef]
21. Lukehurst, C.T.; Frost, P.; Al Seadi, T. Utilisation of Digestate from Biogas Plants as Biofertiliser; International Energy Agency (IEA) Bioenergy; Task 37, Energy from Biogas. 2010. Available online: http://www.iea-biogas.net (accessed on 1 December 2020).

22. Tambone, F.; Genevini, P.; D'Imporzano, G.; Adani, F. Assessing amendment properties of digestate by studying the organic matter composition and the degree of biological stability during the anaerobic digestion of the organic fraction of MSW. *Bioresour. Technol.* **2009**, *100*, 3140–3142. [CrossRef]
23. Nkoa, R. Agricultural benefits and environmental risks of soil fertilization with anaerobic digestates, a review. *Agron. Sustain. Dev.* **2014**, *34*, 473–492. [CrossRef]
24. Béghin-Tanneau, R.; Guérin, F.; Guiresse, M.; Kleiber, D.; Scheiner, J.D. Carbon sequestration in soil amended with anaerobic digested matter. *Soil Till. Res.* **2019**, *192*, 87–94. [CrossRef]
25. Nyberg, K. Impact of Organic Waste Residues on Structure and Function of Soil Bacterial Communities. Ph.D. Thesis, Swedish University of Agricultural Sciences, Uppsala, Sweden, 2006.
26. Alburquerque, J.A.; de la Fuente, C.; Ferrer-Costa, A.; Carrasco, L.; Cegarra, J.; Abad, M.; Bernal, M.P. Assessment of the fertiliser potential of digestates from farm and agroindustrial residues. *Biomass Bioenerg.* **2012**, *40*, 181–189. [CrossRef]
27. Frøseth, R.B.; Bakken, A.K.; Bleken, M.A.; Riley, H.; Pommeresche, R.; Thorup-Kristensen, K.; Hansen, S. Effects of green manure herbage management and its digestate from biogas production on barley yield; N recovery; Soil structure and earthworm populations. *Eur. J. Agron.* **2014**, *52*, 90–102. [CrossRef]
28. Möller, K. Effects of anaerobic digestion on soil carbon and nitrogen turnover; N emissions; and soil biological activity. A review. *Agron. Sustain. Dev.* **2015**, *35*, 1021–1041. [CrossRef]
29. Galvez, A.; Sinicco, T.; Cayuela, M.L.; Mingorance, M.D.; Fornasier, F.; Mondini, C. Short term effects of bioenergy by-products on sol C and N dynamics; Nutrient availability and biochemical properties. *Agric. Ecosyst. Environ.* **2012**, *160*, 3–14. [CrossRef]
30. Grigatti, M.; Boanini, E.; Cavani, L.; Ciavatta, C.; Marzadori, C. Phospohorus in digestate-based compost, chemical speciation and plant availability. *Waste Biomass Valori.* **2015**, *6*, 481–493. [CrossRef]
31. Beni, C.; Servadio, P.; Marconi, S.; Neri, U.; Aromolo, R.; Diana, G. Anaerobic digestate administration, effect on soil physical and mechanical behavior. *Comm. Soil Sci. Plant Anal.* **2012**, *43*, 821–834. [CrossRef]
32. Walsh, J.J.; Rousk, J.; Edward-Jones, G.; Jones, D.L.; Williams, A.P. Fungal and bacterial growth following the application of slurry and anaerobic digestate of livestock manure to temperate pasture soil. *Biol. Fertil. Soils* **2012**, *48*, 889–897. [CrossRef]
33. Bachmann, S.; Gropp, M.; Eichler-Löbermann, E. Phosphorus availability and soil microbial activity in 3 year field experiment amended with digested dairy slurry. *Biomass Bioenerg.* **2014**, *70*, 429–439. [CrossRef]
34. Pulvirenti, A.; Ronga, D.; Zaghi, M.; Tomasselli, A.R.; Mannella, L.; Pecchioni, N. Pelleting is a successful method to eliminate the presence of *Clostridium* spp. From digestate of biogas plants. *Biomass Bioenerg.* **2015**, *81*, 479–482. [CrossRef]
35. FAO. *Guidelines for Soil Description*, 4th ed.; Food and Agriculture Organisation (FAO): Rome, Italy, 2006.
36. Gessa, C.; Ciavatta, C. Complesso di scambio. In *Metodi di Analisi Chimica del Suolo. Ministero per le Politiche Agricole e Forestali*; Angeli, F., Ed.; Osservatorio Nazionale Pedologico e per la Qualità del Suolo: Milano, Italy, 2000; pp. 1–31.
37. Lindsay, W.L.; Norvell, W.A. Development of a DTPA soil test for zinc; iron; manganese and copper. *Soil Sci. Soc. Am. J.* **1978**, *42*, 421–428. [CrossRef]
38. Vance, E.D.; Brookes, P.C.; Jenkinson, D.S. An extraction method for measuring soil microbial biomass C. *Soil Biol. Biochem.* **1987**, *19*, 703–707. [CrossRef]
39. Badalucco, L.; Grego, S.; Dell'Orco, S.; Nannipieri, P. Effect of liming on some chemical; biochemical and micro-biological properties of acid soil under spruce (*Picea abies* L.). *Biol. Fertil. Soils* **1992**, *14*, 76–83. [CrossRef]
40. Pastorelli, R.; Piccolo, R.; Cocco, S.; Landi, S. mRNA recovery and denitrification gene expression in clay-soil bacterial communities under different agricultural managements. *Agrochimica* **2010**, *54*, 179–192.
41. Pastorelli, R.; Piccolo, R.; Landi, S. Changes in active microbial soil communities in agricultural managements, from anthropic to natural. *Agrochimica* **2009**, *53*, 386–397.
42. Lazzaro, L.; Giuliani, C.; Fabiani, A.; Agnelli, A.E.; Pastorelli, R.; Lagomarsino, A.; Benesperi, R.; Calamassi, R.; Foggi, B. Soil and plant changing after invasion, the case of *Acacia dealbata* in a Mediterranean ecosystem. *Sci. Total Environ.* **2014**, *497*, 491–498. [CrossRef]
43. Pastorelli, R.; Landi, S.; Trabelsi, D.; Piccolo, R.; Mengoni, A.; Bazzicalupo, M.; Pagliai, M. Effects of soil management on structure and activity of denitrifying bacterial communities. *Appl. Soil Ecol.* **2011**, *49*, 46–58. [CrossRef]
44. Gee, G.W.; Bauder, J.W. Particle-size analysis. In *Methods of Soil Analysis. Part 1; Physical and Mineralogical Methods*, 2nd ed.; Klute, A., Ed.; Agron Monogr 9; ASA: Madison, WI, USA, 1986; pp. 383–411.
45. Blake, G.R.; Hartge, K.H. Bulk density. In *Methods of Soil Analysis. Part 1; Physical and Mineralogical Methods*, 2nd ed.; Klute, A., Ed.; Agron. Monogr 9; Soil Science Society of America: Madison, WI, USA, 1986; pp. 363–375.
46. Pagliai, M.; Vignozzi, N. Image analysis and microscopic techniques to characterize soil pore system. In *Physical Methods in Agriculture—Approach to Precision and Quality*; Blahovec, J., Kutilek, M., Eds.; Springer: Boston, MA, USA, 2002; pp. 13–38.
47. Greenland, D.J.; Pereira, H.C. Soil damage by intensive arable cultivation, temporary or permanent? *Phil. Trans. R. Soc. B* **1977**, *281*, 193–208. [CrossRef]
48. Kemper, W.D.; Rosenau, R.C. Aggregate stability and size distribution. In *Methods of Soil Analysis. Part 1; Physical and Mineralogical Methods*, 2nd ed.; Klute, A., Ed.; Agron. Monogr. 9; Soil Science Society of America: Madison, WI, USA, 1986; pp. 425–442.
49. Hammer, Ø.; Harper, D.A.T.; Ryan, P.D. PAST: Palaeontological statistics software package for education and data analysis. *Palaeontol. Electron.* **2001**, *4.1*, 9.

50. Wright, E.S. DECIPHER: Harnessing local sequence context to improve protein multiple sequence alignment. *BMC Bioinform.* **2015**, *16*, 322. [CrossRef] [PubMed]

51. Webster, N.S.; Taylor, M.W.; Behnam, F.; Lucker, S.; Rattei, T.; Whalan, S.; Horn, M.; Wagner, M. Deep sequencing reveals exceptional diversity and modes of transmission for bacterial sponge symbionts. *Environ. Microbiol.* **2010**, *12*, 2070–2082. [CrossRef] [PubMed]

52. McLachlan, K.L.; Chong, C.; Vorony, R.P.; Liu, H.W.; Holbein, B.E. Assessing the potential phytotoxicity of digestate during processing of municipal solid waste by anaeroboic digestion; a comparison to aerobic digestion. *Acta Hort.* **2004**, *638*, 225–230. [CrossRef]

53. Pagliai, M. Soil porosity aspects. *Int. Agrophys.* **1988**, *4*, 215–232.

54. Kan, Z.R.; Virk, A.L.; He, C.; Liu, Q.Y.; Qi, J.Y.; Dang, Y.P.; Zhao, X.; Zhang, H.L. Characteristics of carbon mineralization and accumulation under long-term conservation tillage. *Catena* **2020**, *193*, 104636. [CrossRef]

55. Kögel-Knabner, I. The macromolecular organic composition of plant and microbial residues as inputs to soil organic matter. *Soil Biol. Biochem.* **2002**, *34*, 139–162. [CrossRef]

56. Erhart, E.; Siegl, T.; Bonell, M.; Unterfrauner, H.; Peticzka, R.; Ableidinger, C.; Haas, D.; Hartl, W. Fertilization with liquid digestate in organic farming—Effects on humus balance; soil potassium contents and soil physical properties. *Geophys. Res. Abstr.* **2014**, *16*, 4419.

57. Maucieri, C.; Nicoletto, C.; Caruso, C.; Sambo, P.; Borin, M. Effects of digestate solid fraction fertilisation on yield and soil carbon dioxide emission in a horticulture succession. *Ital. J. Agron.* **2017**, *12*, 116–123. [CrossRef]

58. Sikora, L.J.; Stott, D.E. Soil organic carbon and nitrogen. In *Methods for Assessing Soil Quality*; Doran, J.W., Jones, A.J., Eds.; Soil Science Society of America: Madison, WI, USA, 1997; Volume 49, pp. 157–167.

59. Abbruzzini, T.; Oliveira Zenero, M.; de Andrade, P.; Dini Andreote, F.; Campo, J.; Pellegrino Cerri, C. Effects of biochar on the emissions of greenhouse gases from sugarcane residues applied to soils. *Agric. Sci.* **2017**, *8*, 869–886. [CrossRef]

60. Muhayodin, F.; Fritze, A.; Rotter, V.S. A review on the fate of nutrients and enhancement of energy recovery from rice straw through anaerobic digestion. *Appl. Sci.* **2020**, *10*, 2047. [CrossRef]

61. Sharpley, A.N. Effect of soil pH on cation and anion solubility. *Commun. Soil Sci. Plant Anal.* **1991**, *22*, 827–841. [CrossRef]

62. Rivenshield, A.; Bassuk, N. Using organic amendments to decrease bulk density and increase macroporosity in compacted soils. *Arboric. Urban For.* **2007**, *33*, 140–146.

63. Emerson, W.W.; Foster, R.C.; Oades, J.M. Organo-mineral complexes in relation to soil aggregation and structure. In *Interactions of Soil Minerals with Natural Organics and Microbes*; Huang, P.M., Schnitzer, M., Eds.; Soil Science Society of America: Madison, WI, USA, 1986; Volume 17, pp. 521–548. [CrossRef]

64. Pagliai, M.; Vignozzi, N.; Pellegrini, S. Soil structure and the effect of management practices. *Soil Tillage Res.* **2004**, *79*, 131–143. [CrossRef]

65. Bronick, C.J.; Lal, R. Soil structure and management, a review. *Geoderma* **2005**, *124*, 3–22. [CrossRef]

66. Tisdall, J.M.; Oades, J.M. Organic-matter and water-stable aggregates in soils. *J. Soil Sci.* **1982**, *33*, 141–163. [CrossRef]

67. Erktan, A.; Cécillon, L.; Graf, F.; Roumet, C.; Legout, C.; Rey, F. Increase in soil aggregate stability along a Mediterranean successional gradient in severely eroded gully bed ecosystems, combined effects of soil; root traits and plant community characteristics. *Plant Soil* **2015**, *398*, 121–137. [CrossRef]

68. Le Bissonnais, Y.; Arrouays, D. Aggregate stability and assessment of soil crustability and erodibility, II. Application to humic loamy soils with various organic carbon contents. *Eur. J. Soil Sci.* **1997**, *48*, 39–48. [CrossRef]

69. Terhoeven-Urselmans, T.; Scheller, E.; Raubuch, M.; Ludwig, B.; Joergensen, B.G. CO_2 evolution and N mineralization after biogas slurry application in the field and its yield effects on spring barley. *Appl. Soil Ecol.* **2009**, *42*, 297–302. [CrossRef]

70. Herrmann, A. Biogas production from maize, current state; challenges and prospects. 2. Agronomic and environmental aspects. *Bioenerg. Res.* **2013**, *6*, 372–387. [CrossRef]

71. Chen, R.; Blagodatskaya, E.; Senbayram, M.; Blagodatsky, S.; Myachina, O.; Dittert, K.; Kuzyakov, Y. Decomposition of biogas residues in soil and their effects on microbial growth kinetics and enzyme activities. *Biomass Bioenerg.* **2012**, *45*, 221–229. [CrossRef]

72. Girvan, M.S.; Campbell, C.D.; Killham, K.; Prosser, J.I.; Glover, L.A. Bacterial diversity promotes community stability and functional resilience after perturbation. *Environ. Microbiol.* **2005**, *7*, 301–313. [CrossRef] [PubMed]

73. Calbrix, R.L.; Barray, S.; Chabrerie, O.; Fourrie, L.; Laval, K. Impact of organic amendments on the dynamics of soil microbial biomass and bacterial communities in cultivated land. *Appl. Soil Ecol.* **2007**, *35*, 511–522. [CrossRef]

74. Nannipieri, P.; Ascher, J.; Ceccherini, M.T.; Landi, L.; Pietramellara, G.; Renella, G. Microbial diversity and soil functions. *Eur. J. Soil Sci.* **2003**, *54*, 655–670. [CrossRef]

75. Tatti, E.; Decorosi, F.; Viti, C.; Giovannetti, L. Despite long-term compost amendment seasonal changes are main drivers of soil fungal and bacterial population dynamics in a Tuscan vineyard. *Geomicrobiol. J.* **2012**, *29*, 506–519. [CrossRef]

76. Johansen, A.; Carter, M.S.; Jensen, E.S.; Hauggaard-Nielsen, H.; Ambus, P. Effects of digestate from anaerobically digested cattle slurry and plant materials on soil microbial community and emission of CO_2 and N_2O. *Appl. Soil Ecol.* **2013**, *63*, 36–44. [CrossRef]

77. Pivato, A.; Vanin, S.; Raga, R.; Lavagnolo, M.C.; Barausse, A.; Rieple, A.; Laurent, A.; Cossu, R. Use of digestate from a decentralized on-farm biogas plant as fertilizer in soils, An ecotoxicological study for future indicators in risk and life cycle assessment. *Waste Manag.* **2016**, *49*, 378–389. [CrossRef]

78. Negri, M.; Bacenetti, J.; Brambilla, M.; Manfredini, A.; Cantore, A.; Bocchi, S. Biomethane production from different crop systems of cereals in Northern Italy. *Biomass Bioenerg.* **2014**, *63*, 321–329. [CrossRef]
79. Badagliacca, G.; Petrovičovà, B.; Pathan, S.I.; Roccotelli, A.; Romeo, M.; Monti, M.; Gelsomino, A. Use of solid anaerobic digestate and no-tillage practice for restoring the fertility status of two Mediterranean orchard soils with contrasting properties. *Agric. Ecosyst. Environ.* **2020**, *300*, 107010. [CrossRef]
80. Von Cossel, M.; Wagner, M.; Lask, J.; Magenau, E.; Bauerle, A.; Von Cossel, V.; Warrach-Sagi, K.; Elbersen, B.; Staritsky, I.; Van Eupen, M.; et al. Prospects of bioenergy cropping systems for a more social-ecologically sound bioeconomy. *Agronomy* **2019**, *9*, 605. [CrossRef]
81. Springer, U.; Klee, J. Prufung der leistungsfahigkeit von einigen wichtigeren verfahren zur bestimming des kohlemstoffs mittels chromschwefelsaure sowie vorschlag einer neuen schnellmethode. *Pflanz. Dueng Bodenkd* **1954**, *64*, 1. [CrossRef]
82. Bremner, J.M.; Mulvaney, C.S. Nitrogen-total. In *Methods of Soil Analysis. Part 2. Chemical and Microbio-Logical Properties*; Page, A.L., Miller, R.H., Keeney, D.R., Eds.; Soil Science Society of America: Madison, WI, USA, 1982; pp. 595–624.
83. Anderson, J.M.; Ingram, J.S.I. *Tropical Soil Biology and Fertility, A Handbook of Methods*; CAB International: Oxford, UK, 1993.
84. Murphy, J.; Riley, J.P. A modified single solution method for determination of phosphate in natural waters. *Anal. Chim. Acta* **1962**, *27*, 31–36. [CrossRef]
85. Tabatabai, M.; Bremner, J. Arylsulfatase activity of soils. *Soil Sci. Soc. Am. J.* **1970**, *34*, 225–229. [CrossRef]
86. Nübel, U.; Engelen, B.; Felske, A.; Snairdr, J.; Wieshuber, A.; Amann, R.I.; Ludwig, W.; Backhaus, H. Sequence heterogeneities of genes encoding 16S rRNAs in *Paenibacillus polymyxa* detected by temperature gradient gel electrophoresis. *J. Bacteriol.* **1996**, *178*, 5636–5643. [CrossRef]
87. Vainio, E.J.; Hantula, J. Direct analysis of wood-inhabiting fungi using denaturing gradient gel electrophoresis of amplified ribosomal DNA. *Mycol. Res.* **2000**, *8*, 927–936. [CrossRef]
88. Watanabe, T.; Kimura, M.; Asakawa, S. Community structure of methanogenic archaea in paddy field soil under double cropping (rice-wheat). *Soil Biol. Biochem.* **2006**, *38*, 1264–1274. [CrossRef]
89. Hung, C.H.; Cheng, C.H.; Cheng, L.H.; Liang, C.M.; Lin, C.Y. Application of *Clostridium*-specific PCR primers on the analysis of dark fermentation hydrogen-producing bacterial community. *Int. J. Hydrogen Energy* **2008**, *33*, 1586–1592. [CrossRef]
90. Green, T.R.; Popa, R. Turnover of carbohydrate-rich vegetal matter during microaerobic composting and after amendment in soil. *Appl. Biochem. Biotechnol.* **2011**, *165*, 270–278. [CrossRef]
91. Shiratori, H.; Ikeno, H.; Ayame, S.; Kataoka, N.; Miya, A.; Hosono, K.; Beppu, T.; Ueda, K. Isolation and characterization of a new *Clostridium* sp. that performs effective cellulosic waste digestion in a thermophilic methanogenic bioreactor. *Appl. Environ. Microbiol.* **2006**, *72*, 3702–3709. [CrossRef]
92. Zwielehner, J.; Liszt, K.; Handschur, M.; Lassl, C.; Lapin, A.; Haslberger, A.G. Combined PCR-DGGE fingerprinting and quantitative-PCR indicates shifts in fecal population sizes and diversity of *Bacteroides*; *Bifidobacteria* and *Clostridium* cluster IV in institutionalized elderly. *Exp. Gerontol.* **2009**, *44*, 440–446. [CrossRef] [PubMed]
93. Van Dyke, M.I.; McCarthy, A.J. Molecular biological detection and characterization of *Clostridium* populations in municipal landfill sites. *Appl. Environ. Microb.* **2002**, *68*, 2049–2053. [CrossRef] [PubMed]
94. Bagnouls, F.; Gaussen, H. Dry periods and vegetation. *Comptes Rendus Hebd. Seances Acad. Sci.* **1953**, *236*, 1075–1077.

Article

Partial Least Squares Improved Multivariate Adaptive Regression Splines for Visible and Near-Infrared-Based Soil Organic Matter Estimation Considering Spatial Heterogeneity

Xiaomi Wang [1], **Can Yang** [2] **and Mengjie Zhou** [1,*]

[1] College of Resources and Environmental Sciences, Hunan Normal University, 36 Lushan Road, Changsha 410081, China; xiaomiw@yeah.net

[2] Key Laboratory of Metallogenic Prediction of Nonferrous Metals and Geological Environment Monitoring, Ministry of Education, School of Geosciences and Info-Physics, Central South University, Changsha 410083, China; 195011074@csu.edu.cn

* Correspondence: MengjieZhou@hunnu.edu.cn; Tel.: +86-13-47-61-19-070

Abstract: Under the influence of complex environmental conditions, the spatial heterogeneity of soil organic matter (SOM) is inevitable, and the relationship between SOM and visible and near-infrared (VNIR) spectra has the potential to be nonlinear. However, conventional VNIR-based methods for soil organic matter estimation cannot simultaneously consider the potential nonlinear relationship between the explanatory variables and predictors and the spatial heterogeneity of the relationship. Thus, the regional application of existing VNIR spectra-based SOM estimation methods is limited. This study combines the proposed partial least squares–based multivariate adaptive regression spline (PLS–MARS) method and a regional multi-variable associate rule mining and Rank–Kennard-Stone method (MVARC-R-KS) to construct a nonlinear prediction model to realize local optimality considering spatial heterogeneity. First, the MVARC-R-KS method is utilized to select representative samples and alleviate the sample global underrepresentation caused by spatial heterogeneity. Second, the PLS–MARS method is proposed to construct a nonlinear VNIR spectra-based estimation model with local optimization based on selected representative samples. PLS–MARS combined with the MVARC-R-KS method is illustrated and validated through a case study of Jianghan Plain in Hubei Province, China. Results showed that the proposed method far outweighs some available methods in terms of accuracy and robustness, suggesting the reliability of the proposed prediction model.

Keywords: soil organic matter; near-infrared spectroscopy; spatial heterogeneity; multivariate adaptive regression splines; partial least squares regression

Citation: Wang, X.; Yang, C.; Zhou, M. Partial Least Squares Improved Multivariate Adaptive Regression Splines for Visible and Near-Infrared-Based Soil Organic Matter Estimation Considering Spatial Heterogeneity. *Appl. Sci.* **2021**, *11*, 566. https://doi.org/10.3390/app11020566

Received: 10 October 2020
Accepted: 4 January 2021
Published: 8 January 2021

Publisher's Note: MDPI stays neutral with regard to jurisdictional claims in published maps and institutional affiliations.

1. Introduction

Soil organic matter (SOM) content is significantly relevant to soil fertility, biological productivity, and agricultural sustainable development [1,2]. Accurate prediction of SOM content is of great significance for land management [1,3]. Extensive studies have fully affirmed the capability of SOM prediction methods based on visible and near-infrared (VNIR) spectra [4,5]. However, the function relation between SOM content and VNIR spectra should be established because of the strong soil spatial heterogeneity under the influence of complex environmental conditions. In addition, the formation, variation, and decomposition of SOM are generally influenced by various factors, and the VNIR spectra are comprehensively related to various soil properties; hence, the relation of SOM content to VNIR spectra is complex with high nonlinearity [6,7]. Therefore, further exploration of VNIR spectra-based SOM prediction models is needed to improve simulation and prediction accuracies.

An extensive literature review demonstrates that previous methods for predicting soil properties can be categorized into linear and nonlinear prediction methods. For example, multiple linear regression (MLR), partial least squares (PLS) regression, and geographic weighted regression (GWR) are typical linear prediction methods. These methods have been extensively used in many previous studies because they are easily accessible in most software packages [4]. A common characteristic of these prediction methods is that they assume, either explicitly or implicitly, a linear relationship among the analyzed data sets. A nonlinear relationship, however, is prevalent in practice between SOM content and VNIR spectra. Thus, a series of nonlinear prediction methods has been proposed to estimate nonlinear relationships, including, but not limited to, local weighted regression (LWR) [8,9], artificial neural network [10,11], and support vector machine (SVM) methods [12,13]. Although these methods can efficiently determine nonlinear relationships in certain situations, they lead to models based on global optimization and disregard the spatial heterogeneity of the nonlinear relationship. This underlying principle will cause the accuracy of existing methods to easily change with different samples and have difficulty for application to other regions. Therefore, nonlinear estimation methods considering spatial heterogeneity are needed for SOM content prediction.

The multivariate adaptive regression spline (MARS) method is a nonlinear prediction method that considers the effects of local difference. The MARS method is the expansion of splines, thereby providing immense flexibility by automatically determining the splines (i.e., the number of basis functions (BFs) and locations of knots) without human interference. This method has been successfully applied to various fields, such as geotechnical engineering and soil liquefaction assessment [14–16]. The MARS method has proven to be advantageous due to its adaptation to nonlinear relationships and to its adaptive model construction process, while considering the natural local difference in the training data sets. These characteristics of the MARS method indicate its considerable potential to represent the relationships between VNIR spectra and SOM content. Although tracked records of the application of the MARS model to soil property prediction are available in literature [11,17], these records are not well received. The possible reason may be that most of the applications implicitly or explicitly consider the explanatory variables as mutually independent, which is often not the case in reality. In fact, the VNIR spectra of soil samples are the comprehensive representation of soil properties, and thus multicollinearity inevitably exists. As such, multicollinearity should be removed to obtain relatively independent variables when constructing a MARS model. The multicollinearity-removing strategy from PLS regression is verified to be beneficial by reducing the dimensionality, considering the high dimensionality and correlated representations [4,18]. Therefore, this research proposes the utilization of partial least squares regression to remove multicollinearity and obtain extensive information on the explanatory variables; the resultant principal components (PCs) from the PLS regression are used as a replacement for the original explanatory variables to construct a MARS model, thereby overcoming the weakness of the available MARS model for prediction of soil properties. For simplicity, the entire process is referred to as the partial least squares–based multivariate adaptive regression spline (PLS–MARS) method herein. The proposed PLS–MARS method can effectively represent the nonlinear relationship between the data with local difference and numerous variables, as will be discussed in Section 3.

The accuracy of the prediction model relies significantly on calibrated samples. Underrepresentation of calibrated samples will lead to a biased prediction model due to soil spatial heterogeneity. However, preceding studies (e.g., [8,9,11,12]) focused exclusively on developing new prediction models for estimating soil properties, disregarding the influence of the calibration sets on the reliability of the prediction models. In particular, the calibration sets in most of these studies were selected randomly or from several available methods, such as the concentration gradient method (C method) [19,20]. The prediction model may be biased if inappropriate calibration sets are adopted [3]. Hence, reasonable calibration sets should be selected before they are applied to construct prediction models. Accordingly, our newly developed multi-variable analysis method (i.e., regional multi-variable associate rule mining and Rank–Kennard-Stone (MVARC-R-KS) method) [3] was adopted to select calibration sets in the current study (see Subsection III-A for elaboration). The MVARC-R-KS method was used due to the following two reasons. (1) This method is effective in enhancing the accuracy of linear prediction methods, such as PLS regression (Wang, Chen, Guo and Liu [3]), thereby providing the confidence to extend it for non-linear prediction methods (e.g., the proposed PLS–MARS method in this study). (2) The MVARC-R-KS method can consider multiple influential factors (e.g., chemical component, spectrographic information, and environmental factors) and select a representative calibration set, which makes the VNIR-spectrum–based SOM prediction model calibrated from such a calibration set as to be an extensively applicable one [3]. Three commonly used calibration set selection methods, namely, component concentration representative methods (e.g., the C method [19,20]), spectrum representative methods (e.g., the Kennard–Stone (KS) method [21]), and component concentration and spectrum representative methods (e.g., the Rank–Kennard-Stone (Rank–KS) method [22]), are also utilized and compared with the MVARC-R-KS method to investigate their influences on the prediction models and further verify the effectiveness of the proposed method. To the best of our knowledge, such a systematic comparison of the influences of different calibration set selection methods on the performance of SOM prediction models, particularly nonlinear prediction models, such as the proposed PLS–MARS model herein, appears to be original. This comparison is expected to guide the calibration set selection for a specific prediction model in the future.

The remainder of the paper is structured as follows. The study area and materials for this study are introduced in Section 2. The calibration set selection methods and the proposed PLS–MARS prediction method are described in Section 3. Experiments on the simulated data set and real application are discussed in Section 4, illustrating and validating the proposed method. The principal contributions and observations of this study are discussed in Section 5. Section 6 concludes this research.

2. Study Area and Materials

2.1. Study Area

Sampling was done in Jianghan Plain in Hubei Province, China. Jianghan Plain is an important agricultural region because of the humid climate and fertile soil. However, in recent years, the ecological system of the plain has become unstable, and the SOM content has decreased due to long-term agricultural activities. Therefore, estimating the SOM content with high accuracy in this plain is necessary and beneficial to land resource management. To this end, 260 topsoil samples (i.e., 0 cm to 30 cm; see Figure 1) were obtained from this area in June 2014. The samples were taken at least 100 m apart from one another. All samples were partitioned into two parts for chemical study and spectral measurement.

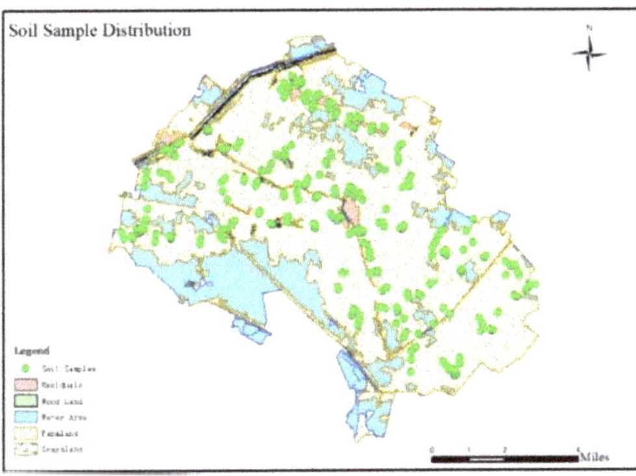

Figure 1. Distribution of the soil samples.

2.2. SOM Content and Spectral Measurement

SOM content was measured by following the analysis and calculation methods in literature [3,23]. An ASD FieldSpec3 portable spectral radiometer was used to measure spectral reflectance, which ranged from 350 nm to 2500 nm with a sampling interval from 1.4 nm to 2 nm. The measurement procedures were carried out in the dark to decrease the interference of external light. The light source was from a standardized white halogen light, which had a 45° incident angle and was at a distance of 45 cm.

3. Methods

VNIR spectra-based SOM content prediction generally consists of the following three parts. Part 1 is the spectral preprocessing. Spectral reflectance is unavoidably impacted by random noises, baseline drift, and scattering effects because of the influence of error from the experimental instruments and ambient noises; thus, the stability of the constructed VNIR-spectrum–based prediction model is affected [3]. Hence, spectral preprocessing is a prerequisite for the construction of the prediction model, which will be detailed in Section 3.1. In part 2, the representative calibration set is selected using methods such as the MVARC-R-KS method [3], which will be detailed in Section 3.2. Part 3 aims to build an SOM prediction model based on the VNIR spectra, such as by using typical prediction methods and the proposed PLS–MARS method, which will be elaborated in Section 3.3.1. Lastly, in Section 3.3.2, the accuracy of the constructed prediction model is evaluated by several assessment indices, as will be described in Section 3.3.2.

3.1. Spectral Preprocessing

According to the characteristics of the spectra, spectral preprocessing was conducted as follows. The spectra ranging from 400 mm to 2350 mm was retained to reduce the interference of noises (Liu et al., 2014b; Liu et al., 2016) (Figure 2a). In addition, continuum removed spectral curves exhibited several diminutive absorption valleys (Figure 2b), which could interfere with the prediction based on the VNIR spectra. Hence, Savitzky–Golay (SG) smoothing was used to denoise this interference. Furthermore, multiplicative scatter correction (MSC) and mean center (MC) operations were utilized to reduce the influence of the unavoidable scattering. In short, the spectra were preprocessed successively through SG smoothing, MSC operation, and MC operation.

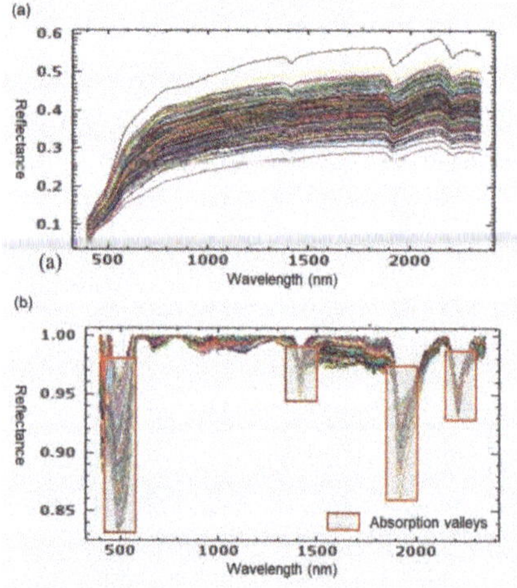

Figure 2. (**a**) Pretreatment spectral curves of the soil samples. (**b**) Continuum removed spectral curves of the soil samples.

3.2. Calibration Set and Validation Set Selection Methods

The MVARC-R-KS method [3] aimed at adaptively selecting the calibration set through multivariate analysis. The procedure of selecting the calibration set using the MVARC-R-KS method comprised two phases [3]. Phase 1 initially mined the clustering distribution zones using the MVARC method, which integrated the a priori algorithm, a density-based clustering algorithm, and the Delaunay triangulation [3]. The association between the SOM content and VNIR spectra may vary according to location and surrounding environment, contributing to the generally strong spatial variability of soil samples. Hence, to open out the spatial heterogeneity of soil samples, environmental variables and SOM content were set as input data for the MVARC method. The spatial heterogeneity of soil samples was checked by Moran's I; results are given in Table 1. Samples in the same cluster mined by the MVARC method were relatively similar with respect to the impact of the surrounding environment. In addition, the influence of environmental variables on SOM in clusters presented significant differences. Phase 2 utilized the Rank–KS algorithm [22] to select the calibration set from the clustering zones. The eventually selected calibration set was an internal component concentration and spectrum representations, as well as an external environment representation. In addition, the size of the calibration set was set as approximately 70% of the soil samples, and the remaining 30% of the samples were used as the validation set, referring to previous studies [23,24].

Table 1. Moran's I index and z-score of data related with prediction models calibrated by different calibration set selection methods.

Prediction Model	Selection Method	Original Data		Predicted Data		Residuals	
		I	Z-Score	I	Z-Score	I	Z-Score
Partial least squares–based geographic weighted regression (PLS–GWR)	Concentration gradient (C) method	1.15	6.33	1.06	5.83	0.01	0.20
	Kennard-Stone (KS) method	1.28	5.87	1.15	5.23	−0.07	−1.27
	Rank–Kennard-Stone (Rank–KS) method	1.33	6.30	1.28	6.03	−0.10	−1.35
	Regional multi-variable associate rule mining and Rank–Kennard-Stone (MVARC-R-KS) method	1.39	6.17	1.24	5.49	−0.07	−0.96
Partial least squares–based multivariate adaptive regression (PLS–MARS)	C	1.15	6.33	1.03	5.69	−0.01	0.06
	KS	1.28	5.87	1.12	5.15	0.07	1.42
	Rank–KS	1.33	6.30	1.12	5.30	−0.01	0.01
	MVARC-R-KS	1.39	6.17	1.17	5.23	0.08	1.33

Apart from the MVARC-R-KS method, three commonly used calibration set and validation set selection methods—C method, KS method, and Rank–KS method—were utilized for comparison to verify the representation of the selected calibration set.

3.3. VNIR-Based Prediction Methods

According to the VNIR-based SOM predicting demands, the PLS–MARS method was proposed to effectively represent the nonlinear relationship between the data with local difference and numerous variables (see Section 3.3.1). To verify the effectiveness of the proposed PLS–MARS method, typical prediction methods, such as MLR, SVM, PLS, and GWR, were used to construct VNIR-based SOM prediction models. To realize a more comparative analysis with the proposed PLS–MARS method, the GWR method was also combined with PLS (named PLS–GWR) to construct a prediction model. Eventually, a series of evaluation indices was used to evaluate the simulation and prediction performance of the prediction models, which is described in detail in Section 3.3.2.

3.3.1. PLS–MARS Method

The PLS–MARS method is a combination of PLS and MARS methods that constructs a nonlinear prediction model with local regression. The flow of PLS–MARS comprises two parts (Figure 3). Part 1 removes the multicollinearity of the explanatory variables by obtaining PCs using a PLS regression method. Part 2 constructs the MARS model based on the predictors and PCs obtained in Part 1. The details of PLS–MARS are described as follows.

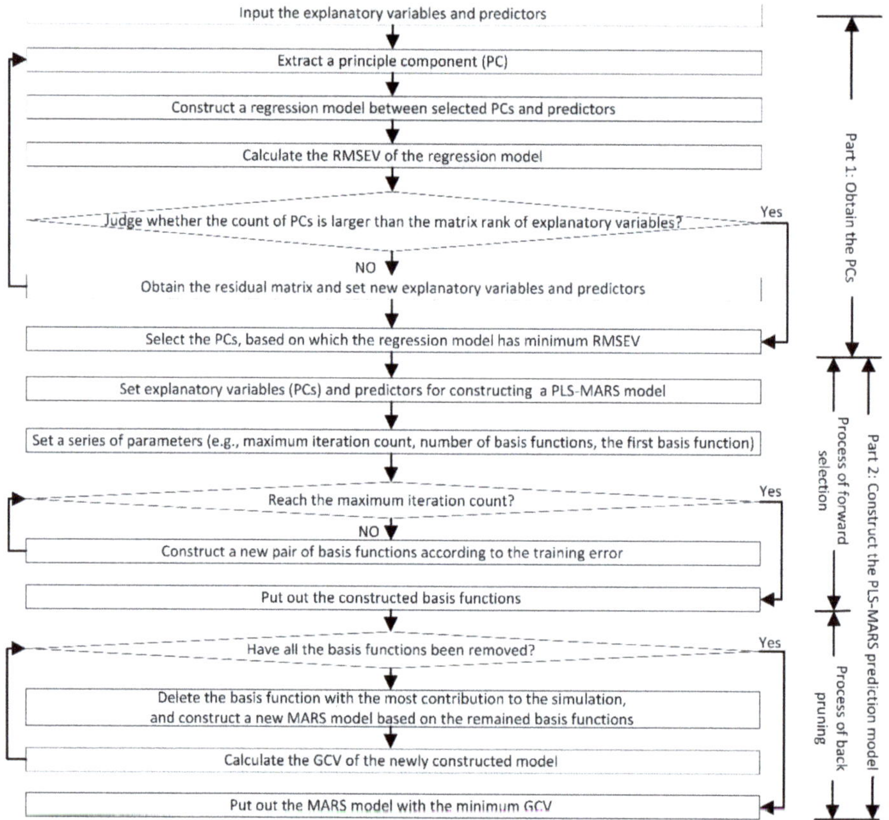

Figure 3. Procedure of constructing the partial least squares–based multivariate adaptive regression spline (PLS–MARS) model.

PCs were abstracted in Part 1. The explanatory variables and predictors were labeled as E_0 and F_0, respectively. The procedure of obtaining PCs was to abstract the PCs first. Thereafter, a series of PLS regression models based on the obtained PCs and predictors was built. Lastly, the accuracy of the models was evaluated. The model with the highest accuracy was selected, and the corresponding PCs were output as the eventual PCs. The steps of this procedure are described in detail as follows.

Step 1: The PC t_1 of the explanatory variables E_0 and PC u_1 of the predictors F_0 were extracted. The estimated values of PCs can be expressed in accordance with Equation (1). The abstracted PCs should represent sufficient information of the variables, and the PCs t_1

and u_1 should be highly correlated. The preceding calculation targets imply that t_1 and u_1 can be calculated based on the conditions in Equation (2) by using the Lagrange multipliers.

$$\begin{cases} t_1 = E_0 w_1 \\ u_1 = F_0 v_1 \end{cases} \tag{1}$$

$$\begin{cases} \langle t_1, u_1 \rangle = \langle E_0 w_1, F_0 v_1 \rangle = w_1^T E_0^T F_0 \Rightarrow Max \\ w_1^T w_1 = 1, \ v_1^T v_1 = 1 \end{cases} \tag{2}$$

Step 2: The regression (i.e., Equation (3)) was constructed based on t_1 and u_1; the model is expressed as follows:

$$\begin{cases} E_0 = t_1 \alpha_1^T + E_1 \\ F_0 = u_1 \beta_1^T + F_1 \end{cases} \tag{3}$$

where α_1 and β_1 are the model parameters to ensure the high correlation between t_1 and u_1, which can be estimated using least squares criterion; E_1 and F_1 are the residual matrixes.

Step 3: The residual matrixes E_1 and F_1 were set as the explanatory variables and predictors, respectively. Step 1 was iteratively repeated until the count of the selected PCs reached h (i.e., the matrix rank of E_0) (Equation (4)). The PCs t of the explanatory variables can be calculated using Equation (2).

$$\begin{cases} E_0 = t_1 \alpha_1^T + \cdots + t_h \alpha_h^T + E_h \\ F_0 = u_1 \beta_1^T + \cdots u_h \beta_h^T + F_h \end{cases} \tag{4}$$

Step 4: The first δ principal component PCs ($0 < \delta \leq h$) were successively selected. In addition, the corresponding PLS regression model between the predictors and extracted PCs of the explanatory variables was constructed.

Step 5: The precision of the PLS model was evaluated using the root-mean-square error (RMSEV) calculated by the cross-validation method (leave-one-out sample). High precision of the PLS regression model led to a small value of RMSEV. The PCs $t = \{t_1, \cdots, t_p\}$ for constructing the PLS regression model with the highest precision were output as PCs for constructing the following the MARS model.

In Part 2, the MARS prediction model was built based on the explanatory variables $\{t_1, \cdots, t_p\}$ and predictors. The MARS model was established based on several BFs in the form of an expansion of splines. The estimation of a true function $f(t)$ utilizing the MARS method [16] based on BFs is expressed as follows:

$$\hat{f}(t_1, \cdots, t_p) = \sum_{m=0}^{M} a_m B_m (t_1, \cdots, t_p) \tag{5}$$

where coefficient a_m is obtained using the least squares method and $B_m(t_1, \cdots, t_p)$ (Equation (6)) is one of the BFs, which is multiplied by several $b_{k,m}$ as shown in Equation (7).

$$B_m(t_1, \cdots, t_p) = \prod_{k=1}^{K_m} b_{k,m}\left(t_{v(k,m)} \middle| P_{k,m}\right) \tag{6}$$

where K_m is the number of $b_{k,m}$ (Equation (7)), which is a bilateral truncation power function decided by the parameters $P_{k,m}$ and explanatory variables $t_v(k, m)$ that correspond to the kth truncated spline BF (SBF) in the mth term of Equation (6).

$$b_{k,m}\left(t_{v(k,m)} \middle| P_{k,m}\right) = \left[S_{k,m} \times \left(t_{v(k,m)} - r_{k.m}\right) \right]_+^q = max(0, \ S_{k,m} \times \left(t_{v(k,m)} - r_{k.m}\right)^q \tag{7}$$

where $P_{k,m} = (S_{k,m}, r_{k.m})$, truncation direction $S_{k,m} = \pm 1$, $r_{k.m}$ is the knot of the BF $b_{k,m}\left(t_{v(k,m)} \middle| P_{k,m}\right)$, and q is the power of SBF reflecting the degree of smoothness of the resulting MARS estimation [16]. In this study, q was set as 1 to simplify the process.

The process of constructing a MARS model consisted of forward selection and backward pruning. In forward selection, four steps were conducted, as follows:

Step 1: BF was set as $B_0(t) = 1$, and the threshold of the number of BFs and the threshold of the model precision were set.

Step 2: Two new adaptive BFs that yielded the minimum training error were iteratively created.

Step 3: Step 2 was repeated until the number of BFs reached the number threshold or the model precision reached the threshold.

Step 4: All BFs and related parameters were provided.

In forward selection, permission was limited to adding BFs. Accordingly, several BFs were used merely for constructing succeeding BFs and contributed insignificantly to the eventual model. The threshold of the number of BFs was generally set as high in the forward selection procession. Hence, the backward pruning for deleting redundant BFs was necessary.

In backward pruning, the model was simplified by deleting one least important BF (based on generalized cross-validation (GCV)) at each step until no more BFs were available to be deleted. A new model was rebuilt at each step; the model with the minimum GCV was selected as the eventual prediction model. GCV is computed in accordance with Equation (8).

$$GCV = \frac{\frac{1}{N}\sum_{i=1}^{N}\left[y_i - \hat{f}(t_i)\right]^2}{\left[1 - \frac{M + d \times \frac{(M-1)}{2}}{N}\right]^2} \tag{8}$$

where N is the size of the calibration set, M is the number of BFs in the model, d is a penalizing factor, which is set as 3 in this study according to a report [16,25], and $(M-1)/2$ denotes the number of the hinge function knots. Consequently, the function penalizes the model for its number of BFs and knots.

In establishing a MARS model, a series of threshold values was preset to realize the adaptive operation and obtain a suitable result. Accordingly, the model with the minimum GCV was selected as the eventual model. The implementation of the PLS–MARS method was realized using MATLAB software.

3.3.2. Fitness Assessment of the VNIR-Based Prediction Model

The prediction model can be evaluated by several assessment indices: the corrected Akaike information criterion (AICc) [26,27], the coefficient of determination of simulation analysis (R_c^2), the coefficient of determination of prediction analysis (R_p^2), root of mean square simulation error (RMSEV), root of mean square prediction error (RMSEP), relative percent deviation (RPD), and Moran's I. AICc estimates the relative amount of information lost by a given model. When comparing a series of models for the same data, the less information a model loses with a lower AICc value, the higher the quality of that model. R_c^2 and RMSEV are used to analyze the simulation precision and stabilization of the model. A low RMSEV value and high R_c^2 value indicate the high stabilization and simulation precision of the model. The prediction precision of the model is estimated by R_p^2, RMSEP, and RPD. If RMSEP is low and R_p^2 is high, then the predictive capability of the model is considered good. In soil spectrographic analysis, if the RRP is less than 1, then the model is not recommended due to poor prediction capability; if the RPD is larger than 1 and less than 1.4, then the prediction capability is still deemed as poor; if the RPD is between 1.4 and 1.8, then the model can be used to perform prediction analysis; if the RPD is larger than 1.8, then the model should have very good prediction capability [28,29]. Moran's I is utilized to test the randomness of residuals. A good model will yield a random series of residuals without autocorrelation [30].

4. Results

4.1. Verification of PLS–MARS Method Based on a Simulated Data Set

An illustrative function (Equation (9)) was set to validate the performance of the proposed PLS–MARS method. The method was proposed mainly for nonlinear prediction applications with numerous explanatory variables. Hence, the size of the explanatory variables in the illustrative function in Equation (9) was set as N ($N = 100$; $N = 500$), and the corresponding relationship between explanatory variables and the predictor was nonlinear. For comparison, MLR, PLS, and SVM methods were also utilized to estimate the illustrative function. Each prediction model was calibrated by 500 samples generated by the Latin hypercube sampling method [16] and tested using 1000 randomly generated samples.

$$f = x_1 \times \sin(x_1) + x_2 \times \sin(x_2) + \cdots + x_{100} \times sin(x_N) \tag{9}$$

The performance of the models was evaluated by fitness assessment indices (e.g., R_c^2, RMSEV, R_p^2, RMSEP, and RPD) (Table 2). The results showed that the MLR model performed well when estimating Equation (9) with 100 explanatory variables, but performed poorly when estimating Equation (9) with 500 explanatory variables. These results suggest the accuracy of the constructed MLR model decreases as the complexity and size of explanatory variables increase. Compared with the MLR model, the PLS method had relatively high accuracy due to the suitability of its equations with numerous explanatory variables. However, SVM and PLS–MARS methods (i.e., as nonlinear prediction methods) had the highest estimation and prediction accuracies when they were used to estimate the relationship in Equation (9).

Table 2. Evaluation results of typical prediction models for the illustrative function.

Illustrative Functions	Explanatory Variable Size	Prediction Model	Coefficient of Determination of Simulation Analysis (R_c^2)	Root of Mean Square Simulation Error (RMSEV (g kg^{-1}))	Coefficient of Determination of Prediction Analysis (R_P^2)	Root of Mean Square Prediction Error (RMSEP (g kg^{-1}))	Relative Percent Deviation (RPD)
f	100	Multiple linear regression (MLR)	0.92	1.08	0.93	0.99	3.9
		partial least squares regression (PLS)	0.92	1.02	0.93	0.93	4.2
		Support vector machine (SVM)	0.96	0.71	0.97	0.61	6.4
		PLS–MARS	0.94	0.89	0.96	0.69	5.7
	500	MLR	0.56	4.29	0.02	31.2	0.2
		PLS	0.64	3.33	0.75	2.98	1.9
		SVM	0.69	3.09	0.78	2.79	2.2
		PLS–MARS	0.65	3.21	0.76	2.93	2.1

In summary, the comparison experiments on the simulated data set verified that the PLS–MARS method exhibits a high performance for estimating nonlinear relationships to numerous explanatory variables.

4.2. Case Study of PLS–MARS Method

In this subsection, the PLS–MARS method was used to construct a prediction model between VNIR spectra and SOM content based on the representative calibration set in the study area introduced in Section 2. The study area was utilized as a demonstration zone to validate the effectiveness and accuracies of the PLS–MARS method. Three missions were carried out as follows.

Mission 1 selected the representative calibration set by utilizing the MVARC-R-KS method [3], which is specified in Section 4.2.1. Mission 2 (in Section 4.2.2) constructed the PLS–MARS method, which was calibrated by the selected calibration set using the MVARC-R-KS method [3]. Typical prediction methods based on typical calibration sample selection methods were utilized to predict SOM content in the research area to verify the effectiveness of the proposed PLS–MARS method and the influence of the calibration set. These comparison experiments are elaborated in Section 4.2.3.

4.2.1. Calibration Set Selected Using MVARC-R-KS Method

The procedure of selecting the calibration set using the MVARC-R-KS method comprised two phases [3]. Phase 1 consisted of obtaining clustering zones with similar influences of the environment on SOM content. Phase 2 utilized the Rank–KS method to select representative samples from the clustering zones. The selected samples were eventually merged as the calibration set (Figure 4). Typical methods, such as C, KS, and Rank–KS methods, were used for comparison to verify the representation of the calibration set. If the calibration set was representative, then the distribution characteristics of the calibration set would be similar to the entire samples. The statistical values of SOM content and the spectral reflectance information of the calibration sets selected by the aforementioned methods were calculated and are shown in Figure 5 and Table 3. The results showed that the mean and median values of the SOM content in the calibration set selected by the KS method were lower than those in the entire samples. In contrast to the KS method, the C, Rank–KS, and MVARC-R-KS methods obtained calibration sets having a considerably similar distribution to the entire samples for SOM content. Compared with the calibration sets selected by the KS, Rank–KS, and MVARC-R-KS methods, the calibration set selected by the C method had less similar spectral reflectance information to that of the entire samples. Compared with the C, KS, and Rank–KS methods, the MVARC-R-KS method further took into consideration the spatial variation and impacts of the environmental variables [3].

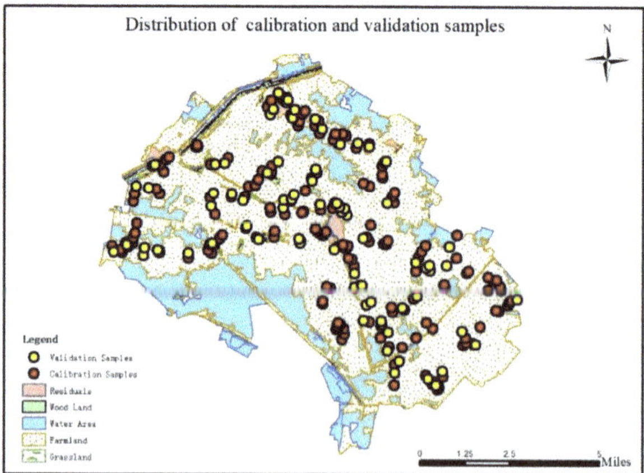

Figure 4. Calibration and validation samples selected utilizing the regional multi-variable associate rule mining and Rank–Kennard-Stone (MVARC-R-KS) method.

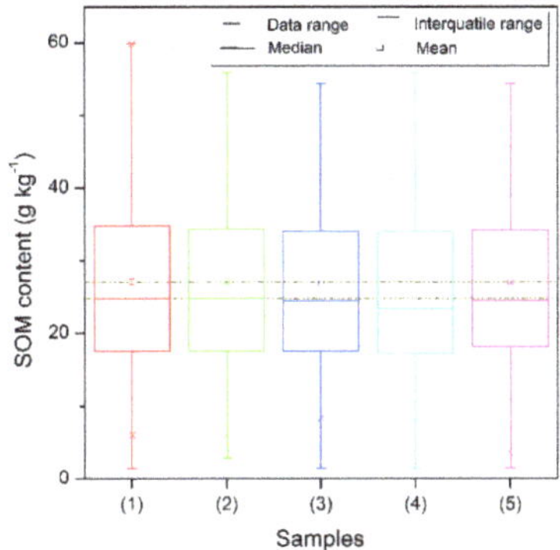

Figure 5. Statistics of the soil organic matter (SOM) content in the following samples: (1) all samples; (2) samples selected utilizing the concentration gradient (C) method; (3) samples selected utilizing the Rank–Kennard-Stone (Rank–KS) method; (4) samples selected using the Rank method; (5) samples selected utilizing the MVARC-R-KS method.

Table 3. Statistical values of the spectral reflectance information in the samples.

Statistic Values	All Samples	Samples Selected by Typical Methods			
		C	Rank–KS	KS	MVARC-R-KS
Mean	0.3542	0.3495	0.3563	0.3561	0.3541
Variation	0.0369	0.0360	0.0404	0.0401	0.0405

In summary, the calibration set selected by the MVARC-R-KS method was the SOM content, spectrum representative, and environment representative [3]. Hence, the MVARC-R-KS method seemed to be reasonable to select a representative calibration set.

4.2.2. Performance of the PLS–MARS Model Calibrated by the Calibration Set Selected Utilizing the MVARC-R-KS Method

The PLS–MARS method was used to build the SOM prediction model based on the VNIR spectra in accordance with the selected calibration set (Figure 4). PCs were first calculated and selected. The count of PCs was set at 10, with the highest precision (i.e., RMSEV) of the constructed regression model (Figure 6) based on the PC selection procedure described in Phase 1 of Section 3.2. Thereafter, the PLS–MARS model was constructed based on the PCs and SOM content. The PLS–MARS model was eventually evaluated. Table 4 shows the fit assessments of the prediction model. The result indicated that the PLS–MARS model calibrated by the samples selected utilizing the MVARC-R-KS method [3] could estimate SOM content with high stability and prediction precision.

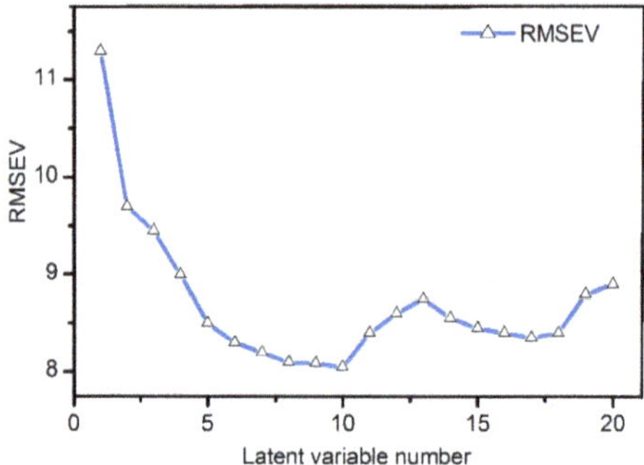

Figure 6. Root-mean-square error (RMSEV) of the prediction model constructed using the latent variables (i.e., principal components (PCs)).

Table 4. Evaluation results of prediction models calibrated by different calibration set selection methods.

Prediction Model	Selection Method	Corrected Akaike Information Criterion (AICc)	R_c^2	RMSEV (g kg^{-1})	R_p^2	RMSEP (g kg^{-1})	RPD
MLR	C	1005.54	0.54	8.9	0.11	28.0	0.44
	KS	973.4	0.62	8.19	0.11	26.0	0.47
	Rank–KS	948.65	0.50	9.04	0.22	23.1	0.52
	MVARC-R-KS	937.79	0.54	8.9	0.09	30.8	0.39
PLS	C	877.64	0.73	6.39	0.53	8.21	1.48
	KS	844.87	0.78	5.87	0.50	8.82	1.38
	Rank–KS	829.80	0.71	6.51	0.56	8.48	1.51
	MVARC-R-KS	785.50	0.79	5.83	0.70	6.98	1.61
SVM	C	996.75	0.54	8.70	0.50	8.78	1.38
	KS	972.02	0.58	8.16	0.39	9.47	1.29
	Rank–KS	912.02	0.55	8.17	0.46	9.34	1.38
	MVARC-R-KS	935.36	0.50	8.84	0.55	8.11	1.56
PLS–GWR	C	1187.90	0.75	5.80	0.54	8.18	1.49
	KS	1227.43	0.79	5.19	0.50	8.76	1.42
	Rank–KS	1190.26	0.76	5.55	0.60	8.02	1.60
	MVARC-R-KS	1213.80	0.74	6.12	0.66	6.83	1.78
PLS–MARS	C	1156.61	0.80	5.45	0.53	8.20	1.48
	KS	1213.20	0.83	5.12	0.55	8.13	1.50
	Rank–KS	1166.12	0.79	5.50	0.62	7.88	1.63
	MVARC-R-KS	1135.50	0.84	5.14	0.71	6.52	1.94

4.2.3. Comparison of the PLS–MARS Model with Typical Prediction Models

The MLR, PLS, SVM, and PLS–GWR methods were utilized to construct the SOM prediction model based on the VNIR spectra in the research region to verify the availability of the PLS–MARS method. The precision of the prediction method was substantially correlated with the selected calibration set. Hence, the C, KS, Rank–KS, and MVARC-R-KS methods were introduced to obtain the representative calibration sets.

The performance of the models was evaluated using fitness assessment indices (e.g., AICc, R_c^2, RMSEV, R_p^2, RMSEP, and RPD) (Table 4). Two conclusions could be drawn. One was that the performance of the prediction models highly relied on the selected calibration sets; such performance was achieved for the following reasons. The results imply that the accuracies of the models based on different calibration sets fluctuated heavily. For example, the prediction models calibrated using the calibration sets selected by the C and KS methods had poor prediction capabilities. The Rank–KS method selected the calibration set for building prediction models with relatively higher accuracies than those of the C and KS methods. The prediction models calibrated by the samples obtained utilizing the MVARC-R-KS method had the highest prediction accuracies. These results sufficiently verified the dependency of the performance of the prediction models on the selected calibration sets. Because the PLS–GWR model was a local prediction model, the calibration sets had less effect on the prediction results. Another conclusion was that the PLS–MARS method outperformed the typical prediction models for estimating VNIR-spectrum–based SOM content in the study area. The following results support this conclusion. The results in Table 4 show that the MLR model had poor prediction accuracies and thus could not be used to predict SOM content in the study area. The possible reason is due to the inapplicability of these methods for complex high-dimension data sets. Although the PLS and SVM models had relatively higher simulation and prediction accuracies, their model accuracies were lower than the PLS–GWR and PLS–MARS models that considered the local difference. The Moran's I indexes in Table 1 indicate that both the PLS–MARS model and PLS–GWR model could maintain the spatial heterogeneity distribution characteristics of data. In addition, random spatial residuals without autocorrelation could further verify the effectiveness of the PLS–GWR model and the PLS–MARS model. Compared with the PLS–GWR model, the PLS–MARS model, which considered the nonlinear relationship between SOM content and VNIR spectra, had the highest simulation and prediction accuracies among the prediction models. Since PLS–GWR and PLS–MARS were constructed based on the same PCs, their AICc values were comparable. Nevertheless, the AICc values indicated that the proposed PLS–MARS model had higher model quality than the PLS–GWR model. In summary, the PLS–MARS model calibrated by the samples obtained utilizing the MVARC-R-KS method indicated immense potential for building a prediction model of SOM content based on the VNIR spectra in the riverside region.

5. Discussion

Intrinsic strategies for the prediction method and selected calibration set were significant factors that affected the performance of the constructed prediction model. The influence of the calibration set on model performance, and the good performance of the MVARC-R-KS method and the proposed PLS–MARS prediction method will be discussed in Sections 5.1 and 5.2, respectively.

5.1. Influence of the Calibration Set on the Model Performance

If the calibration set can well represent the relation of the SOM content with VNIR spectra, then the prediction model calibrated using this calibration set will have a good prediction performance. Table 4 shows that the prediction models calibrated using the calibration sets selected by the classical C and KS methods lacked good prediction performance, whereas the prediction models based on the Rank–KS method could build models with considerably high R_p^2 and RPD in the study area. This phenomenon was mainly due to the fact that the calibration set obtained utilizing the C or KS method was representative of either the SOM content or the spectrographic information. By contrast, the calibration set obtained utilizing the Rank–KS method was both SOM content and spectral information representative. These results also indicated that the component concentrations and spectrographic information should be simultaneously taken into account to obtain the representative calibration set to accurately build a VNIR-spectrum–based SOM prediction model. However, in comparison with the Rank–KS method, the MVARC-R-KS method could select the more representative calibration sets, which could calibrate the prediction models with the highest accuracy. This result could be attributed to the consideration of the impact of the surrounding environment and spatial heterogeneity on the samples in the MVARC-R-KS method, which were substantially consistent with real situations.

In summary, the prediction models depended substantially on the selected calibration sets. A representative calibration set instead of a randomly selected calibration set should be obtained in advance to construct a prediction model with good performance. Hence, calibration set selection should be taken seriously during the construction of the prediction models. This study suggests that the MVARC-R-KS method [3] can be set as the optimal choice in the riverside region. This is because this method simultaneously considers the influence of multiple influential factors (i.e., chemical components, spectrographic information, and environmental factors), thereby making the VNIR-spectrum–based SOM prediction model calibrated by the selected calibration set an extensively applicable one.

5.2. Strategies of the PLS–MARS Method

The performance of the PLS–MARS method was analyzed using the simulated data set and real application. In addition, the effectiveness and accuracy of PLS–MARS were compared with those of several commonly used prediction methods (i.e., MLR, PLS, SVM and PLS–GWR methods) in literature. The results in the simulated application implied that the PLS, SVM, and PLS–MARS methods could efficiently estimate the relationship between numerous explanatory variables and their responses. However, if the relationship exhibited significantly non-linear features and spatial heterogeneity, such as the relation of the SOM content with VNIR spectra in the study area, then the PLS–MARS method produced more accurate results than the rest of the prediction methods tested. This superiority is generally due to the consideration of the natural local difference of data sets, and the construction of a local non-linear optimization model in the PLS–MARS method rather than the global non-linear optimization characteristics of the other prediction methods. The adaptive strategy of the PLS–MARS method also facilitated the automatic estimation of natural relationships and supported their promotion. Thus, the PLS–MARS method has immense potential for estimating the complex nonlinear relationships of geographical data with numerous explanatory variables, and the PLS–MARS method calibrated by the samples obtained utilizing the MVARC-R-KS method shows great potential in estimating significantly nonlinear relationships in spatial and non-spatial domains.

6. Conclusions

The PLS–MARS method was proposed to construct a nonlinear model for SOM content prediction, in which the MVARC-R-KS method was adopted to select the optimal calibration set. Several commonly used calibration set selection methods were applied to calibrate the PLS–MARS model, and their influences on the prediction precision of the established model were investigated. The implementation procedure for the proposed method was described in detail. The proposed approach was illustrated and validated through a simulated data set and practical application in the Jianghan Plain. A comparative study with conventional prediction methods indicated that the proposed PLS–MARS method presents the following advantages: (1) the proposed method can efficiently estimate the nonlinear relationship underlying data with numerous variables; (2) the proposed method can accurately construct the highly nonlinear relationship of the SOM content with VNIR spectra, with consideration of the spatial heterogeneity of the relationship; (3) the proposed method can automatically construct a prediction model without substantial prior knowledge of the data set.

The following new findings were obtained based on the application of the PLS–MARS method to estimate SOM content in the Jianghan Plain in China: (1) the accuracy of prediction models, including the PLS–MARS model, depends significantly on the selected calibration sets; (2) the combined PLS–MARS and MVARC-R-KS methods have immense stability and prediction capabilities when estimating the relation of SOM content with VNIR spectra in the study area.

In general, the proposed PLS–MARS method can accurately estimate the nonlinear relationship with local differences. The PLS–MARS method calibrated by the samples obtained utilizing the MVARC-R-KS method has considerable potential for estimating SOM. Furthermore, the PLS–MARS method is conceptually simple and easily executed via a programming language, thereby ensuring easy application.

Future studies will focus on further practical applications of the PLS–MARS method. For example, a PLS–MARS model calibrated by the calibration set selected utilizing

MVARC-R-KS method can serve as a potential prediction model to estimate other soil properties (e.g., iron content, copper content, and soil organic C) based on the VNIR spectra in the study area or other areas. The proposed PLS–MARS method can also be used to characterize the nonlinear relationships of other geographic phenomena.

Author Contributions: X.W. and M.Z. conceived and designed the experiments; X.W. and C.Y. performed the experiments; all the authors analyzed the data; X.W. wrote the paper; all authors contributed to the revision of the manuscript. All authors have read and agreed to the published version of the manuscript.

Funding: This research was funded by [the Scientific Research Program of the Hunan Education Department] grant number [19C1113], [the Natural Science Foundation of Hunan Province] grant number [2020JJ5363], and the National Natural Science Foundation of China grant number [41901314].

Institutional Review Board Statement: Not applicable.

Informed Consent Statement: Not applicable.

Data Availability Statement: Data available on request due to restrictions privacy. The data presented in this study are available on request from the corresponding author. The data are not publicly available because the data is the result of the whole research group's hard work.

Conflicts of Interest: The authors declare no conflict of interest.

References

1. Singh, M.; Sarkar, B.; Bolan, N.S.; Ok, Y.S.; Churchman, G.J. Decomposition of soil organic matter as affected by clay types, pedogenic oxides and plant residue addition rates. *J. Hazard. Mater.* **2019**, *374*, 11–19. [CrossRef]
2. Tian, P.; Mason-Jones, K.; Liu, S.; Wang, Q.; Sun, T. Form of nitrogen deposition affects soil organic matter priming by glucose and cellulose. *Biol. Fertil. Soils* **2019**, *55*, 383–391. [CrossRef]
3. Wang, X.; Chen, Y.; Guo, L.; Liu, L. Construction of the Calibration Set through Multivariate Analysis in Visible and Near-Infrared Prediction Model for Estimating Soil Organic Matter. *Remote Sens.* **2017**, *9*, 201. [CrossRef]
4. Guo, L.; Zhang, H.; Chen, Y.; Qian, J. Combining Environmental Factors and Lab VNIR Spectral Data to Predict SOM by Geospatial Techniques. *Chin. Geogr. Sci.* **2019**, *29*, 258–269. [CrossRef]
5. Moura-Bueno, J.M.; Dalmolin, R.S.D.; ten Caten, A.; Dotto, A.C.; Demattê, J.A.M. Stratification of a local VIS-NIR-SWIR spectral library by homogeneity criteria yields more accurate soil organic carbon predictions. *Geoderma* **2019**, *337*, 565–581. [CrossRef]
6. Liu, Y.; Chen, Y. Estimation of total iron content in floodplain soils using VNIR spectroscopy—A case study in the Le'an River floodplain, China. *Int. J. Remote Sens.* **2012**, *33*, 5954–5972. [CrossRef]
7. Liu, Y.; Chen, Y. Feasibility of Estimating Cu Contamination in Floodplain Soils using VNIR Spectroscopy—A Case Study in the Le'an River Floodplain, China. *Soil Sedim. Contam. An Int. J.* **2012**, *21*, 951–969. [CrossRef]
8. Liu, Y.; Guo, L.; Jiang, Q.; Zhang, H.; Chen, Y. Comparing geospatial techniques to predict SOC stocks. *Soil Tillage Res.* **2015**, *148*, 46–58. [CrossRef]
9. Liu, Y.; Song, Y.; Guo, L.; Chen, Y.; Lu, Y.; Liu, Y. Geostatistical models of soil organic carbon density prediction based on soil hyperspectral reflectance. *Trans. Chin. Soc. Agric. Eng.* **2017**, *33*, 183–191.
10. Tekin, Y.; Tümsavas, Z.; Mouazen, A.M. Comparing the artificial neural network with parcial least squares for prediction of soil organic carbon and pH at different moisture content levels using visible and near-infrared spectroscopy. *Rev. Bras. Ciência Solo* **2014**, *38*, 1794–1804. [CrossRef]
11. Liess, M.; Schmidt, J.; Glaser, B. Improving the Spatial Prediction of Soil Organic Carbon Stocks in a Complex Tropical Mountain Landscape by Methodological Specifications in Machine Learning Approaches. *PLoS ONE* **2016**, *11*, e0153673.
12. Huang, N.; Wang, L.; Guo, Y.; Niu, Z. Upscaling plot-scale soil respiration in winter wheat and summer maize rotation croplands in Julu County, North China. *Int. J. Appl. Earth Obs. Geoinf.* **2017**, *54*, 169–178. [CrossRef]
13. Laamrani, A.; Berg, A.A.; Voroney, P.; Feilhauer, H.; Blackburn, L.; March, M.; Dao, P.D.; He, Y.; Martin, R.C. Ensemble Identification of Spectral Bands Related to Soil Organic Carbon Levels over an Agricultural Field in Southern Ontario, Canada. *Remote Sens.* **2019**, *11*, 1298.
14. Zheng, G.; Zhang, W.B.; Zhou, H.Z.; Yang, P.B. Multivariate adaptive regression splines model for prediction of the liquefaction-induced settlement of shallow foundations. *Soil Dyn. Earthq. Eng.* **2020**, *132*, 10.
15. Huang, H.; Ji, X.L.; Xia, F.; Huang, S.H.; Shang, X.; Chen, H.; Zhang, M.H.; Dahlgren, R.A.; Mei, K. Multivariate adaptive regression splines for estimating riverine constituent concentrations. *Hydrol. Process.* **2020**, *34*, 15. [CrossRef]
16. Liu, L.-L.; Cheng, Y.-M. Efficient system reliability analysis of soil slopes using multivariate adaptive regression splines-based Monte Carlo simulation. *Comput. Geotech.* **2016**, *79*, 41–54. [CrossRef]
17. Brillante, L.; Bois, B.; Mathieu, O.; Lévêque, J. Electrical imaging of soil water availability to grapevine: A benchmark experiment of several machine-learning techniques. *Precis. Agric.* **2016**, *17*, 637–658. [CrossRef]

18. Guo, L.; Zhao, C.; Zhang, H.; Chen, Y.; Linderman, M.; Zhang, Q.; Liu, Y. Comparisons of spatial and non-spatial models for predicting soil carbon content based on visible and near-infrared spectral technology. *Geoderma* **2017**, *285*, 280–292. [CrossRef]
19. De Jong, E.; Schappert, H.J.V. Calculation of soil respiration and activity from CO_2 profiles in the soil. *Soil Sci.* **1972**, *113*, 328–333.
20. Tang, J.; Baldocchi, D.D.; Qi, Y.; Xu, L. Assessing soil CO_2 efflux using continuous measurements of CO_2 profiles in soils with small solid-state sensors. *Agric. Forest Meteorol.* **2003**, *118*, 207–220.
21. Technometrics. Index to contents, Volume 11, 1969. *Technometrics* **1969**, *11*, 848–851. [CrossRef]
22. Liu, W.; Zhao, Z.; Yuan, H.; Song, C.; Li, X. An optimal selection method of samples of calibration set and validation set for spectral multivariate analysis. *Spectrosc. Spectr. Anal.* **2014**, *34*, 947–951.
23. Liu, Y.; Lu, Y.; Guo, L.; Xiao, F.; Chen, Y. Construction of Calibration Set Based on the Land Use Types in Visible and Near-InfRared (VIS-NIR) Model for Soil Organic Matter Estimation. *Acta Pedol. Sin.* **2016**, *53*, 332–341.
24. Rana, P.; Gautam, B.; Tokola, T. Optimizing the number of training areas for modeling above-ground biomass with ALS and multispectral remote sensing in subtropical Nepal. *Int. J. Appl. Earth Obs. Geoinf.* **2016**, *49*, 52–62. [CrossRef]
25. Friedman, J.H.; Roosen, C.B. An introduction to multivariate adaptive regression splines. *Stat. Methods Med. Res.* **1995**, *4*, 197–217. [CrossRef] [PubMed]
26. Brunsdon, C.; Fotheringham, S.; Charlton, M. Geographically Weighted Regression. *J. R. Statal Soc. Ser. D (Statian)* **1998**, *47*, 431–443. [CrossRef]
27. Hurvich, C.M.; Simonoff, J.S.; Tsai, C.L. Smoothing parameter selection in nonparametric regression using an improved Akaike information criterion. *J. R. Statal Soc.* **1998**, *60*, 271–293. [CrossRef]
28. Shi, T.; Cui, L.; Wang, J.; Fei, T.; Chen, Y.; Wu, G. Comparison of multivariate methods for estimating soil total nitrogen with visible/near-infrared spectroscopy. *Plant Soil* **2012**, *366*, 363–375. [CrossRef]
29. Viscarra Rossel, R.A.; McGlynn, R.N.; McBratney, A.B. Determining the composition of mineral-organic mixes using UV–vis–NIR diffuse reflectance spectroscopy. *Geoderma* **2006**, *137*, 70–82.
30. Chen, Y. Spatial Autocorrelation Approaches to Testing Residuals from Least Squares Regression. *PLoS ONE* **2016**, *11*, e0146865. [CrossRef]

 *applied sciences*

Article

Orchard Floor Management Affects Tree Functionality, Productivity and Water Consumption of a Late Ripening Peach Orchard under Semi-Arid Conditions

Pasquale Losciale [1,*], Liliana Gaeta [2], Luigi Manfrini [3], Luigi Tarricone [4] and Pasquale Campi [2]

1   Department of Soil, Plant and Food Sciences–DiSSPA-, University of Bari "Aldo Moro", 70126 Bari, Italy
2   Research Centre for Agriculture and Environment, CREA-Council for Agricultural Research and Economics, 70125 Bari, Italy; liliana.gaeta@crea.gov.it (L.G.); pasquale.campi@crea.gov.it (P.C.)
3   Department of Agricultural and Food Sciences–DISTAL, Alma Mater Studiorum, University of Bologna, 40126 Bologna, Italy; luigi.manfrini@unibo.it
4   Research Centre for Viticulture and Enology, CREA-Council for Agricultural Research and Economics, 70010 Turi, Italy; luigi.tarricone@crea.gov.it
*   Correspondence: pasquale.losciale@uniba.it

Received: 23 October 2020; Accepted: 13 November 2020; Published: 17 November 2020

Abstract: Semi-arid conditions are favorable for the cultivation of late ripening peach cultivars; however, seasonal water scarcity and reduction in soil biological fertility, heightened by improper soil management, are jeopardizing this important sector. In the present two-year study, four soil managements were compared on a late ripening peach orchard: (i) completely tilled (control); (ii) mulched with reusable reflective plastic film; (iii) mulching with a Leguminosae cover-crop flattened after peach fruit set; (iv) completely tilled, supplying the water volumes of the plastic mulched treatment, supposed to be lower than the control. Comparison was performed for soil features, water use, tree functionality, fruit growth, fruit quality, yield and water productivity. Even receiving about 50% of the regular irrigation, reusable reflective mulching reduced water loss and soil carbon over mineralization, not affecting (sometimes increasing) net carbon assimilation, yield, and fruit size and increasing water productivity. The flattening technique should be refined in the last part of the season as in hot and dry areas with clay soils and low organic matter, soil cracking increased water evaporation predisposing the orchard at water stress. The development and implementation of appropriate soil management strategies could be pivotal for making peach production economically and environmentally sustainable.

Keywords: mulching; flattening; irrigation; photosynthesis; transpiration; soil quality; water stress integral; fruit growth; water use efficiency; productivity

---

1. Introduction

Fruit growing is a key sector for the Mediterranean economy, society and environment. It is the highest value among the agricultural productions, representing 17% of the total EU agricultural turnover (FAO—Food and Agriculture Organization of the United Nations, 2018). Furthermore, orchards contribute to land preservation via stewardship and climate regulation through evapotranspiration [1] and represent one of the most typical fruit crops of the Mediterranean Basin [2], thus being at the basis of both its economy and dietary culture.

Fruit growing has a double-faced nature. On one side, the great demand of high-quality products. The fruits, most of which are delivered to the fresh market, are asked to meet the consumer demand with very high-quality standards [3] as fruit consumption improves health and well-being [4]. This, at world level, makes the fruit sector very competitive, creating concerns and often pushing the fruit growers more on the yield than in quality production. Orchard intensification techniques are one of the results from this request. On the other side, society has expressed concern about the exploitation of agricultural inputs because of their dramatic impact on natural resources and ecosystem functioning [5].

Peach is among the most representative and valued fruit species in the Mediterranean Basin. The southern Italy environment is usually hot and dry until the beginning of autumn, therefore is particularly suitable for early and late ripening peach cultivars. However, late cultivars are more water demanding due to the long-lasting persistence of fruits on the plant [6].

Climate change and foreseeable future resource limitations (mainly water) threaten Mediterranean fruit production. The secure supply of high-quality fruit is under jeopardy because of increased heat and water stress conditions [7], reduced productive land [8], decreased soil fertility due to intensive management practices, reduced water availability and competition for water with other productive sectors and human activities [8]. In Mediterranean countries, the reduction in organic matter, which impacts soil fertility, is a common occurrence, for climatic reasons [9]. This is often aggravated by improper management of the residues of agricultural products, the low use of organic amendments and the rapid mineralization of the organic compounds due also to intensive tillage practices [10]. A multi-year life cycle assessment study showed that fertilizers and energy consumption (i.e., electricity, fossil fuel) and water were the main impact factors in peach cultivation [11]. Fertilizers and energy consumption are indeed indicated to be the most impactful on emissions and climatic disorders. Climate change, with its increased temperatures, can have negative effects on tree productivity if scheduling irrigation is not applied properly. However, agriculture is already the main user of water, consuming about 70% of freshwater, and it must achieve savings, rather than further increases, in its water needs [12]. Water scarcity can be tackled by improving water saving techniques. In particular the rational management of soil can increase water use efficiency either through decreasing soil evaporation, using artificial [13–15] or natural [16,17] mulching material, and increasing soil water holding capacity, via decompaction and organic matter enrichment. Recent experiences in horticultural crops [18,19] reported that water and N recycling at agroecosystem level can be enhanced by cover crop practice and natural mulching covering techniques, independently from the soil management strategy. The authors pointed out that the improved nitrogen surplus was not sufficiently retained in the agroecosystem without cover crop. On the contrary, the practices adopted in the treatments with the cover crop or temporary intercropping considerably improved the N self-sufficiency of the system. Beneficial effects have been found in grapes [20,21]. Grape ecosystem services provided by Mediterranean vineyards are particularly threatened, because soil functions are often impaired by yearly repeated intensive agricultural practices or weed and pest management. The authors demonstrated that the potential of soil management practices to enhance soil functioning, can be promoted by the presence of a cover crop, even temporarily, in the inter row [20,21]. Moreover, Almagro found that improved soil management in rainfed Mediterranean agroecosystems can be a powerful strategy to mitigate the current atmospheric CO_2 increase, through soil carbon sequestration and stabilization [22]. However, few publications on the effect of orchard floor modification on fruit trees, in general, and on peach tree development and functioning, including fruit production, in particular, are available [23]. These authors found that changed soil management practices such as zero tillage, supply of organic amendments, understory mowing, retention of crop residues, can result in worthwhile gains within a long-term period. In a Mediterranean peach orchard, these gains can include increased production and also increased sustainability with higher level of soil organic carbon and litter carbon pools [23]. However, to date, orchard soil and water management often rely only on grower and extension service experience or, in the most advanced cases, are driven by data on soil water content and/or climate conditions and plant status [24–26]. Real-time tree performance, as well as the inter-relation among the different chemical, physical and microbiological variables affecting soil fertility, is little considered. Solutions to the major threats encountered by the fruit production sector may be found through approaches that consider the soil–plant system as a whole and, therefore, address improving the entire orchard performance to cope with water scarcity and climate change [27,28]. Since orchards are particularly complex systems, different methods and approaches should be adopted under different, even contrasting, pedo-climatic, economic and social conditions [29]. Even lesser explored in horticulture, the use of different mulching material in the fruit orchard is deserving of particular interest. The use of plastic mulch was adopted with the aim of increasing the water use efficiency in a dryland rainfed area. However, the plastic adopted in the mulching was dark in color with negligible results on light diffusion and probably enhancing soil temperature [30,31]. Recent studies pointed out the positive effect of mulching with high-reflective biodegradable plastic film on productivity and water use efficiency on peach [15].

The aim of the present study was to investigate the effect of four different orchard floor managements on tree functionality, productivity and water consumption of a late ripening peach cultivar orchard under semi-arid conditions. The comparison was performed among the following treatments: (i) completely tilled, (ii) mulched with reusable and reflective plastic film, (iii) mulching with a Leguminosae cover crop flattened after peach fruit set, with the aim of increasing the organic matter and water holding soil capacity; (iv) completely tilled and reducing irrigation at the same volumes supplied to the plastic mulched treatment.

2. Materials and Methods

2.1. Experimental Set-Up and Pedo-Climatic Conditions

The trial was carried out in 2015 and 2016 at the Experimental farm of the Council for Agricultural Research and Economics (Research Centre for Agriculture and Environment), in southern Italy (Rutigliano, lat.: 40°59′ N, long.: 17°01′ E, alt.: 147 m asl) on 3-year-old peach (*Prunus persica* (L.) Batsch var. *laevis*) trees of a late ripening cultivar "Calred" [32] grafted on Missour rootstock, trained as slender spindle and spaced 4.0 × 2.5 m. The experimental site is under the Mediterranean climate, characterized by warm and dry summers. The average air temperature throughout the year and during the vegetative–reproductive season is 15.5 and 20 °C, respectively, and the annual rainfall is about 535 mm, mainly concentrated in the autumn and late winter periods and usually greatly reduced, or absent, in the spring–summer period [33].

Four different orchard managements were tested: soil tilled (T); inter-row mulching with a reusable and machine-resistant reflective plastic film (C/820 Black Silver Orchard; thickness: 100 µm; Ginegar Plastic product Ltd., Ginegar, Israel) to reduce soil evaporation and to increase the diffuse light (M); inter-row mulching with horse bean (*Vicia faba.* L.) sown in November and flattened after peach fruit set forming a natural mulching on the soil (F). The last treatment (S) was established on tilled soil supplying the same irrigation volume of M that was supposed to be lower than the control (T) as evaporation was limited by the plastic mulching. Since horse bean contributed to Nitrogen fixation, N supply on F treatment was halved in comparison to T, M and S, while the Phosphorus and Potassium supply was increased by 25% in order to feed the service crop.

In order to verify the homogeneity of soil characteristics at the beginning of the trial, as well as to evaluate the evolution of the soil conditions as a function of the four treatments, three soil samples per treatment were collected at the beginning of the trial (before flattening period), after the harvest of the first year and at the end of the second year (just before the winter pruning). Soil was evaluated for its physico-chemical traits: soil texture by hydrometer method, total carbon organic content (TOC, %) by the dry-combustion procedure with a TOC Vario Select analyzer (Elementar, Germany), pH, electric conductibility (EC, dS m^{-1}), N (g kg^{-1}, Kjeldahl procedure) and P (mg kg^{-1}, Olsen method) content. Soil texture was similar among the four treatment and it was classified as clay loam [34]. Soil water content in volume at field capacity (FC, −0.03 MPa) and wilting point (WP, −1.5 MPa) were 0.34 m^3 m^{-3} and 0.21 m^3 m^{-3}, respectively (measured using the Richards chambers). At 0.6 m of depth, the parent rock is present; this reduces the capacity of the root systems to expand beyond this layer. At the beginning of the trial also, the evaluated chemical trait results were not statistically different among the treatments (Table 1).

Table 1. Soil chemical traits at the beginning of the trial (14 April 2015) after the harvest of the first year of the study (24 September 2015) and before the winter pruning of the second year (16 December 2016). Within each date and for each variable, different letters indicate a statistical difference at $p \leq 0.05$.

Date	Treatment	N (g kg^{-1})	P (mg kg^{-1})		pH	EC (dS m^{-1})	TOC (%)	
14/04/2015	T	1.06	34.40		8.43	0.17	1.05	a
	M	1.00	28.32		8.38	0.17	0.98	b
	F	0.98	44.70		8.42	0.17	1.03	a
	S	1.20	41.48		8.40	0.14	1.05	a
	F-value	*3.08*	*1.45*		*0.47*	*4.12*	*10.72*	
	p-value	*0.129*	*0.334*		*0.715*	*0.081*	*0.013*	
24/09/2015	T	1.21	42.05		8.33	0.18	1.22	
	M	1.15	35.87		8.26	0.18	1.16	
	F	1.00	53.00		8.32	0.19	1.18	
	S	1.17	51.46		8.21	0.20	1.22	
	F-value	*1.30*	*1.49*		*0.94*	*0.05*	*0.95*	
	p-value	*0.372*	*0.325*		*0.489*	*0.981*	*0.484*	
16/12/2016	T	0.68	37.97	ab	8.31	0.19	1.11	a
	M	0.66	39.83	ab	8.38	0.12	1.20	a
	F	0.94	52.36	a	8.27	0.17	1.18	a
	S	0.75	27.14	b	8.33	0.13	0.99	b
	F-value	*1.07*	*5.53*		*0.28*	*0.43*	*8.57*	
	p-value	*0.392*	*0.010*		*0.838*	*0.736*	*0.002*	

2.2. Water Supply and Soil Water Content

Water was supplied with drip irrigation system having 2 drippers per tree, and a flow rate of 8 l h^{-1} per dripper. Volumetric soil water content (SWC) was measured by capacitive probes (10HS, Decagon Devices Inc., Pullman, WA, USA) linked to dataloggers (Grillobee, TecnoEL, Italy). For each treatment, three points were monitored. Capacitive probes were installed horizontally into the soil profile, on the row at 0.3 m from peach trees and at −0.1, −0.3 and −0.5 m from the soil surface, in order to intercept the dynamics of soil water content below the dripping lines. For each treatment, water content in the whole soil profile was calculated averaging the values of the three depts in each of the three points. Probes were previously calibrated in order to measure the volumetric soil water content (SWC) and identify the intervention threshold (IT). The IT corresponded to the SWC at which the readily available water was completely used. The IT of 0.26 m^3 m^{-3} was adopted; this value was obtained considering a depletion fraction (fraction of available soil water that can be depleted from the root zone before moisture stress) of 0.5 [35]. When the IT was reached, the amount of water necessary to return at FC was supplied [36]. T, M and F were irrigated monitoring the SWC while S received the same water volume of M.

2.3. Leaf Functionality and Tree Water Relations

Three plants similar in canopy size and potential crop load were selected for each treatment. At fruit cell division, pit hardening, fruit cell expansion and close to the harvest stages, leaf net photosynthesis (Pn, μmol m^{-2} s^{-1}), stomatal conductance (gs, mol m^{-2} s^{-1}) and transpiration (Tr, mol m^{-2} s^{-1}) were measured on well-exposed leaves placed on the east and west side of the canopy, 4 times during the day (9.00–17.00 h), with an open circuit infrared gas analyzer fitted with an LED light source (Li-COR 6400XT, LI-COR, Lincoln, Nebraska, USA). At each time of the day and canopy side, light intensity was maintained constant across the 4 treatments, setting the LED light source at the natural irradiance experienced by the leaf immediately before the measurement. The values obtained on the west and east side of the canopy were averaged for each tree. At the same time of measure, stem water potential (Ψs, MPa) was measured on the same trees belonging to the 4 treatments according to [37]. Pn and Tr, collected during the day, were integrated [38], providing the specific amount of CO_2 (ΣPn, mol m^{-2}) and water (ΣTr, mol m^{-2}) fixed and transpired by a square meter of leaf during the time of measure. To take into account the amount of energy used by trees to raise water from the soil during the time of measure, the water stress integral (S_Ψ, GPa) was calculated as the difference between the integral by time of stem water potential and the integral by time of the minimum stem water potential measured in the same time range [39,40].

2.4. Fruit Growth and Productivity

The fruit growth pattern was monitored during the season by means of a digital caliper implemented with a datalogger able to store the data (HK—Horticultural Knowledge s.r.l. Bologna, Italy) on twelve fruit per tree. The fruit volume (V, cm^3) and the absolute growth rate (AGR, cm^3 day^{-1}) were calculated assuming the shape of the peach as a spheroid and measuring the three axes of each peach [15].

At harvest (ready for hand selection: when fruit flesh firmness was around 3–5 kg cm^{-2}), the number of fruits per tree (NF), the average fruit weight (FW, g), the yield (Y, t ha^{-1}) and the irrigation water productivity (WPi, kg) of fresh fruit per cubic meter supplied with irrigation [41] were evaluated on the same trees monitored for fruit growth and leaf functionality. The total soluble solids content (TSS, °Brix), flesh firmness (FF, kg cm^{-2}) and the percentage of fruit skin red overcolor (RC, %) were measured on 10 fruit per tree.

2.5. Statystical Analysis

For each period of measure soil data, ΣPn and ΣTr were subjected to ANOVA. Fruit growth data (V and AGR) used for the statistical analysis were obtained averaging the measures taken on each tree. A by-time repeated ANOVA was performed analyzing separately the data of fruit cell division, pit hardening, fruit cell expansion and ripening stages, respectively. Three productivity, water use efficiency and fruit quality variables were tested by means of an ANCOVA considering the number of fruits as the covariate variable.

3. Results

3.1. Pedo-Climatic Conditions

The two-year study's thermic patterns during the experiment period (1 June–10 September) were almost comparable with an average minimum, maximum and average temperature of about 20.6, 29.5 and 25.0 °C, respectively. The year 2015 was less rainy than 2016 with a cumulative rainfall during the period of 126 mm. The year 2016 showed a cumulative rainfall of 206 mm till 31 August and an additional 147 mm of rain fallen in the first fortnight of September (Figure 1).

Soil measurements were performed on 14 April 2015, 24 September 2015 and 16 December 2016, before peach's full bloom, after the harvest and before the winter pruning of the second season, respectively. At the beginning of the trial, the soil chemical traits evaluated resulted in being not statistically different among the four treatments (Table 1). N, P, pH, conductibility and total organic carbon content were about 1.0 g kg^{-1}, 37.2 mg kg^{-1}, 8.4, 0.16 dS m^{-1} and 1.0 g kg^{-1}, respectively. After the first harvest, the treatments continued to be similar for soil chemical traits. At the end of the two-year trial, S showed the lowest value of P and TOC while the highest P content was recorded for F, followed by T and M (Table 1).

3.2. Water Supply and Soil Water Content

In order to have all the treatments at the same soil moisture conditions at the beginning of the experiment, three and two full irrigations were provided regardless of the treatments in 18 May, 29 May and 5 June in 2015, as well as in 6 June and 14 June, in 2016. In 2015, the seasonal water supply was 1557 m^3 ha^{-1} for T and F, and 815 m^3 ha^{-1} for M, S. The irrigation season lasted about

4 months (Figure 2A). In 2016, T and F received 1810 m³ ha⁻¹ while M and S 1023 m³ ha⁻¹ and the duration of the irrigation season was about 3 months (Figure 2B). M and S received about 50% less water than T and F in both years. Soil water content remained between the field capacity (FC) and the intervention threshold (IT) till the end of July in T, M and F during the two seasons. In the same period, S showed SWC lower than IT for several days (Figure 3, June–July). From the beginning of the irrigation season till the end of July, M and F showed the highest SWC values (average SWC of 0.32 and 0.33 m³ m⁻³ in 2015 and 2016, respectively, for M; 0.31 and 0.32 m³ m⁻³ for F). In the same period, T revealed SWC similar to M and F, in 2015 (average of 0.31 m³ m⁻³), while in 2016, it decreased at an average value of 0.29 m³ m⁻³. The average SWC of S in the same period was 0.29 and 0.26 m³ m⁻³ in 2015 and 2016, respectively (Figure 3). Due to local watershed restrictions occurring during the period August–September for both the years, the irrigation frequency (number of peaks in Figure 3) was reduced for all the treatments. In this period, the SWC of T, F and S fell below the intervention threshold several times and in S it reached the wilting point; M showed the highest soil water content, rarely below IT (Figure 3). In this period, the comparison between T and F, receiving almost the same water volume in each irrigation, showed that in August–September, after water supply SWC declined faster in F than in T, reaching values lower than T and close to WP at the end of the irrigation season (Figure 3).

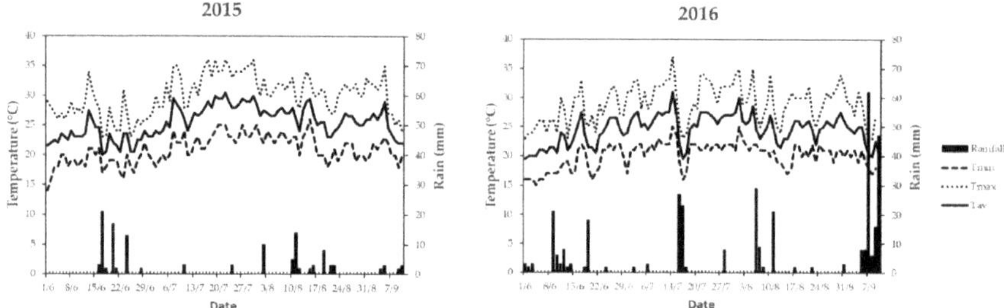

Figure 1. Air temperature (lines) and rainfall (bars) recorded for the peach orchard under investigation during the two years of study.

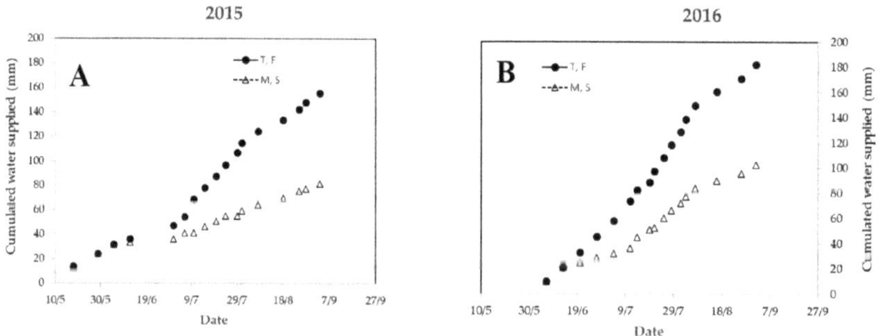

Figure 2. Cumulated water volumes supplied to T, F (closed circles) and to M, S (open triangles) recorded in 2015 (**A**) and 2016 (**B**).

3.3. Leaf Functionality and Tree Water Relations

Full bloom occurred on 7 April and 14 March in 2015 and 2016, respectively. Leaf gas exchange and stem water potential measures were performed 55, 86, 87, 88 and 120 days after full bloom (DAFB) in 2015 (1/6, 2/7, 3/7, 4/7 and 5/8), and 92, 107, 120, 135, and 163 DAFB (14/6, 29/6 12/7, 27/7 and 24/8) in 2016.

2015

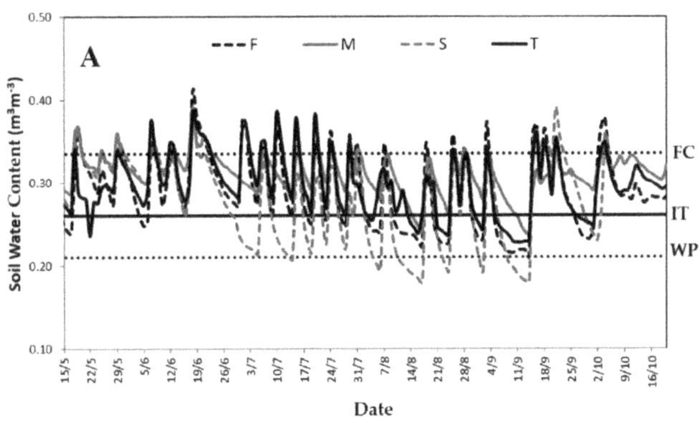

2016

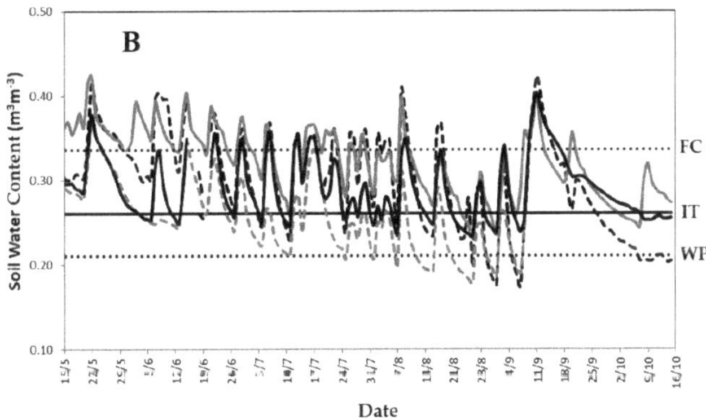

Figure 3. Soil water content (m^3 m^{-3}) pattern recorded on the 4 treatments in 2015 (**A**) and 2016 (**B**). Horizontal dotted and continuous lines represent the field capacity (FC), the wilting point (WP) and the intervention threshold (IT), respectively.

3.3.1. Season 2015

At 55 DAFB, when water supply differentiation was not established yet, cumulative photoassimilation (ΣPn), transpiration (ΣTr) and the integral water stress (S_Ψ) were similar among the four treatments (Figure 4; Table S1). The average air temperature (T_{air}) and vapor pressure deficit (VPD) during the measure period (9:00–17:00) were 29.9 °C and 2.3 kPa, respectively; the midday stem water potential was around −0.9 MPa and the average stomatal conductance was about 0.164 mol m^{-2} s^{-1}. At 86–87 and 88 DAFB, trees were at the end of the pit hardening stage and water supply was differentiated among the four treatments. At 86 DAFB, the average VPD and air temperature were 3.5 kPa and 36.4 °C, respectively. A slight reduction in ΣPn and ΣTr were observed in S, while the water stress integral S_Ψ was statistically lower in S and F in comparison with T and M (Figure 4; Table S1). The midday stem water potential was around −1.2 MPa in M and T and −1.5 MPa in F and S. The average gs recorded during the measure period was about 0.102 mol m^{-2} s^{-1} for M and T, 0.094 mol m^{-2} s^{-1} for F and 0.085 mol m^{-2} s^{-1} for S. At 87 DAFB, the average VPD was about 4.5 kPa and the air temperature recorded during the time of measure was about 37.5 °C. S showed the cumulative photoassimilation and transpiration to be lower than the remaining treatments (0.142 and 72.5 mol m^{-2}, respectively). The highest S_Ψ was recorded in M (−3.6 GPa), followed by T (−7.0 GPa); F and S showed the lowest values of the water stress integral of about −10 GPa (Figure 4; Table S1). The midday stem water potential was −1.1, −1.25, −1.3 and −1.4 MPa in M, T, F and S, respectively, while the average gs was around 0.105 mol m^{-2} s^{-1} in M, T, F and 0.064 mol m^{-2} s^{-1} in S. At 88 DAFB, the average VPD and the air temperature of the period of measure (9:00–17:00) were 3.8 kPa and 37.1 °C, respectively. The highest cumulative photoassimilation was recorded on M (0.25 mol m^{-2}), followed by T and F (0.198 and 0.188 mol m^{-2}, respectively). S revealed the lowest ΣPn (0.125 mol m^{-2}) and the same trend was observed for the cumulative transpiration (Figure 4A,B; Table S1). The average gs recorded during the period was 0.134 mol m^{-2} s^{-1} for M followed by T and F, with an average gs of about 0.096 mol m^{-2} s^{-1} and S (gs, 0.057 mol m^{-2} s^{-1}). M had the highest S$_\Psi$ (−2.5 GPa) followed by T (−6.7 GPa); the lowest S$_\Psi$ values were recorded in F and S with values of −14.2 and −17.2 GPa, respectively (Figure 4C). At 120 DAFB, trees were in the full fruit cell expansion stage. The average (9:00–17:00) air temperature and VPD were 33.5 °C and 2.5 kPa, respectively. The lowest ΣPn and ΣTr were recorded in S (0.166 and 48.6 mol m^{-2}, respectively) while the remaining treatments were similar (Figure 4A,B; Table S1). The same trend was observed for S$_\Psi$ reaching the lowest levels of the season (Figure 4C; Table S1). The midday stem water potential was −1.4, −1.5, −1.6 and −1.8 MPa for M, T, F and S, respectively, and the average stomatal conductance recorded within the measure period (9:00–17:00) was around 0.106 mol m^{-2} s^{-1} for M, T, F, and 0.066 mol m^{-2} s^{-1} for S.

3.3.2. Season 2016

At 92 DAFB, when water supply was the same for all the treatments, no differences were recorded in terms of cumulative photoassimilation, transpiration and water stress integral; the average VPD and air temperature during the period of measure (9:00–17:00) were about 2.0 kPa and 29.4 °C, respectively; the midday stem water potential was −0.6 MPa and the average gs was about 0.182 mol m^{-2} s^{-1}. At pit hardening (107 DAFB), ΣPn, ΣTr and S$_\Psi$ were similar among the treatments (Figure 5; Table S1). Average air temperature and VPD were 29.6 °C and 2.4 kPa, respectively; the midday stem water potential was −0.8 MPa and the average gs was 0.115 mol m^{-2} s^{-1}. At the beginning of fruit cell expansion (12 July, 124 DAFB), the average VPD was about 4.5 kPa and air temperature 37.6 °C. T and F showed the highest ΣPn (around 0.23 mol m^{-2}), while the lowest one was recorded on S (~0.15 mol m^{-2}). M revealed an intermediate ΣPn of 0.18 mol m^{-2} (Figure 5A; Table S1). The highest cumulative transpiration was recorded on F (146.1 mol m^{-2}), followed by T (127.0 mol m^{-2}); the lowest ΣTr was observed in M and S with an average cumulative transpiration of 89.8 mol m^{-2} (Figure 5B). The average gs during the period of measure was about 0.107 mol m^{-2} s^{-1} for T and F and 0.073 mol m^{-2} s^{-1} for M and S. F and T had a quite similar water stress integral (−3.9 GPa), higher than S$_\Psi$ recorded in S and M of about −12 GPa (Figure 5C). The midday stem water potential was −1.0 MPa for T and F, and −1.3 MPa for S and M. At the fruit cell expansion stage (135 DAFB), the average air temperature and vapor pressure deficit recorded during the period of measure were about 32.5 °C and 2.8 kPa, respectively. M and F showed ΣPn and ΣTr higher than T and S (Figure 5A,B; Table S1). The average gs recorded from 9:00 to 17:00 was about 0.118 mol m^{-2} s^{-1} in M and F, and 0.098 mol m^{-2} s^{-1} in T and S. The water stress integral was −7.8 GPa in T, followed by M (−10.6 GPa); S and R had the lowest S$_\Psi$ of about −14.8 GPa (Figure 5C; Table S1). The midday stem water potential at 135 DAFB was about −1.0 MPa in T, −1.1 MPa in M and −1.3 MPa in F and S. Close to the harvest (163 DAFB), the average T$_{air}$ and VPD were 29.1 °C and 1.9 kPa, respectively. M showed a cumulative net photoassimilation and transpiration higher than T and S and the lowest values were observed in F (Figure 5A,B; Table S1). The average stomatal conductance followed the same trend with gs of 0.126 mol m^{-2} s^{-1} in M, about 0.099 mol m^{-2} s^{-1} in T and S, and 0.064 mol m^{-2} s^{-1} in F. S$_\Psi$ in M was −4.4 GPa, higher than T and S (~−9.0 GPa); the lowest water stress integral was observed in F with S$_\Psi$ of −13.4 GPa (Figure 5C). The midday stem water potential followed the same trend with values of −1.4, −1.6, −1.6 and −1.8 MPa recorded in M, T, S and F, respectively.

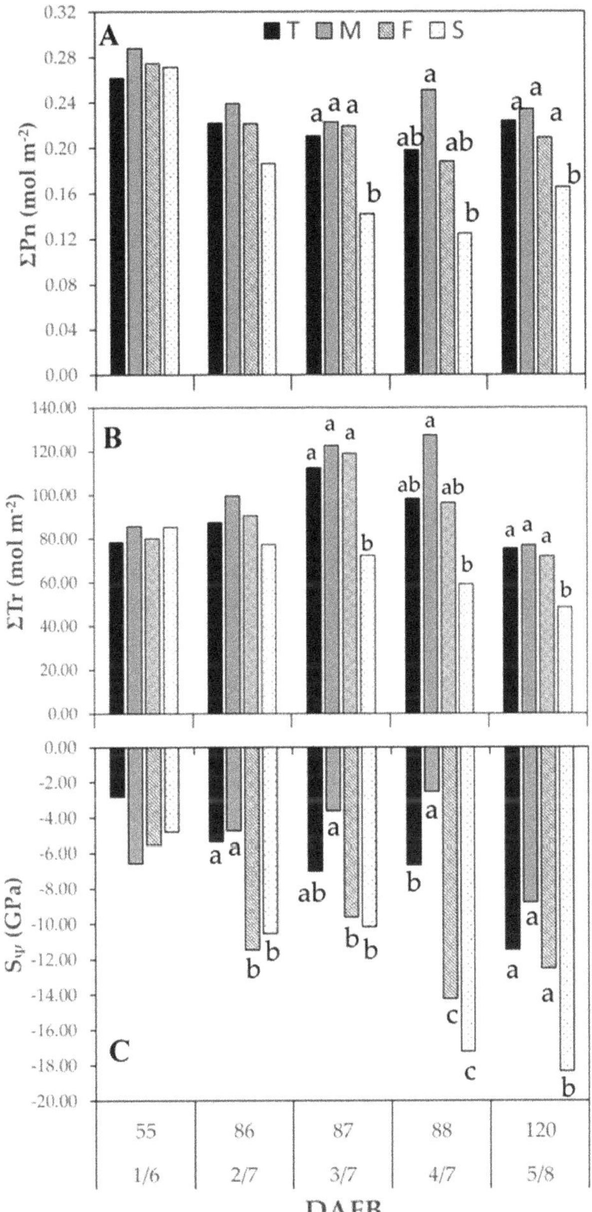

Figure 4. Cumulative leaf net photosynthesis (ΣPn) (**A**), transpiration (ΣTr) (**B**) and water stress integral (S_Ψ) (**C**) calculated for T (black bars), M (grey bars), F (dashed bars) and S (dotted bars) during the time of measure (9:00–17:00 h) of each day of measurement in 2015. Within the same date different letters indicate a statistical difference at $p \leq 0.05$.

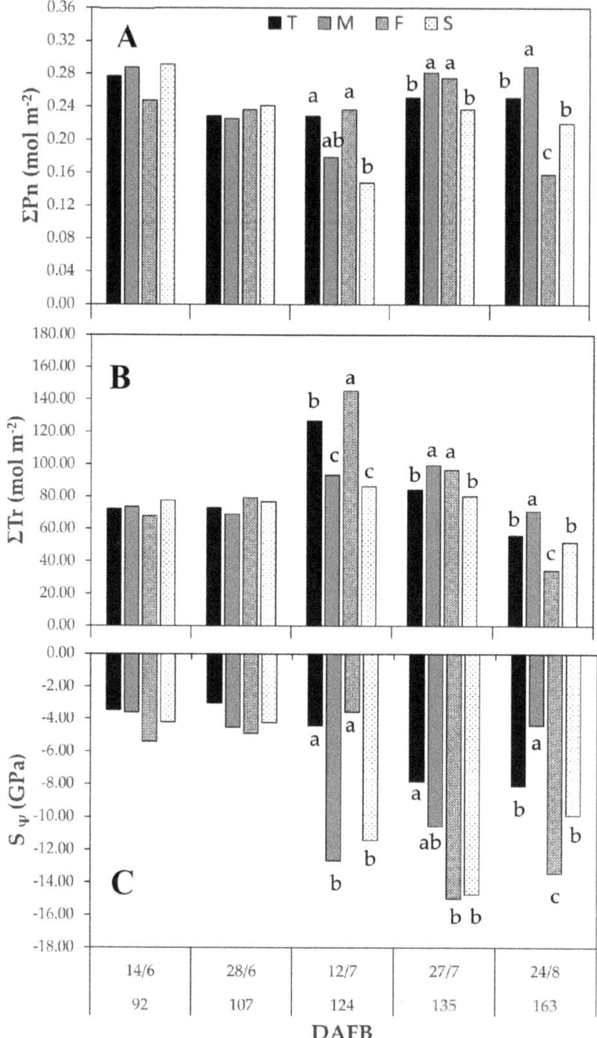

Figure 5. Cumulative leaf net photosynthesis (ΣPn) (**A**), transpiration (ΣTr) (**B**) and water stress integral (S$_\Psi$) (**C**) calculated for T (black bars), M (grey bars), F (dashed bars) and S (dotted bars) during the time of measure (9:00–17:00 h) of each day of measurement in 2016. Within the same date different letters indicate a statistical difference at $p \leq 0.05$.

3.4. Fruit Growth and Productivity

3.4.1. Season 2015

At the end of the fruit cell division stage (38–77 DAFB), no differences for fruit volume were recorded among the four treatments while the average AGR values observed in M and S were lower than those measured on T and F (Table 2; Figure 6). During the pit hardening stage (77–105 DAFB), the fruit volume was similar among the treatments (Table 2), and the absolute growth rate was higher in T, M and F (~0.59 cm^3 day^{-1}) than in S with an AGR value of 0.43 cm^3 day^{-1} (Table 2). The reduced AGR in S was observed starting from 94 DAFB (Figure 6B). In the full fruit cell expansion stage

(105–125 DAFB), T, M and F continued to have fruits bigger than S (Table 2) with a difference between S and the remaining treatments growing progressively (Figure 6A). The average AGR recorded between 105 and 125 DAFB was higher in M, T and F (~1.07 cm^3 day^{-1}) than in S with an AGR of 0.73 cm^3 day^{-1} (Table 2). F maintained an AGR similar to M and T till 115 DAFB; afterwards it decreased, reaching values closer to S (Figure 6B). During the last days before the harvest (125–148 DAFB), M and T showed an average fruit volume (~105.8 cm^3) higher than F and S with a value of about 88 cm^3 (Table 2). The same behavior was observed for the absolute growth rate with values of about 2.34 cm^3 day^{-1}, for M and T and 1.86 cm^3 day^{-1} for S and F (Table 2).

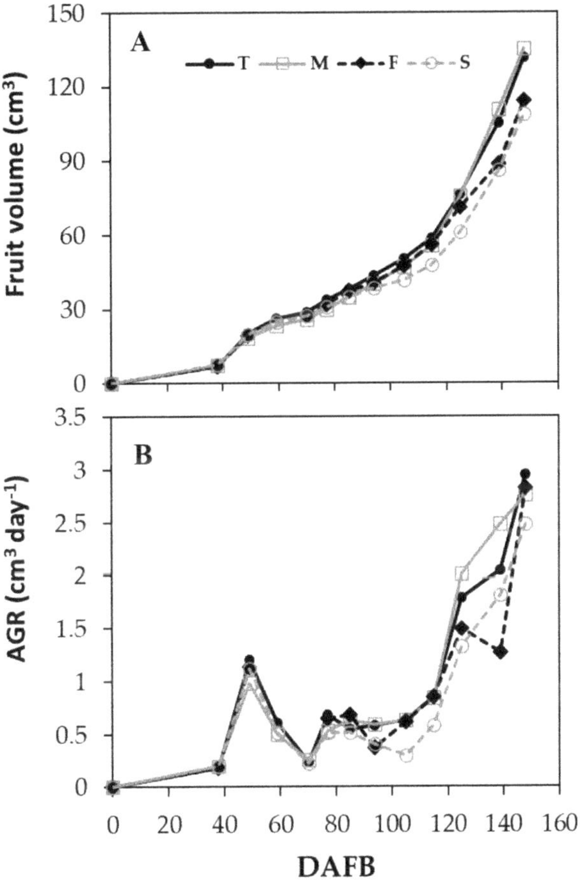

Figure 6. Fruit growth (**A**) and absolute growth rate (**B**) pattern recorded in 2015 for T, M, F and S treatments.

Table 2. Average fruit volume (V) and absolute growth rate (AGR) recorded for T, M, F and S treatments in different fruit growth stages (Days After Full Bloom range, DAFB Range), in 2015. Within each stage, a by-time repeated ANOVA was performed. For each variable, different letters indicate a statistical difference at $p \leq 0.05$.

DAFB Range	Treatment	V (cm^3)		AGR (cm^3 day^{-1})	
38–77	T	23.18		0.58	b
	M	20.93		0.49	c
	F	21.46		0.65	a
	S	21.00		0.51	c
	F-value	*3.14*		*5.25*	
	p-value	*0.108*		*0.041*	
77–105	T	41.35		0.61	a
	M	37.95		0.60	a
	F	39.42		0.58	a
	S	36.56		0.43	b
	F-value	*3.56*		*5.83*	
	p-value	*0.088*		*0.033*	
105–125	T	61.77	a	1.09	a
	M	59.42	a	1.16	a
	F	58.13	a	0.98	a
	S	50.13	b	0.73	b
	F-value	*5.08*		*9.14*	
	p-value	*0.044*		*0.012*	
125–148	T	104.46	a	2.26	a
	M	107.10	a	2.42	a
	F	91.26	b	1.86	b
	S	85.28	b	1.87	b
	F-value	*5.30*		*5.06*	
	p-value	*0.040*		*0.044*	

In 2015, the third leaf season for the peach orchard, the harvest was performed in two picks: on 3 and 7 September, 149 and 153 DAFB, respectively. A late fruit drop occurred during the season modifying the initial crop load imposed. The ANCOVA revealed the significative effect of the number of fruit as the covariate variable for the yield (F = 63.26; $p < 0.001$), fruit weight (F = 19.98; $p = 0.001$) and water productivity (F = 30.98; $p < 0.001$), whose values have been adjusted accordingly. The yield was similar among the treatments (~11.00 t ha^{-1}), while the fruit fresh weight was higher in M and T (166.9 and 149.4 g, respectively) than in F and S with values of 112.9 and 93.6 g, respectively (Table 3). Water productivity was higher in M and S (average of 12.01 kg m^{-3}) than in T and F with an average value of 6.93 kg m^{-3} (Table 3). No differences for the total soluble solid content was observed, recording an average value of about 18 °Brix. The percentage of fruit skin over color was higher in M (~92.3%) than in the remaining treatments (~56.7%) while the flesh firmness was higher in F, followed by T, S and M with values of 3.7, 3.0, 2.8, and 2.2 kg cm^{-2}, respectively (Table 3).

Table 3. Yield (Y), fruit fresh weight (FW), water productivity (WPi), sugar content (TSS) flesh firmness (FF) and fruit skin red overcolor, measured on the 4 treatments under investigation in 2015. Data were subjected to ANCOVA analysis, considering the number of fruits per tree as a covariate variable and were adjusted accordingly. For each variable, different letters indicate a statistical difference at $p \leq 0.05$.

Treatment	Y (t ha^{-1})	FW (g)		WPi (kg m^{-3})		TSS ($^\circ$Brix)	FF (kg cm^{-2})		RC (%)	
T	11.91	149.45	a	7.22	b	18.07	3.01	ab	58.33	b
M	10.82	166.94	a	12.23	a	17.34	2.18	b	92.31	a
F	10.96	112.91	b	6.64	b	18.73	3.70	a	55.00	a
S	10.44	93.61	b	11.79	a	18.37	2.85	ab	56.67	a
F-value	*0.70*	*17.59*		*17.32*		*2.02*	*5.12*		*7.48*	
p-value	*0.571*	*<0.001*		*<0.002*		*0.169*	*0.019*		*0.005*	

3.4.2. Season 2016

During the pit hardening (57–107 DAFB) and the first part of fruit cell expansion stages (107–140 DAFB), no differences for fruit volume and AGR were recorded among the four treatments (Table 4; Figure 7). Afterwards (140–164 DAFB), the average fruit volume for the period was similar among the treatments, while the absolute growth rate was higher in M and T (5.89 and 5.16 cm^3 day^{-1}, respectively) than in S and F, with values of 4.63 and 4.31 cm^3 day^{-1}, respectively (Table 4). Starting from 150 DAFB, a divergent pattern for AGR was observed comparing T and M versus F and S: while in the former treatments AGR continued to increase, in the latter ones it decreased (Figure 7B). Close to the harvest, the fruit volume of M and T was higher than F and S (Figure 7A).

Table 4. Average fruit volume (V) and absolute growth rate (AGR) recorded on T, M, F and S treatments in different fruit growth stages, in 2016. Within each stage, a by-time repeated ANOVA was performed. For each variable, different letters indicate a statistical difference at $p \leq 0.05$.

DAFB Range	Treatment	V (cm^3)	AGR (cm^3 day^{-1})	
57–107	T	35.99	0.73	
	M	33.19	0.59	
	F	35.45	0.73	
	S	35.11	0.69	
	F-value	*0.44*	*2.59*	
	p-value	*0.731*	*0.149*	
107–140	T	77.81	1.94	
	M	73.17	1.64	
	F	80.54	2.02	
	S	73.17	1.77	
	F-value	*1.85*	*1.67*	
	p-value	*0.239*	*0.271*	
140–164	T	190.61	5.16	a
	M	180.13	5.89	a
	F	182.93	4.63	b
	S	180.27	4.31	b
	F-value	*0.36*	*8.51*	
	p-value	*0.785*	*0.014*	

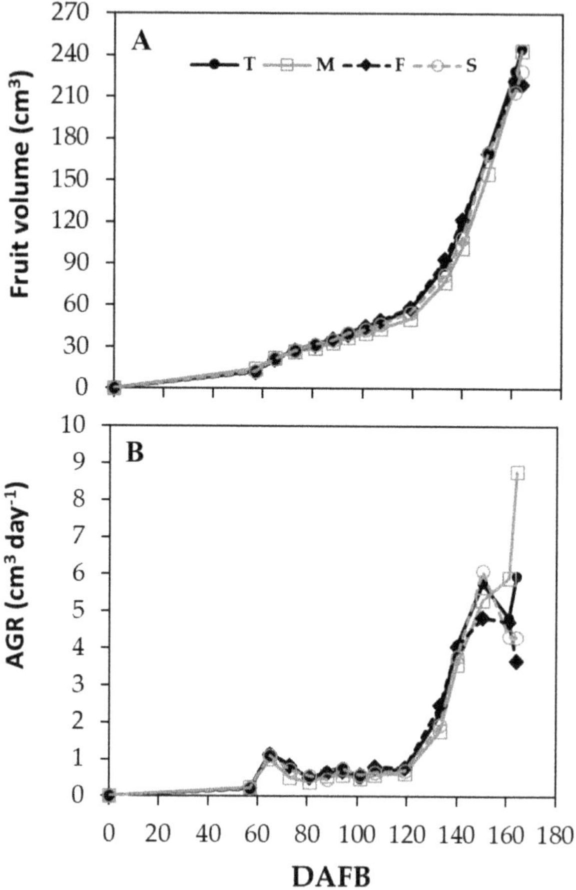

Figure 7. Fruit growth (**A**) and absolute growth rate (**B**) pattern recorded in 2016 on T, M, F and S treatments.

In 2016, the harvest occurred on 26 August (165 DAFB) with a single pick. Even in this season a late fruit drop affected the imposed crop load. The ANCOVA showed a significant effect of the number of fruits as a covariate variable for yield ($F = 126.95$; $p < 0.001$) and WPi ($F = 76.84$; $p < 0.001$), whose values have been adjusted accordingly. Y was similar for the four treatments, with values ranging from 16.78 t ha^{-1} (M) to 13.56 t ha^{-1} (F). Fruit fresh weight was higher in M (256.65 g) than in T (234.46 g) and the lowest values were observed in S and F with FW of 216.92 and 211.65 g, respectively (Table 5). The highest WPi was recorded in M (16.2 kg m^{-3}), followed by S (14.03 kg m^{-3}), T (8.38 kg m^{-3}) and F with WPi of 7.45 kg m^{-3} (Table 5). No difference in terms of sugar content, flesh firmness and percentage of red overcolor was observed among the treatments (Table 5).

Table 5. Yield (Y), fruit fresh weight (FW), water productivity (WPi), sugar content (TSS) flesh firmness (FF) and fruit skin red overcolor, measured on the 4 treatments under investigation in 2016. Data were subjected to ANCOVA analysis, considering the number of fruits per tree as a covariate variable and were adjusted accordingly. For each variable, different letters indicate a statistical difference at $p \leq 0.05$.

Treatment	Y (t ha^{-1})	FW (g)		WPi (kg m^{-3})		TSS (°Brix)	FF (kg cm^{-2})	RC (%)
T	15.25	234.46	ab	8.39	b	14.16	4.58	84.07
M	16.78	256.65	a	16.22	a	13.75	4.96	93.89
F	13.56	211.65	b	7.45	b	14.13	4.26	93.33
S	14.35	216.92	b	14.03	a	13.78	3.58	84.63
F-value	*0.19*	*3.49*		*4.56*		*0.73*	*0.76*	*2.38*
p-value	*0.899*	*0.052*		*0.026*		*0.553*	*0.539*	*0.125*

4. Discussion

The thermic pattern during the two years of study was quite similar and in line with the average of the site while year 2015 was less rainy than 2016 (Figure 1). The similar chemical–physical conditions of the soil recorded at the beginning of the trial suggested that the starting point for the four treatments was the same (Table 1). At the end of 2016, the different floor and water managements affected the soil chemical features (Table 1). S, subjected to water restriction and tillage, showed a decrease in TOC and assimilable P. As observed in other studies, under semi-arid conditions the high temperature and evapo-transpirative demand of the environment, jointly with the low soil moisture lasting for long time during the year, may have increased the mineralization processes, reducing the already low organic carbon in the soil [10,42,43]. An additional, related consequence could have been the reduction in P in its assimilable form in the tilled treatment subjected to water shortage [44]. A decrease in P in its assimilable form was observed in an evergreen forest, in the Mediterranean area, subjected to drought stress [45]. M, experiencing the same water supply of S, did not show a decrease in TOC suggesting the positive role of artificial, reflective mulching in preventing organic matter over degradation (Table 1). The use of organic mulching (F) did not affect TOC, while it was not possible to verify if the increase in P occurring in this treatment was given by the floor management per se or to the increased amount of P supplied as fertilizer to feed the service crop (Table 1). Artificial reflective mulching reduced soil evaporation behaving as a physical barrier against the water loss but also reducing solar radiation absorption by the soil and the wind speed at the surface [46]. Even when receiving about 50% less water than the control (T), M maintained SWC higher than T for most of the irrigation season in both years (Figure 3). These preliminary results confirmed what was observed in other studies on rain-fed peach orchards where mulching with plastic film reduced water loss of about 15% in comparison with tilled soil [14]. The flattening technique (F) ameliorated SWC conditions in the rainier year (2016) till the end of July. Afterward, its SWC dropped faster than T after irrigation, in both the years (Figure 3). This rapid water depletion might have been attributable to the disruption of the natural mulching and to the onset of relevant soil cracking occurring in clay soils, low in organic matter [47]. August was characterized by a general water limitation caused by watershed restriction usually occurring in this area (water demand for crop and for civil use increases in this period). Under this stressful condition, M continued to maintain SWC higher than the remaining treatments (Figure 3); only in a few cases its soil water content dropped below the IT, assuring an adequate soil moisture for all the season long [13,15]. When the soil evaporation was not contrasted, the reduction in water supply (S), produced a progressive consumption of the readily available water with values of SWC very low and below the IT (Figure 3).

Soil and water management strategies affected leaf functionality and water relations in peach trees. In both the years, when water supply was not yet differentiated among the four treatments and SWC was within the readily available water range (Figure 3), T, M, F and S behaved similarly (Figures 4 and 5) for carbon assimilation (ΣPn), water transpiration (ΣTr) and plant water status, expressed as water stress integral (S_Ψ). Reflective mulching, preventing the excessive soil evaporation and increasing the diffuse light, maintained the pedo-climatic conditions favorable to photosynthetic activity of leaves [13,14] for the entire vegetative–reproductive season in the two years of study (Figures 4A and 5A). Excluding the measure performed in 2016 at 124 DAFB (12 July), ΣPn, ΣTr and S_Ψ in M were similar, and in some cases higher than T receiving double the amount of water (Figures 4 and 5). When water shortage was associated to tillage (S), tree water status (S_Ψ) was affected and net photosynthesis was subjected to stomatal limitation (reduction in average gs and ΣTr) and probably to non-stomatal limitation as well (Figures 4 and 5). Previous research in pear, apple, peach and grapevine suggested that stomatal closure could affect leaf thermoregulation inducing the increase in leaf temperature and raising the activity of photorespiration [48–50]. This process is a photoprotective strategy for the plant, but it means a loss of carbon in terms of biomass accumulation [51–53]. The use of natural mulching associated with full irrigation (F) did not affect leaf functionality (ΣPn and ΣTr) in comparison with T till the end of July, in both years (Figures 4 and 5). This suggested the absence of such competition between the main and the service crop and the positive effect of the flattening technique in controlling weeds. This positive effect was already observed in vegetable crops [54,55]. In August, when the flattened crop was disrupted and the soil cracking caused a rapid water depletion (Figure 5, 163 DAFB), S_Ψ, ΣPn and ΣTr values in F were the lowest of the four treatments. From a leaf functioning point of view the use of the flattening technique for all of the dry season appeared to be detrimental under particular pedoclimatic conditions such as hot summer, clay and poor of organic matter soils.

The differences among the treatments appeared more evident when high temperature and VPD were associated with soil moisture limitation. The comparison between the days 86 and 87 DAFB of year 2015 revealed that SWC did not change markedly within each treatment as well as S_Ψ (Figure 4C). However, at 87 DAFB, the environment was more water demanding than the previous day: the average T_{air} and VPD passed from 36.4 to 37.5 °C and from 3.5 to 4.5 kPa, respectively. As a consequence, soil with an adequate SWC allowed leaves to maintain the stomata opened, increasing ΣTr and ΣPn. On the other hand, S, having a low SWC, reduced gs and leaf transpiration [56], thus carbon assimilation (Figure 4A). Trees, being in the middle of the Soil Plant Air Continuum (S.P.A.C.) were strongly influenced by the status of rhizosphere and air. The comparison between F and S at 86, 87 and 88 DAFB revealed that, even if the two treatments had the same S_Ψ and midday stem water potential, S showed a cumulative net photosynthesis and leaf transpiration lower than F (Figure 4). The same behavior was observed in 2016, comparing M and S at 124 DAFB and F and S at 135 DAFB. This late ripening peach cultivar seemed to have a "pessimistic" (also called conservative or near iso-hydric) behavior. At low soil water content and high vapor pressure deficit, S sustained its stem water potential, at the same level of F, above a safe threshold to prevent embolism [57,58]. This defense strategy was at the expense of CO_2 fixation since it was regulated by stomatal closure [59,60]. These findings are in contrast with those described by Xiloyannis et al. (1980) on another late ripening peach cultivar considered aniso-hydric [61], suggesting the needing to deepen this issue.

During the first year of study, the pattern of fruit growth of the four treatments was similar till the end of fruit cell division stage (38–77 DAFB); however, M and S, receiving less water, showed an average absolute growth rate lower than T and F (Table 2). This difference was completely recovered in M during the pit hardening stage (77–105 DAFB), while S continued to have an average AGR lower than the remaining treatments (Table 2). Passing from pit hardening to the fruit cell expansion stage (105–125 DAFB), peach fruit became more water and carbon demanding [62]. Water shortage, jointly with the reduction in net photosynthesis, led to fruit growth limitation. As revealed by the average V and AGR in the period 105–125 DAFB (Table 2; Figure 6), the reduced AGR initially observed in S during the pit hardening (77–105 DAFB) resulted in fruit size lower than the remaining treatments (Table 2). In August, during the last part of fruit cell expansion and close to the harvest (125–148 DAFB), fruit growth was limited even in F, suggesting that the rapid water depletion occurring in this period affected fruit volume and growth rate (Table 2, Figure 6). All the advantage of F on S was lost, and, at the end of the season, the two treatments showed the smallest fruits (Figure 6, Table 3). The same behavior was observed in 2016. The rainier season alleviated the effect of water shortage and till 140 DAFB (first part of fruit cell expansion) no differences were recorded among the treatments for fruit volume and AGR (Table 4, Figure 7). During the last part of fruit cell expansion (140–164 DAFB), even in this season, F and S showed the lowest average AGR (Table 4), fruit volume (Figure 6) and fruit size (Table 5).

Fruit yield was generally low considering that Calred is a late ripening cultivar. However, it should be taken into account that this cultivar was subjected to a late fruit drop in both the years and that, probably, trees were not at a fully mature productive status, as in 2015 and 2016, they were at third and fourth leaf, respectively. The different orchard floor management and water supply did not affect yield, but the fruit size with biggest fruit picked on T and M (Tables 3 and 5). Water productivity of M was about 70 and 90% higher than T suggesting that the use of artificial reflective mulching could be considered a water friendly strategy in rainfed [14] as well as in irrigated peach orchards (Tables 3 and 5). Although in 2015 peaches seemed to ripe earlier in M in comparison with the remaining treatments (Table 3), fruit quality (TSS, FF and RC) was generally not affected by the different managements (Tables 3 and 4).

5. Conclusions

Under semi-arid conditions and where water supply is limited, the choice of an appropriate orchard floor management could be of pivotal importance for getting the peach production both economic and environment-friendly. Even receiving about 50% of the regular irrigation, reusable reflective mulching reduced water loss and soil carbon over mineralization, not affecting (sometimes increasing) net carbon assimilation, yield, and fruit size. As a consequence, water productivity was drastically increased. These first results suggested that the reflective mulching strategy could be considered to be water and soil "friendly". This management technique is firstly described and explored on a peach orchard, thus the studies on the development and the use of alternative material for mulching should be explored. The flattening technique as mulching strategy should be refined for the final part of the irrigation season, especially in those hot and dry areas with clay soils, low in organic matter, thus predisposed to cracking.

Supplementary Materials: The following are available online at http://www.mdpi.com/2076-3417/10/22/8135/s1, Table S1: Anova results for Cumulative leaf net photosynthesis (ΣPn), transpiration (ΣTr) and water stress integral (SΨ) calculated in each day of measurement in 2015 and 2016. For each variable, within the same date the asterisk indicates $p \leq 0.05$.

Author Contributions: Conceptualization, P.L.; data curation, L.G., L.M., L.T. and P.C.; formal analysis, L.G. and P.C.; investigation, L.G., L.T. and P.C.; methodology, P.L. and P.C.; supervision, P.L.; validation, P.L. and L.G.; writing—original draft, P.L. and L.M.; Writing—review and editing, P.L., L.G. and L.M. All authors have read and agreed to the published version of the manuscript.

Funding: This research received no external funding.

Acknowledgments: The authors would like to thank Di Gennaro D. and Amendolagine A.M. for their contribution in data collecting, and Introna P. and Volpicella M. for their valuable operative effort in conducting the orchard.

References

1. Morandi, B.; Manfrini, L.; Zibordi, M.; Corelli-Grappadelli, L.; Losciale, P. From fruit anatomical features to fruit growth strategy: Is there a relationship? *Acta Hortic.* **2016**, *1130*, 185–191. [CrossRef]

2. Tous, J.; Ferguson, L. Mediterranean fruits. In *Progress in New Crops*; Janick, J., Ed.; ASHS Press: Arlington, VA, USA, 1996; pp. 416–430.

3. Charles, M.; Corollaro, M.L.; Manfrini, L.; Endrizzi, I.; Aprea, E.; Zanella, A.; Corelli Grappadelli, L.; Gasperi, F. Application of a sensory-instrumental tool to study apple texture characteristics shaped by altitude and time of harvest. *J. Sci. Food Agric.* **2018**, *98*, 1095–1104. [CrossRef] [PubMed]

4. Corollaro, M.L.; Manfrini, L.; Endrizzi, I.; Aprea, E.; Demattè, M.L.; Charles, M.; Bergamaschi, M.; Biasioli, F.; Zibordi, M.; Corelli Grappadelli, L.; et al. The effect of two orchard light management practices on the sensory quality of apple: Fruit thinning by shading or photo-selective nets. *J. Hort. Sci. Biotech.* **2015**, *90*, 99–107. [CrossRef]

5. Demestihas, C.; Plénet, D.; Génard, M.; Raynal, C.; Lescourret, F. Ecosystem services in orchards. A review. *Agron. Sustain. Dev.* **2017**. [CrossRef]

6. Morandi, B.; Manfrini, L.; Losciale, P.; Zibordi, M.; Corelli Grappadelli, L. The positive effect of skin transpiration in peach fruit growth. *J. Plant. Phys.* **2010**, *167*, 1033–1037. [CrossRef]

7. Lopez, G.; Boini, A.; Manfrini, L.; Torres-Ruiz, J.M.; Pierpaoli, E.; Zibordi, M.; Losciale, P.; Morandi, B.; Corelli-Grappadelli, L. Effect of shading and water stress on light interception, physiology and yield of apple trees. *Agr. Water Manag.* **2018**, *210*, 140–148. [CrossRef]

8. Perulli, G.D.; Bresilla, K.; Manfrini, L.; Boini, A.; Sorrenti, G.; Corelli Grappadelli, L.; Morandi, B. Beneficial effect of secondary treated wastewater irrigation on nectarine tree physiology. *Agr. Water Manag.* **2019**, *221*, 120–130. [CrossRef]

9. González-Hidalgo, J.C.; Pena-Monné, J.L.; de Luis, M. A review of daily soil erosion in Western Mediterranean areas. *Catena* **2007**, *71*, 193–199. [CrossRef]

10. Ventrella, D.; Stellacci, A.M.; Castrignanò, A.; Charfeddine, M.; Castellini, M. Effects of crop residue management on winter durum wheat productivity in a long term experiment in Southern Italy. *Eur. J. Agron.* **2016**, *77*, 188–198. [CrossRef]

11. Vinyes, E.; Asin, L.; Alegre, S.; Muñoz, P.; Boschmonart, J.; Gasol, C.M. Life Cycle Assessment of apple and peach production, distribution and consumption in Mediterranean fruit sector. *J. Clean. Prod.* **2017**, *149*, 313–320. [CrossRef]

12. Water for Sustainable Food and Agriculture FAO Report 2017. Available online: http://www.fao.org/3/a-i7959e.pdf (accessed on 6 November 2020).

13. Yaghi, T.; Arslan, A.; Naoum, F. Cucumber (*Cucumis sativus* L.) water use efficiency (WUE) under plastic mulch and drip irrigation. *Agric. Water Manag.* **2013**, *128*, 149–157. [CrossRef]

14. Wang, C.; Wang, H.; Zhao, X.; Chen, B.; Wang, F. Mulching affects photosynthetic and chlorophyll a fluorescence characteristics during stage III of peach fruit growth on the rain-fed semiarid Loess Plateau of China. *Sci. Hortic.* **2015**, *194*, 246–254. [CrossRef]

15. Campi, P.; Gaeta, L.; Mastrorilli, M.; Losciale, P. Innovative Soil Management and Micro-Climate Modulation for Saving Water in Peach Orchards. *Front. Plant Sci.* **2020**, *11*. [CrossRef] [PubMed]

16. Fan, J.; Gao, Y.; Wang, Q.; Sukhdev, S.M.; Li, Y. Mulching effects on water storage in soil and its depletion by alfalfa in the Loess Plateau of northwestern China. *Agric. Water Manag.* **2014**, *138*, 10–16.

17. Diacono, M.; Fiore, A.; Farina, R.; Canali, S.; Di Bene, C.; Testani, E.; Montemurro, F. Combined Agro-ecological Strategies for Adaptation of Organic Horticultural Systems to Climate Change in Mediterranean Environment. *Ital. J. Agron.* **2016**, *11*, 85–91. [CrossRef]

18. Antichi, D.; Sbrana, M.; Martelloni, L.; Abou Chehade, L.; Fontanelli, M.; Raffaelli, M.; Mazzoncini, M.; Peruzzi, A.; Frasconi, C. Agronomic Performances of Organic Field Vegetables Managed with Conservation Agriculture Techniques: A Study from Central Italy. *Agronomy* **2019**, *9*, 810. [CrossRef]

19. Tosti, G.; Benincasa, P.; Farneselli, M.; Guiducci, M.; Onofri, A.; Tei, F. Processing Tomato–Durum Wheat Rotation under Integrated, Organic and Mulch-Based No-Tillage Organic Systems: Yield, N Balance and N Loss. *Agronomy* **2019**, *9*, 718. [CrossRef]

20. Salomé, C.; Coll, P.; Lardo, E.; Metay, A.; Villenave, C.; Marsden, C.; Blanchart, E.; Hinsinger, P.; Le Cadre, E. The soil quality concept as a framework to assess management practices in vulnerable agroecosystems: A case study in Mediterranean vineyards. *Ecol. Indic.* **2016**, *61*, 456–465. [CrossRef]

21. Tarricone, L.; Debiase, G.; Masi, G.; Gentilesco, G.; Montemurro, F. Cover Crops Affect Performance of Organic Scarlotta Seedless Table Grapes Under Plastic Film Covering in Southern Italy. *Agronomy* **2020**, *10*, 550. [CrossRef]

22. Almagro, M.; Garcia-Franco, N.; Martínez-Mena, M. The potential of reducing tillage frequency and incorporating plant residues as a strategy for climate change mitigation in semiarid Mediterranean agroecosystems. *Agric. Ecosyst. Environ.* **2017**, *246*, 210–220. [CrossRef]

23. Montanaro, G.; Dichio, B.; Briccoli Bati, C.; Xiloyannis, C. Soil management affects carbon dynamics and yield in a Mediterranean peach orchard. *Agric. Ecosyst. Environ.* **2012**, *161*, 46–54. [CrossRef]

24. Losciale, P.; Manfrini, L.; Morandi, B.; Pierpaoli, E.; Zibordi, M.; Stellacci, A.M.; Salvati, L.; Corelli Grappadelli, L. A multivariate approach for assessing leaf photoassimilation performance using the IPL index. *Physiol. Plant.* **2015**, *154*, 609–620. [CrossRef] [PubMed]

25. Manfrini, L.; Pierpaoli, E.; Zibordi, M.; Morandi, B.; Muzzi, E.; Losciale, P.; Corelli Grappadelli, L. Monitoring strategies for precise production of high-quality fruit and yield in apple in Emilia-Romagna. *Chem. Eng. Trans.* **2015**, *44*, 301–306.

26. Manfrini, L.; Corelli Grappadelli, L.; Morandi, B.; Losciale, P.; Taylor, J.A. Innovative approaches to orchard management: Assessing the variability in yield and maturity in a 'Gala' apple orchard using a simple management unit modeling approach. *Eur. J. Hortic. Sci.* **2020**, *84*, 1–8. [CrossRef]

27. Jordan, M.O.; Vercambre, G.; Gomez, L.; Pagès, L. The early spring N uptake of young peach trees (Prunus persica) is affected by past and current fertilizations and levels of C and N stores. *Tree Physiol.* **2014**, *34*, 61–72. [CrossRef]

28. Morandi, B.; Boselli, F.; Boini, A.; Manfrini, L.; Corelli Grappadelli, L. The fruit as a potential indicator of plant water status in apple. *Acta Hortic.* **2017**, *1150*, 83–90. [CrossRef]

29. Tozzi, F.; van Hooijdonk, B.M.; Tustin, D.S.; Corelli Grappadelli, L.; Morandi, B.; Losciale, P.; Manfrini, L. Photosynthetic Performance and Vegetative Growth in a New Red Leaf Pear: Comparison of Scion Genotypes Using a Complex, Grafted-Plant System. *Front. Plant Sci.* **2018**. [CrossRef]

30. Ham, J.M.; Kluitenberg, G.J.; Lamont, W.J. Optical properties of plastic mulches affect the field temperature regime. *J. Amer. Soc. Hort. Sci.* **1993**, *118*, 188–193. [CrossRef]

31. Zheng, W.; Wen, M.; Zhao, Z.; Liu, J.; Wang, Z.; Zhai, B.; Li, Z. Black plastic mulch combined with summer cover crop increases the yield and water use efficiency of apple tree on the rainfed Loess Plateau. *PLoS ONE* **2017**, *12*. [CrossRef]

32. Okie, W.R. *Handbook of Peach and Nectarine Varieties—Agriculture Handbook number 714*; U.S. Department of Agriculture; Agricultural Research Service: Washington, DC, USA, 1998; p. 42.

33. Campi, P.; Palumbo, A.D.; Mastrorilli, M. Evapotranspiration estimation of crops protected by windbreak in a Mediterranean region. *Agr. Water Manag.* **2012**, *104*, 153–162. [CrossRef]

34. U.S. Department of Agriculture; Soil Conservation Service. *Soil Taxonomy: A Basic System of Soil Classification for Making and Interpreting Soil Surveys; Soil Survey Staff, 1975 Department of Agriculture Handbook 436*; U.S. Government Printing Office: Washington, DC, USA, 1975.

35. Allen, R.G.; Pereira, L.S.; Raes, D.; Smith, M. *Crop Evapotranspiration. Guide-Lines for Computing Crop Water Requirements. FAO Irrigation and Drainage Paper No. 56*; FAO—Food and Agriculture Organization of the United Nations: Rome, Italy, 1998.

36. Campi, P.; Navarro, A.; Palumbo, A.D.; Modugno, F.; Vitti, C.; Mastrorilli, M. Energy of biomass sorghum irrigated with reclaimed wastewaters. *Eur. J. Agron.* **2016**, *76*, 176–185. [CrossRef]

37. Naor, A.; Klein, I.; Doron, I. Stem water potential and apple size. *J. Am. Soc. Hortic. Sci.* **1995**, *120*, 577–582. [CrossRef]

38. Losciale, P.; Chow, W.S.; Corelli Grappadelli, L. Modulating the light environment with the peach "asymmetric orchard": Effects on gas exchange performances, photoprotection, and photoinhibition. *J. Exp. Bot.* **2010**, *61*, 1177–1192. [CrossRef] [PubMed]

39. Myers, B.J. Water stress integral. A link between short-term stress and long-term growth. *Tree Physiol.* **1988**, *4*, 315–323. [CrossRef]

40. Fernández, M.D.; Hueso, J.J.; Cuevas, J. Water stress integral for successful modification of flowering dates in 'Algerie' loquat. *Irrig. Sci.* **2010**, *28*, 127–134. [CrossRef]

41. Fernández, J.E.; Alcon, F.; Diaz-Espejo, A.; Hernandez-Santana, V.; Cuevas, M.V. Water use indicators and economic analysis for on-farm irrigation decision: A case study of a super high density olive tree orchard. *Agr. Water Manag.* **2020**, *237*. [CrossRef]

42. Davidson, E.A.; Janssens, I.A. Temperature sensitivity of soil carbon decomposition and feedbacks to climate change. *Nature* **2006**, *440*, 165–173. [CrossRef]

43. Abdelhafez, A.A.; Abbas, M.H.H.; Attia, T.M.S.; El Bably, W.; Mahrous, S.E. Mineralization of organic carbon and nitrogen in semi-arid soils under organic and inorganic fertilization. *Environ. Technol. Innov.* **2018**, *9*, 243–253. [CrossRef]

44. Cramer, M.D., Hawkins, H.J., Verboom, G.A. The importance of nutritional regulation of plant water flux. *Oecologia* **2009**, *161*, 15–24. [CrossRef]

45. Sardans, J.; Peñuelas, J. Increasing drought decreases phosphorus availability in an evergreen Mediterranean forest. *Plant. Soil* **2004**, *267*, 367–377. [CrossRef]
46. Eberbach, P.L.; Humphreys, E.; Kukal, S.S. The effect of rice straw mulch on evapotranspiration, transpiration and soil evaporation of irrigated wheat in Punjab, India. *Agric. Water Manag.* **2011**, *98*, 1847–1855.
47. Ritchie, J.T.; Adams, J.E. Field measurement of evaporation from soil shrinkage cracks. *Soil Sci. Soc. Am. Proc.* **1974**, *38*, 131–134. [CrossRef]
48. Flexas, J.; Bota, J.; Escalona, J.M.; Sampol, B.; Medrano, H. Effects of drought on photosynthesis in grapevines under field conditions: An evaluation of stomatal and mesophyll limitations. *Funct. Plant. Biol.* **2002**, *29*, 461–471. [CrossRef]
49. Losciale, P.; Zibordi, M.; Manfrini, L.; Morandi, B.; Bastias, R.M.; Corelli Grappadelli, L. Light management and photoinactivation under drought stress in peach. *Acta Hortic.* **2011**, *922*, 341–347. [CrossRef]
50. Losciale, P.; Manfrini, L.; Morandi, B.; Novak, B.; Pierpaoli, E.; Zibordi, M.; Corelli Grappadelli, L.; Anconelli, S.; Galli, F. Water restriction effect on pear rootstocks: Photoprotective processes and the possible role of photorespiration in limiting carbon assimilation. *Acta Hortic.* **2014**, *1058*, 237–244. [CrossRef]
51. Galmés, J.; Abadia, A.; Cifre, J.; Medrano, H.; Flexas, J. Photoprotection processes under water stress and recovery in Mediterranean plants with different growth forms and leaf habits. *Physiol. Plant.* **2007**, *130*, 495–510. [CrossRef]
52. Foyer, C.H.; Harbinson, J. Oxygen metabolism and the regulation of photosynthetic electron transport. In *Causes of Photooxidative Stress and Amelioration of Defence Systems in Plants*; Foyer, C.H., Mullineaux, P.M., Eds.; CRC Press: Boca Raton, FL, USA, 1994; pp. 1–42.
53. Osmond, C.B. Photorespiration and photoinhibition: Some implication for the energetics of photosynthesis. *Biochim. Et Biophys. Acta* **1981**, *639*, 77–98. [CrossRef]
54. Canali, S.; Campanelli, G.; Ciaccia, C.; Leteo, F.; Testani, E.; Montemurro, F. Conservation tillage strategy based on the roller crimpertechnology for weed control in Mediterranean vegetable organic cropping systems. *Eur. J. Agron.* **2013**, *50*, 11–18. [CrossRef]
55. Ciaccia, C.; Testani, E.; Campanelli, G.; Sestili, S.; Leteo, F.; Tittarelli, F.; Riva, F.; Canali, S.; Trinchera, A. Ecological service providing crops effect on melon-weed competition and allelopathic interactions. *Org. Agr.* **2015**, *5*, 199–207. [CrossRef]
56. Jarvis, P.G. The interpretation of the variations in leaf water potential and stomatal conductance found in canopies in the field. *Philos. Trans. R. Soc. Lond. B* **1976**, *273*, 593–610.
57. Lauri, P.É.; Barigah, T.S.; Lopez, G.; Martinez, S.; Losciale, P.; Zibordi, M.; Manfrini, L.; Corelli Grappadelli, L.; Costes, E.; Regnard, J.L. Genetic variability and phenotypic plasticity of apple morphological responses to soil water restriction in relation with leaf functions and stem xylem conductivity. *Trees* **2016**, *30*, 1893–1908. [CrossRef]
58. Tardieu, F.; Simonneau, T. Variability among species of stomatal control under fluctuating soil water status and evaporative demand: Modelling isohydric and anisohydric behaviours. *J. Exp. Bot.* **1998**, *49*, 419–432. [CrossRef]
59. Gollan, T.; Turner, N.C.; Schulze, E.D. The responses of stomata and leaf gas exchange to vapour pressure deficits and soil water content. *Oecologia* **1985**, *65*, 356–362. [CrossRef] [PubMed]
60. Socías, X.; Correia, M.J.; Chaves, M.; Medrano, H. The role of abscisic acid and water relations in drought responses of subterranean clover. *J. Exp. Bot.* **1997**, *48*, 1281–1288. [CrossRef]
61. Xiloyannis, C.; Uriu, K.; Martin, G.C. Seasonal and diurnal variations in absicic acid, water potential, and diffusive resistance in leaves from irrigated and non-irrigated peach trees. *J. Am. Soc. Hortic. Sci.* **1980**, *105*, 412–415.
62. Morandi, B.; Rieger, M.W.; Corelli Grappadelli, L. Vascular flows and transpiration affect peach (Prunus Persica Batsch.) fruit daily growth. *J. Exp. Bot.* **2007**, *58*, 3941–3947. [CrossRef]

Publisher's Note: MDPI stays neutral with regard to jurisdictional claims in published maps and institutional affiliations.

Review

Bioremediation of PAH-Contaminated Soils: Process Enhancement through Composting/Compost

Tahseen Sayara [1] and Antoni Sánchez [2,*] 

[1] Department of Environment and Sustainable Agriculture, Faculty of Agricultural Sciences and Technology, Palestine Technical University-Kadoorie, 7 Tulkarm, Palestine; tsayara@yahoo.com

[2] Departament d'Enginyeria Química, Biològica i Ambiental, Escola d'Enginyeria, Universitat Autònoma de Barcelona, 08193 Bellaterra, Barcelona, Spain

* Correspondence: antoni.sanchez@uab.cat

Received: 22 April 2020; Accepted: 22 May 2020; Published: 26 May 2020

Abstract: Bioremediation of contaminated soils has gained increasing interest in recent years as a low-cost and environmentally friendly technology to clean soils polluted with anthropogenic contaminants. However, some organic pollutants in soil have a low biodegradability or are not bioavailable, which hampers the use of bioremediation for their removal. This is the case of polycyclic aromatic hydrocarbons (PAHs), which normally are stable and hydrophobic chemical structures. In this review, several approaches for the decontamination of PAH-polluted soil are presented and discussed in detail. The use of compost as biostimulation- and bioaugmentation-coupled technologies are described in detail, and some parameters, such as the stability of compost, deserve special attention to obtain better results. Composting as an ex situ technology, with the use of some specific products like surfactants, is also discussed. In summary, the use of compost and composting are promising technologies (in all the approaches presented) for the bioremediation of PAH-contaminated soils.

Keywords: bioremediation; composting; PAHs; organic co-substrates; soil

1. Introduction

Globally, different anthropogenic activities have resulted in increasing environmental pollution, and its consequences has injured almost all components of the ecosystem [1–3]. Soil, as a vital component of the terrestrial ecosystem, is prone to pollution from different sources, including industrial and agricultural activities [4–7]. Wide verities of pollutants entering the soil posing a huge threat and risk to human health and natural ecosystem [8–12]. Polycyclic aromatic hydrocarbons (PAHs), petroleum, and related derivatives represent the main sources of soil contamination [13–17]. Indeed, these organic pollutant groups are listed as priorities and receive considerable attention, owing to their toxic, genotoxic, mutagenic, and potentially cancer-causing properties [18,19].

To deal with this problem, several treatment technologies are used, including chemical, physical, and biological, as well as thermal for remediation of these contaminated soils. Among the best approaches is the bioremediation technology, which is categorized as a promising approach that continues to gain more attention due to its efficiency, cost-effectiveness, and environmental-friendly byproducts [20–22]. The process mainly relies on the activity of a wide spectrum of microorganisms to degrade the target contaminants to lower toxic levels. Bioremediation of PAH-contaminated soil has been performed utilizing distinctive approaches [7]. In any case, composting as a remediation approach has been considered a reasonable strategy in this field, because it provides nutrients for indigenous microorganisms to degrade the target contaminants; simultaneously, applying this approach is a great opportunity for feasible and sustainable reuse of the natural biodegradable fraction of wastes. Additionally, the process is cost-effective compared with other approaches—for instance, composting costs between \$50–\$140 per ton, while applying slurry or biopiling treatments cost \$170 per ton and \$130–\$260 per cubic meter, respectively [23–27]. Bioremediation of PAH-contaminated soil through composting could be implemented through incorporating PAH-contaminated soils to the composting process, or by adding compost to contaminated soils. Also, bioaugmentation or surfactant application might be included to achieve the final set objectives [16,25,28–34]. Biodegradation of PAHs intrinsically depends on microbial activity, where bacteria and fungi are considered the foremost vital variables governing the bioremediation process [35–37]. However, the functionality of these microorganisms is affected by different factors within the composting mixture, including biotic and abiotic factors. In this context, the environmental condition (pH, temperature, moisture,), nutrient availability, oxygen presence, and bioavailability of the contaminants are essential parameters for process control and performance [38].

This review focuses on the application of composting and compost addition for the bioremediation of soils contaminated with PAHs. In this regard, the impact of different controlling factors like temperature, PAH structure and concentration, co-substrate stability, co-substrate mixing ration, and bioaugmentation are discussed. Moreover, other issues, such as bioavailability, surfactant application, and the degradation pathways of PAHs are illustrated, in order to provide an insight into the process that is necessary for new development.

2. Soil Contamination with PAHs

Soil represents a vital component of all terrestrial ecosystems. However, it is subjected to degradation or decline in its quality as a result of different anthropogenic activities that have resulted in increasing the rate of contamination [4,5,7]. Therefore, polluted land is a worldwide concern, and can be viewed as major obstruction to sustainable development and modern environmental protection [39]. Soil contamination has been recognized as one of the major dangers to soil function in Europe by the Communication from the European Commission "Towards a Thematic Strategy for soil Protection" [40,41]. The issue has expanded with expanding public awareness and concern about the presence of chemicals in the environment, particularly due to their different unfavorable impacts on the ecosystem and human health. Polycyclic aromatic hydrocarbons (PAHs) have been recorded as pollutants of priority importance due to their properties and ubiquitous occurrence, as well as their recalcitrance [18,42,43]. Consequently, great efforts worldwide have been directed toward remediating these pollutants from the environment.

2.1. PAHs: Properties and Sources

PAHs are a group of ubiquitous organic pollutants with at least two aromatic rings (Figure 1), and are poorly soluble in water (Table 1). Due to their chemical structure, PAHs have hydrophobic properties, which refers to their ability to accumulate on the surface of solid materials like soil, sediment, sewage sludge, and solid wastes. The dangers emerging from the presence of PAHs in soil are related to the toxic nature of those pollutants [42,43]. It is noteworthy that some substances in this group have been recognized as mutagenic, carcinogenic, and teratogenic [18,19].

Sources of PAHs are categorized into natural as well as anthropogenic sources: hydrothermal process volcanoes, forest fires, and waste burning are natural sources of PAHs. Anthropogenic sources include waste incinerators, burning of fossil fuels during heating processes, incomplete combustion of organic matter, petrochemical spills on land, wood burning, petrol and diesel oil combustion, gasification, and plastic waste incineration [13–15,17]. Globally, 16 to 32 PAH compounds are subjected to mandatory control, due to their harmful properties [44]. PAH persistence and hydrophobicity in environmental components are the main factors that exacerbate the pollution problem, taking into account that soils receives a considerable share of this pollution (sink), due to their complex matrix structure that facilitates the sorption of these pollutants. Soil organic matter is a decisive factor in determining the degree of PAH sorption into the soil, along with the physicochemical properties of the PAHs themselves [18,19,23,45,46]. Therefore, the remediation of soils polluted by aged PAHs has become a major issue for environmental scientists in recent years [12–14,47,48].

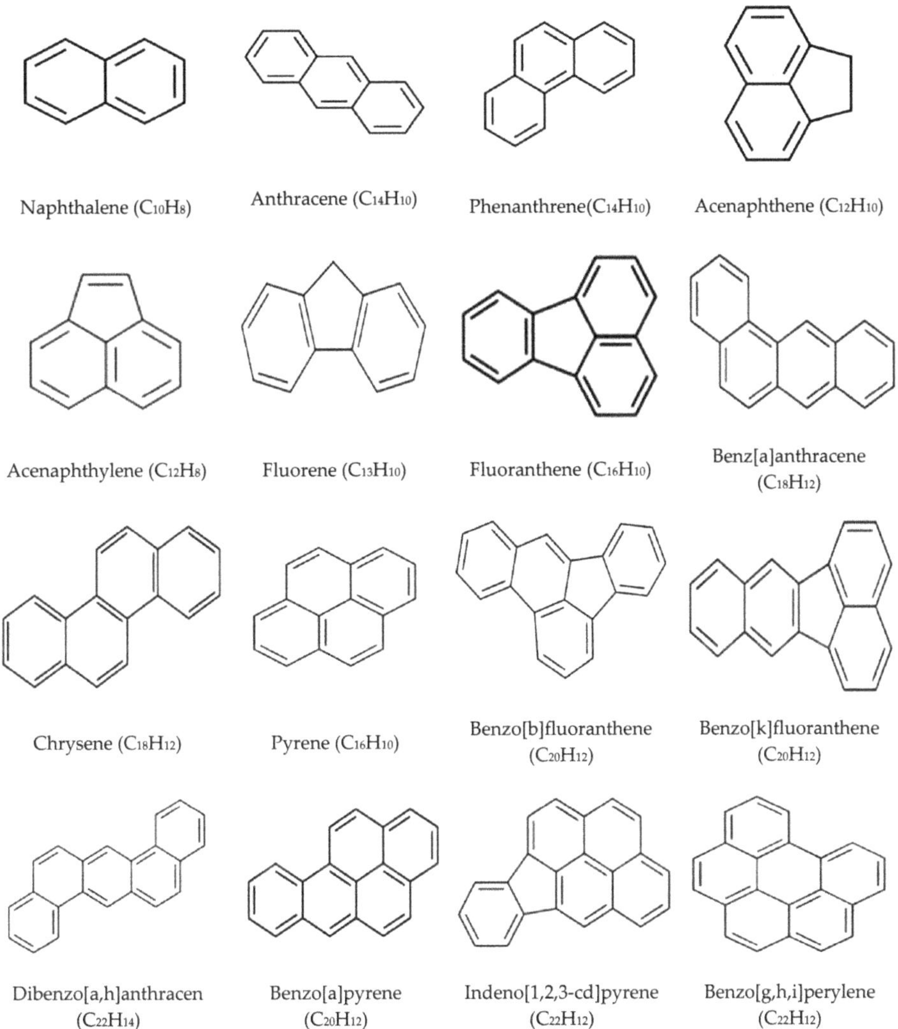

Figure 1. Structure and chemical formula of the 16 polycyclic aromatic hydrocarbons (PAHs) listed as priority pollutants by the United States Environmental Protection Agency (USEPA).

2.2. Bioremediation of PAH-Contaminated Soils

When natural biodegradation processes cannot achieve the desired goals, in this case, human intervention becomes necessary to stimulate the process above naturally occurring microbial process [49]. Accordingly, several approaches have been used to enhance bioremediation efficiency. These approaches, which could be used separately or in combination (two or more) include, but are not limited to, biostimulation (providing nutrients for increasing the microbial activity), bioaugmentation (introducing a consortium of indigenous or exogenous microorganisms), using surfactants, and co-metabolism [50,51]. Recently, various studies have been done in attempt to understand the process hierarchy and to provide solutions for different process limitations. For instance, much research has been carried out to better understand the microbial behavior and its interaction with the contaminants during the bioremediation process, whereas others have focused on introducing exogenic and genetically engineered microbes for process enhancement [52].

Table 1. Selected properties of the 16 USEPA PAHs.

PAH	Number of Rings	Molecular Weight	Aqueous Solubility (mg/L)	Vapor Pressure. (Pa)	Log K_{ow}
Naphthalene	2	128	31	1.0×10^2	3.37
Acenaphthylene	3	152	16	0.9	4.00
Acenaphthene	3	154	3.8	0.3	3.92
Flourene	3	166	1.9	9.0×10^{-2}	4.18
Phenanthrene	3	178	1.1	2.0×10^{-2}	4.57
Anthracene	3	178	0.045	1.0×10^{-3}	4.54
Pyrene	4	202	0.13	6.0×10^{-4}	5.18
Flouranthene	4	202	0.26	1.2×10^{-3}	5.22
Benzo(a)anthracene	4	228	0.011	2.8×10^{-5}	5.91
Chrysene	4	228	0.006	5.7×10^{-7}	5.91
Benzo(b) flouranthene	5	252	0.0015	-	5.80
Benzo(k) flouranthene	5	252	0.0008	5.2×10^{-8}	6.00
Benzo(a)pyrene	5	252	0.0038	7.0×10^{-7}	5.91
Dibenzo(a,h) anthracene	5	278	0.0006	3.7×10^{-10}	6.75
Indeno(1,2,3-cd)pyrene	6	276	0.00019	-	6.50
Benzo(ghi)perylene	6	276	0.00026	1.4×10^{-8}	6.50

* K_{ow}: octanol–water partition.

3. Composting Technology

Composting is defined as an aerobic process, which fundamentally requires oxygen, optimal moisture content, and porosity to stabilize the organic waste, and its common control variables are temperature, oxygen, and moisture [53]. Thus, composting bioremediation is the adaptation and application of the composting technology for wastes and contaminant treatments.

In order to achieve optimum results within a reasonable time during any composting treatment, process-controlling parameters have to be adjusted within the optimum values, and the process passes through two main stages. First is the decomposition/active stage, which is characterized by extensive microbial activity that leads to a steadily increase in the temperature, passing from the mesophilic ranges (25–45 °C) to reach the thermophilic ones (more than 45 °C). To maintain aerobic conditions for effective microbial activity during this stage, a high rate of aeration is needed. Second is the curing stage; this take place at a lower temperature, and microbial activity is relatively low, as the nutrients pool has been depleted. Material humification is an important characteristic occurring in this stage [54], which gives an interesting value to the produced compost, especially for soil bioremediation, as will be discussed later in this work.

4. Bioremediation of PAH-Contaminated Soil by Composting

Composting technology is categorized as ex situ technology, which has been used for the treatment of contaminated soils. During the last few decades, the process received more attention, as it has proved its high efficiency in degrading various organic contaminants like, among others, PAHs, pesticides, explosives, and chlorophenols [25,55–60]. Essentially, the process relies on the addition of compost or organic co-substrates/amendments to the contaminated soil, and while the co-substrate matures, due to the action of various microbial populations within the mixture, the target pollutants are degraded [57,61]. Thus, treatment of PAH-contaminated soil combined with composting of organic waste could be an interesting option and a sustainable method with much increasing attention. It would enable eco-friendly disposal of such waste and enhance the biodegradation rate of PAHs [7,60,62]. The biodegradation process efficiency depends fundamentally on the bioavailability of the substrates, environmental conditions (pH, moisture, temperature), the presence of oxygen, and the availability of nutrients [38]. Remarkably, the bioremediation of PAH-contaminated soils through composting has confirmed this technique's capability to overcome most obstacles that might hinder reaching its goal, which is the removal of contaminants [10,63–67].

As the process is based on mixing the contaminated soil with organic co-substrates, any failure may result in producing much greater quantity of contaminated material, and this is recognized as the main concern of using such approach. This weakness, and the general scarcity of information on the toxicity, distribution, and bioavailability of such contaminants in compost-amended soils, may therefore result in the drawing up of excessively stringent soil assessment measures with remediation cost implications [68].

4.1. Effect of PAH Characteristics and Concentrations

The physical and chemical properties of PAHs have a considerable effect on their biodegradation rate. Microbial assimilation and biodegradation of these compounds basically depends on their solubility. Nevertheless, most of compounds belonging to this group are characterized as poorly soluble in water, especially with their increasing molecular weight and angularity (Table 1, Figure 1), which thus increase their hydrophobicity [69–72]. This was obvious in many studies dealing with the biodegradation of different PAHs. For instance, Han et al. [73] investigated the application of different agricultural waste on the biodegradation of aged PAHs in soil microcosms over 90 days. The initial concentration of total PAHs in the soil was 36.1 mg kg^{-1} dry soil, where four-ring PAHs comprised 41.7% of the total PAHs. The results demonstrated higher degradation rates of 40.7–61.2% for PAHs with low molecular weight (LMW), compared to 18.7–33.1% for those with high molecular weight (HMW) in all soil microcosms. Similarly, Lukić et al. [74] showed that LMW-PAH removal was more favorable in the mesophilic phase, with 11% and 15% residues in the soil, than in the thermophilic phase, with 29% and 31% residues. Additionally, more resistance to degradation was observed for HMW PAHs, resulting in a decrease in the total removal, which was less than 50% for both benzo[a]pyrene and benzo[k]flouranthene, in all treatments [75,76]. In this regard, even though both compounds have the same number of benzene rings (five) and molecular weights, the higher octanol–water partition coefficient (log K_{ow}) of benzo[k]flouranthene increased its hydrophobic properties and consequently its degradation rate under the same conditions. Indeed, higher log K_{ow} leads to a higher potential of bioaccumulation, which is the main factor responsible for the lower biodegradability of such compounds [77]. Obviously, and according to the obtained result in different studies, there is a consistent relationship between the persistence of PAHs in the environment and increasing their numbers of benzene rings, which ultimately affects their biodegradation rate.

The concentrations of PAHs also have a substantial influence on the microbial activity in such treatments, since high concentrations would lead to toxic or inhibition conditions. Meanwhile, low concentrations could be below the rate needed to stimulate microbial cultures to degrade these contaminants [78,79]. This was obvious in the study conducted by Sayara et al. [78], in which the PAH concentrations had a crucial effect. Low concentrations were found to be less than the rates that are assumed to initiate the degradation process, since microbial communities prefer the utilization of readily available nutrients, which are consumed quickly before initiating biodegradation of the target PAHs. The same results were obtained by Zappi et al. [80], where low concentrations of PAH did not degrade, even when the system was supplanted with additional carbon sources. Wu et al. [66] showed that compost addition is an effective approach for enhancing PAH removal from soils, but increasing the ratio of added compost does not necessarily help to increase removal. Nevertheless, enhanced removal by compost addition seems more effective for higher initial PAH concentrations. In this regard, Jorgensen et al. [81] demonstrated that the degradation rate of a compound is proportional to its concentration, especially for highly soluble compounds, and argued that the degradation of hydrocarbons is governed by first-order kinetics. However, this argument may be validated to some extent, as high concentration may become detrimental to microbial activity and disturb nutrient balance, especially when LMW PAHs are present [78].

4.2. Effect of Temperature

Providing optimum temperature is an intrinsic factor for the successful biodegradation of PAH. The importance of this factor stems from its influence on the metabolic activity, bioavailability, solubility, and diffusion rate of the target contaminate [82]. It is noteworthy that the solubility of PAHs increases with temperature, which ultimately increases the bioavailability of the PAH molecules. However, increasing temperature is associated with decreasing oxygen solubility, which on turn reduces the metabolic activity of aerobic microorganisms. Furthermore, and to a certain extent, the specified temperature range will determine the types of dominant microorganisms and their enzymatic activity that will undertake the degradation [73].

The successive stages during the normal composting process (mesophilic phase, thermophilic phase and curing phase) are expected to be accompanied by specific populations of bacteria, and different effects on contaminants are found with different stages of compost product. The biodegradation of PAHs occurs over a wide temperature range, and microorganisms have found to be adapted to biodegrade PAHs at extreme temperature conditions. Under mesophilic and thermophilic temperatures ranges, it has been found that the enzymatic activity of microorganisms increases, which helps in increasing the rate of hydrocarbon degradation. However, it should be underlined that a great amount of research has been directed to focus on the process behavior under mesophilic conditions, as it is believed that a wide spectrum of microbial communities could be present in active roles under these temperatures, and thus reasonable degradation rates could be achieved [78,83–86].

During in-vessel composting of pyrene-contaminated soil, composting temperature affected the prevailing of some microbial groups over others, and the predominant bacterial community changed over time. The degradation of pyrene was dominated by α-, β-, and γ-Proteobacteria, as well as *Actinobacteria*, at 38 °C during 14 days of composting, and then *Streptomyces* at 55 °C. Later, at 70 °C and after 42 days of composting, *Acinetobacter* and *Thermobifida* occupied leading position. Finally, *Thermobifida* and *Streptomyces* flourished after 60 days of composting at 38 °C [87]. Concerning the temperature effect, Lukić et al. [74] claimed that degradation rates of 89% and 59% for three-ring and four-ring PAHs, respectively, were achieved in reactors under mesophilic temperatures. In contrast, reactors displaying a thermophilic range ended with 71% and 41% removal for the same pollutants, respectively, during the bioremediation process. The addition of compost significantly promoted the removal of PAHs and alkanes up to 88% after 50 days of incubation under mesophilic temperatures (28 °C), compared to the natural biodegradation of hydrocarbons in soils without compost [67]. Additionally, the composting of PAH-contaminated soils under different conditions and different organic substrates were found to perform better under mesophilic conditions [23,25,78,88,89]. LMW-PAH concentrations, such as naphthalene, acenaphthylene, acenaphthene, fluorene, anthracene, and phenanthrene, were decreased by an average of 89% at 38 °C, which is twice that compared to the concentration reduction at 55 °C, which was an average of 45%. Simultaneously, no big difference was observed concerning HMW PAHs, including fluoranthene, pyrene, benzo[a]anthracene, and chrysene, where the removal rate was by an average of 67% at 38 °C, compared to 69% at 55 °C. Nevertheless, a high temperature was considered adverse to microbial activity, and volatilization was the leading mechanism of PAH removal [88]. Under these conditions, it is assumed that a longer incubation period under the mesophilic phase could facilitate PAH removal, due to the richest microbial diversity and possible increased microbial activity [27]. According to these studies, and others in the literature, mesophilic temperatures demonstrated their viability and were found to be more favorable for degrading LMW PAHs, with great success in many cases due to the large microbial diversity;

however, these temperatures were not found to be so efficient in the degradation of recalcitrant PAHs [69,75,88,90].

On the other hand, thermophilic ranges have been documented as enhancing PAH degradation. During the composting of hydrocarbon-polluted sediments (total petroleum hydrocarbons (TPH) = 40.3 gkg^{-1} dw) with different organic co-substrates, Alves et al. [91] point out that fish sludge achieved higher temperatures and was able to maintain thermophilic temperature longer than other amendments, which ultimately led to greater TPH removal rates (39.5%). It was assumed that such conditions are conducive to develop fungal communities and exert a surfactant effect, thus promoting the degradation rates. Similarly, Zhu et al. [92] proposed that enhanced solubility under thermophilic conditions could explain the higher removal rate (46%) of benzo[a]pyrene in composting treatments compared to 29% under mesophilic ones. However, whether the increased solubility or microbial community changes contribute to the high-temperature impacts needs further investigation. Viamajala et al. [82] further demonstrated that the elevated temperature during the thermophilic phase of composting enhanced the solubilization rates of phenanthrene, and hence its degradation. Based on the aforementioned observations, it is clear that the impact of composting temperature is correlated to the physiochemical properties of the targeted PAH, as the corresponding degrading microorganism are specific to temperature [93]. Generally, and despite of the different observations, mesophilic temperatures and the dominant microorganisms under these conditions are believed to be more preferable for the degradation of such compounds [69,78,83–86,94].

4.3. Effect of Organic Co-Substrate Stability

Even though various organic co-substrates/amendments have demonstrated their viability in the composting of PAH-contaminated soils, composition of these materials varies significantly in the sources and stages of decomposition [25,59,78,95,96], which as a consequence influences the removal rate in different ways [59,69,78,97]. The selected organic co-substrates for the bioremediation process should contribute in improving and overcoming any deficiencies or limitations that influence the process performance and efficiency. Accordingly, selection of the most suitable organic co-substrates represents as a major challenge in such studies [59,95].

In the bioremediation of PAH-contaminated soils, organic matter stability is of particular importance, as this parameter is directly correlated to the organic substrates' composition and biological activity [98]. Various studies have pointed out that the fate of PAHs is dependent on the quality and nature of the amended organic matter [25,69,78,93]. Bioremediation of PAH-contaminated soils with more stable compost has proved to be more effective than with less stable or fresh organic amendments [25,78,97]. In this context, the preference of these substrates related to the presence of humic substances was found to form a considerable part of stable compost and was proportional with its degree of stability [25,78]. This humic matter was documented to enhance the organic compounds' bioavailability [78,97]. During the composting process, organic co-substrates provided nutrients for microorganisms [99]; meanwhile, humic matter evolution is expected to facilitate the microbial accessibility to PAHs. This behavior is established as a result of decreasing humic matter binding affinity and increasing of the heterogeneity of binding sites, closer to soil humic matter, which is conducive to microbial accessibility to PAHs [97,100]. Additionally, stable compost contains low biodegradable organic matter content and a higher concentration of organic macromolecules [101], which are believed to enhance the biodegrading of the contaminant. The presence of easily degradable organic matter is assumed to reduce the process efficiency, as microbial cultures prefer to use easily degradable organic matter and thus decrease or retard utilization of the contaminant. Another important point in this item is represented by effect of the potential working surface area. In this regard, less degraded organic matter generally has coarse fractions (>5 mm), whereas most humified organic matter is generally presented in fine fraction [102]. The finest compost size fraction (<3 mm) with a higher surface area ratio provides more accessibility to microorganisms and releases more nutrients compared to coarse compost fraction [101,103].

During the bioremediation process, an increase in the content of humic matter from 0.23% to 0.70% was observed, and these changes resulted from the structural changes that occurred in the material composition. Potentiometric titrations of humic acid solution showed increases in the buffering and redox capacities of humic acids [104]. Plaza et al. [97] reported that the composting process caused a structural conversion of humic acids from an organic substrate by reducing the aliphatic fraction and increasing the polarity and aromatic polycondensation in a PAH-contaminated soil. This conversion decreased the PAH binding affinity of humic acids, and thus improved PAH-degrading microbial accessibility. Similar results were observed in other studies, which supports the application of stable organic co-substrates [24,66,73,78,105,106].

It is worth mentioning that when the same ratio of inorganic fertilizer (Nitrogen (N) .Phosphorus (P) .Potassium (K)) was compared with organic compost on the bioremediation of diesel-polluted agricultural soil over a two-month period, the results revealed that total petroleum hydrocarbon removal from polluted soil was 71.40 ± 5.60% and 93.31 ± 3.60% for N.P.K. and compost-amended options, respectively [107]. Also, after 30 weeks, the removal efficiencies of TPH in the soils were 29.3% under natural attenuation, 82.1% when nutrients (NH_4NO_3 and K_2HPO_4) were added, and 63.7% when the mixture was supplemented with 20% (*w/w*, dry weight basis) of aged refuse. However, a removal efficiency of 90.2% was recorded when nutrient and aged refuse were combined together. Nutrients plus aged refuse made the TPH concentration decrease to below the threshold level of commercial use required for Chinese soil quality for TPH (<3000 mg/kg) in 30 weeks. It was also found that dehydrogenase activity, bacterial counts, and degrader abundance in the soil were remarkably enhanced by the addition of aged refuse (20% *w/w*) [108]. All these results confirm the suitability of stable compost over other organic and inorganic substrates. Therefore, one can conclude that introducing an adequate organic co-substrate is usually more efficient in enhancing the bioremediation process, as observed in different studies. This advantage presumably corresponds to the capacity of the compost to perform simultaneously for both bioaugmentation and biostimulation.

4.4. Effect of the Mixing Ratio

The suitability of different substrates based on their physiochemical properties is recognized as an important factor in the composting of PAH-contaminated soils. Determining the appropriate quantity to be added to the mixture is also of great importance, since an inappropriate ratio may hamper or inhibit microbial activity and bioavailability [78,93]. It has been determined that even though microbial metabolism may be temporarily increased using a certain mixing ratio, the long-term inhibition of functionally important organisms may result in the failure of the bioremediation of high-molecular-weight PAHs [78]. The amount of various nutrients, and the ratio of nutrients like C, N, and P in particular, are quite conceivable as being involved in the success of the bioremediation process. Furthermore, determining the minimum quantity of the amendment that could support and maintain the desired activity with a high degradation rate is directly related to process economics [78].

As reported in the study conducted by Wang et al. [109], a microorganism's selection of nutrients could delay the degradation of pollutants, as normally microorganisms prefer easily degradable materials over resistant ones. This study revealed that that treatments with amendment ratios of 1:1 and 2:1 had average TPH removal rates of 30.7% and 33.3%, respectively, but the amendment ratio of 3:1 had a slower net degradation rate of between 11.6% and 26.8%. An excess of readily degradable carbon might overtake the TPH and act as substrate for the metabolism of microbial degraders. Therefore, the proper amount of amendments should be taken into account in composting to balance the motivating effect on microorganisms and the competing effect with pollutants [109]. Similarly, Hickman and Reid [110] concluded that the compost additions combined with earthworms at a ratio of 1:0.5–1:1 (soil/compost, *w/w*) were efficient in enhancing the removal of extractable petroleum hydrocarbons and PAHs. However, when higher ratios of compost (1:2 and 1:4) were used, PAH losses were not advanced, which may indicate that the activity of earthworms were restricted by a higher addition of compost. Wu et al. [66] showed that compost addition is an effective approach for enhancing PAH removal from soils, especially for higher initial PAH concentrations, but increasing the ratio of compost added does not necessarily help to increase removal.

Experiments with different ratios of contaminated soil to green waste from 0.6:1 to 0.9:1 have demonstrated that in general, PAH removal is significantly enhanced in reactors increased with green waste until a maximum mixing ratio of 0.7:1 [89]. The same observation was found in Sayara et al. [78]. These results imply that low mixing ratios were not sufficient to stimulate the microbial growth; on the other hand, excessive amounts could eventually inhibit the targeted contaminants, as microbial communities prefer to use more available and easily degradable nutrients. Furthermore, co-composting of sediments (S) polluted by PAHs with urban green waste (GW) was performed using two mixing ratios (1:1 and 3:1; S/GW). In the first six months of treatment, the PAH concentrations in the 1:1 and 3:1 ratio scenarios was reduced by 57% and 26%, respectively. Despite the fact that only two mixing ratios were tested, the results again demonstrate that the low mixing ratio (3:1) was not sufficient to enhance the degradation process [94]. When different corn straw ratios (1%, 2%, 4%, or 6% w/w) were investigated for the remediation of aged PAHs in soils, removal rates were significantly ($p < 0.05$) enhanced under the 6% ratio, mainly for HMW PAHs. This indicates that the high amendment of corn straw was a potential option for the remediation of PAH-contaminated soils [111].

4.5. Bioaugmentation

When the indigenous microbial activity is not sufficient, or does not have the potential to achieve the set goals for bioremediation [112,113], it appears imperative to accelerate the process using different approaches. Among these approaches is bioaugmentation. The mechanism of this approach fundamentally depends on introducing exogenous microorganism strains that are characterized by their high capacity and diverse metabolic profiles in degrading the target contaminants [16,25,29,31,32]. However, and as concluded in many studies, the application of this approach has not always been effective in enhancing biodegradative capacity, mainly during the composting of contaminated soils [25,88,114,115]. For instance, 84% of petroleum hydrocarbon was degraded when *Candida catenulate CM1* was used as an inoculant, while a removal rate of only 48% was obtained without inoculation, indicating a positive impact of bioaugmentation [29]. On the other hand, treatments using different substrates (mixing ratio = 1:1) were performed at the laboratory and field scales, and incubated with/without fungal inoculum (*Phanerochaete velutina*). Laboratory scale treatment showed that HMW PAHs were degraded significantly in the fungal-inoculated microcosms, such that 96% of four-ring PAHs and 39% of five- and six-ring PAHs were removed in three months, whereas 55% of four-ring PAHs and only 7% of five- and six-ring PAHs were degraded in non-inoculated ones. However, the field scale achieved similar degradation rates. Importantly, the number of gram-positive, PAH-ring, hydroxylating dioxygenase genes in the field scale experiment was found to increase 1000-fold, indicating that bacterial PAH degradation played a major role [116]. Wu et al. [117] compared bioaugmentation using *Acinetobacter SZ-1* strain and biostimulation using $(NH_4)_2SO_4$ and KH_2PO_4 in a petroleum-contaminated soil. It was found that the dissipation of total petroleum hydrocarbons (TPH) and the amounts of cultivable TPH, alkane, and PAH-degrading microorganisms were higher for biostimulation than for bioaugmentation. Similarly, Canet et al. [118] demonstrated that fungal inoculation, including four well-known PAH-degrading microorganisms (*P. chrysosporium* IMI 232175, *Coriolus versicolor* IMI 210866, *Pleurotus ostreatus* IMI 341687, and Wye isolate #7) in a mixture composed of non-sterile, coal-tar-contaminated soil and wheat straw, was unsuccessful in improving PAH removal. Sayara et al. [25] reported that the introduction of the white-rot fungi *T. versicolor* ATCC 42530 was not able to improve the decomposition of PAHs; on the contrary, organic substrates were capable of achieving significant degradation rates. Furthermore, inoculation with *P. chrysosporium* in a soil composting system was ineffective at enhancing the removal of benzo[a]pyrene [119].

Actually, several biotic and abiotic barriers have been documented to be behind the failure of bioaugemtation, mainly during field application of this mechanisms [51,120–122]. Biotic factors, including competition between indigenous and exogenous microorganisms for nutrients and the biodiversity of indigenous microorganisms, could act as a barrier to the invasion of exogenous microorganisms, in addition to antagonistic interactions and predation by protozoa and bacteriophages. Abiotic factors include all the physicochemical properties of pollutants and soils, such as pH, contaminant concentration, soil type, temperature, humidity aeration, nutrient content, and redox potential.

5. Bioavailability of PAHs

In some cases, when optimal conditions for microbial degradation are provided but low or even no degradation take place, the bioavailability of the pollutant would be the most probable reason for disabling the process from proceeding forward. Actually, the bioavailability of PAHs is directly linked to the intrinsic relationship between physicochemical and microbiological factors within the composting matrix. In particular, this factor determines the fraction of the chemical compound in the soil that can be utilized or transformed by living microorganisms [68,123–126].

Bioremediation is governed by PAH sorption onto the soil matrix in such a way that gradual sorption diminishes the possibility of desorption, and thus the PAH overstates its persistency within the soil organic matrix. This would explain the biphasic behavior of contaminants during bioremediation processes, which are associated with high removal rates in the initial phase, which is primarily limited by microbial degradation kinetics; in the second phase, though, the removal rate is low and generally limited by slow desorption. PAHs with low bioavailability are characterized with low desorption mainly in the second phase of bioremediation [127,128].

5.1. Factors Affecting PAH Bioavailability

PAHs are characterized by their high hydrophobicity, consequently increasing their affinity for being adsorbed into soil organic matter and ultimately being less available for biological uptake. Different studies [70–72,129,130] have highlighted that the following factors have an essential role in determining the bioavailability of PAHs. First is contamination time (ageing): the irreversible sorption of PAH is exponentially proportional with contact time, thus decreasing the bioavailability of pollutants to microorganisms and therefore the rate and extent of biodegradation. For instance, the removal efficiency of anthracene from freshly- and age-spiked agricultural soil was investigated. The results revealed that 72% of anthracene was removed in freshly-spiked soil, while only 34% was degraded in aged soil [131]. Nevertheless, it is worth mentioning that recently contaminated soil would exhibit toxicity to or even inhibit indigenous microorganisms until they adapted to the new environment [132,133]. The second factor for determining the bioavailability of PAHs is their physicochemical properties: PAHs' water solubility is considered a crucial factor regarding their bioavailability. It is inversely proportional with PAHs' molecular weight, which in turn reduces their accessibility to microorganisms (Section 2.1). The last factor is the physicochemical properties of the soil: organic matter, particle size, and shape have a major influence on PAHs' bioavailability. Mineral surfaces (i.e., clays) and organic matter of the soil matrix are characterized by their high affinity to adsorb PAHs

The addition of organic co-substrates to the composting mixture is believed to enhance the bioavailability of PAHs, which consequently increases the biodegradation rate [25,50,71]. Kobayashi et al. [106] demonstrated that water-extractable organic matter (WEOM) from cow manure compost was observed to increase the apparent solubility of phenanthrene, pyrene, and benzo[a]pyrene to 8.4, 34, and 89 times higher than their measured water values, respectively, thus promoting their solubility and biodegradation. Additionally, in a diesel-spiked soil, Wu et al. [66] showed that compost addition initially decreased PAH removal by up to 89% because of the decreased bioavailability resulting from strong sorption. However, as time increased, compost amendment enhanced PAH removal by more than two-fold compared with unamended soil, to which 30% was contributed by desorption and 70% by degradation. In coal tar- and coal ash-contaminated soils, compost addition was beneficial overall for enhancing PAH removal up to 94%, and 40% of the total loss was due to the enhanced desorption [66].

The stability of the used co-substrates plays a major role in stimulating bioavailability and biodegradation of PAHS (as discussed in Section 4.3). This type of substrates was found to have more humic matter [71]. In this context, humic matter was able to increase the microbial activities much more than those developed in humin (aged organic matter), demonstrating that humin is able to sequester organic contaminants in a stronger way [70]. An important observation is that the bioavailability of the more readily degradable or LMW PAHs was decreased due to competitive inhibition of the enzymes, which is associated with biodegradation when the enzymes present in a multiple-PAH mixture. However, the bioavailability of those usually more recalcitrant PAHs (HMW PAHs) was increased by producing inducible enzymes for catabolism [134].

5.2. Surfactant

As mentioned in the previous sections, some of PAHs are characterized by their high hydrophobicity as well as low solubility, as they have the ability to be strongly adhere to soil particles and be slowly released into the water phase [135]. Among the different alternatives to overcome the problems of low bioavailability during bioremediation of PAHs is the application of surfactants. The functionality of these additives basically depends on reducing interfacial surface tension, and thus increasing their solubility [30,33,34,136,137]. The efficiency of these surfactants is influenced by many factors, including surfactant type and concentration, PAH hydrophobicity, temperature, pH, salinity, dissolved organic matter, and microbial community. An imperative and crucial element for effective PAH remediation is the selection of the optimum ratio of mixed surfactants to avoid the inhibition of microbial activities [33,34,137]. Nowadays, various groups of surfactants are available, and each one is being used under certain conditions to be compatible with the available environment. Furthermore, biosurfactants that are produced by microorganisms are receiving more favor, as they considered more environmentally friendly [33,138]. Interestingly, Both Tween-80 and rhamnolipid were found to improve the bioremediation fluoranthene [139]; however, it should be considered that the application of surfactants may not always lead to enhanced PAH biodegradation or removal. In fact, if the surfactant is preferentially used as an easier carbon substrate than PAHs for soil microorganisms, it may actually inhibit PAH biodegradation. Selection of surfactant types is therefore crucial for the effectiveness of surfactant-aided bioremediation of PAH-contaminated soils [140].

6. PAH Biodegradation Pathway

As illustrated in the literature, a wide spectrum of microorganisms has been classified, and these microorganisms are known for their high potential in degrading PAHs. These microorganisms include, but not limited to, bacteria, fungus, actinomycetes, protozoa, and algae [69,87]. Actually, the biodegradation of PAHs has the possibility of being undertaken either under aerobic or anaerobic conditions. However, aerobic conditions are more preferable, due to their documented efficiency [141]. As a result, composting as an aerobic technique has received more attention for treatment for such types of pollution [25,78,79,123,132,141].

Fortunately, it has been documented that a wide variety of bacterial cultures have the potential to biodegrade LMW PAHs directly, using them as the sole carbon and energy source [142–145]. Otherwise, PAHs (like HMW PAHs) have to proceed through the accumulation of these compounds in the body of microorganisms, and then be decomposed through sequential steps and multiple routes (Figure 2) into a bioavailable form that could be metabolized by microorganisms [123,141,146,147]. Hydroxylation of the aromatic ring via a di- or monooxygenase enzymes or dehydrogenase is the first step in the degradation process, with the formation of a *cis*-dihydrodiol, which gets rearomatized to a diol intermediate by the action of a dehydrogenase. These diol intermediates may then be cleaved by intradiol or extradiol ring-cleaving dioxygenases through either an ortho-cleavage or meta-cleavage pathway, leading to intermediates such as catechols and protocatechol acid that are ultimately converted to tricarboxylic acid cycle intermediates, which could be considered as the end of the biodegradation [123,144,146–149]. Bacteria can also degrade PAHs via the cytochrome P450-mediated pathway, with the production of trans-dihydrodiols.

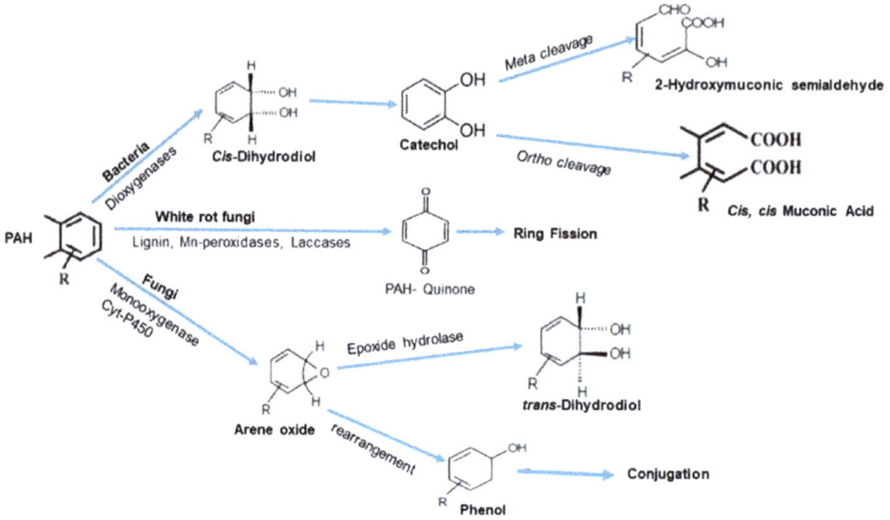

Figure 2. Bacterial and fungal biodegradation pathways of PAHs.

It should be noted that HMW PAH degradation pathways still need more investigation, as few bacterial isolates were found to be capable of degrading them. Also, their biodegradation in some cases is complicated and passes through different routes, or even proceeds via co-metabolism, like that of benzo[a]pyrene [150–152].

Fungal enzymatic activity also has a key role in the biodegradation of PAHs. Lignolytic and non-lignolytic fungi have the capability of oxidizing PAHs utilizing cytochrome P-450 monooxygenase and a lignin-degrading enzyme system for oxidizing aromatic rings. Usually, an oxygen atom is incorporated into the aromatic nucleus, whereas the remaining atom is reduced to water to yield cis-transdihydrodiols. The formed arene oxide, though non-enzymatic, can undergo some rearrangement to form a phenol, which can be further conjugated with glucose, xylose, gluconeric acid, and sulfate. On the other hand, ligninolytic fungi, which are usually known as white rot fungi, have been characterized by their capability to degrade PAHs through ligninolytic and non-ligninolytic culture conditions. Ligninolytic enzymes oxidizes the PAH ring by producing hydroxyl free radicals by the donation of one electron; consequently, PAH–quinones and acids are formed instead of dihydrodiols. [123,153–155]. Extracellular enzymes of white rot fungi, which include laccase, LiP, and MnP, have a key role in the degradation of PAHs [153,156–159].

7. Conclusions

The main conclusion of this review is that the use of compost and composting in several strategies significantly improves the removal of PAHs in contaminated soils. However, this strategy should be well studied and tested. For instance, future studies are required on compost stability, as it is an important parameter for considering the removal of PAHs. Composting also needs to be optimized to improve PAH removal. This could include, for instance, the use of some additives like surfactants, which can be of help for the desorption and further removal of PAHs. Furthermore, more investigations are still needed regarding the biodegradation of PAHs combined with other hydrocarbons in mixtures, biodegradation

HMW PAHs, and the microbial interactions within PAH-degrading consortia. In summary, composting and compost opens a wide number of strategies to improve the bioremediation of PAH-contaminated soils. However, it is important to define this strategy and to test its efficiency before full-scale application.

Author Contributions: Writing—original draft preparation, T.S.; writing—review and editing, A.S. All authors have read and agreed to the published version of the manuscript.

Funding: This research received no external funding.

Acknowledgments: Tahseen Sayara would like to thank Palestine Technical University for its administrative support.

Conflicts of Interest: The authors declare no conflict of interest.

Abbreviations

PAHs	Polycyclic aromatic hydrocarbons
HMW	High molecular weight
LMW	Low molecular weight
USEPA	United States Environmental Protection Agency

References

1. Levillain, J.; Cattan, P.; Colin, F.; Voltz, M.; Cabidoche, Y.M. Analysis of environmental and farming factors of soil contamination by a persistent organic pollutant, chlordecone, in a banana production area of French West Indies. *Agric. Ecosyst. Environ.* **2012**, *159*, 123–132. [CrossRef]
2. Yuan, G.L.; Qin, J.X.; Li, J.; Lang, X.X.; Wang, G.H. Persistent organic pollutants in soil near the Changwengluozha glacier of the Central Tibetan Plateau, China: Their sorption to clays and implication. *Sci. Total Environ.* **2014**, *472*, 309–315. [CrossRef]
3. Zeng, G.; Wan, J.; Huang, D.; Hu, L.; Huang, C.; Cheng, M.; Xue, W.; Gong, X.; Wang, R.; Jiang, D. Precipitation, adsorption and rhizosphere effect: The mechanisms for Phosphate-induced Pb immobilization in soils—A review. *J. Hazard. Mater.* **2017**, *339*, 354–367. [CrossRef] [PubMed]
4. Kavamura, V.N.; Esposito, E. Biotechnological strategies applied to the decontamination of soils polluted with heavy metals. *Biotechnol. Adv.* **2010**, *28*, 61–69. [CrossRef] [PubMed]
5. Xu, P.; Zeng, G.M.; Huang, D.L.; Feng, C.L.; Hu, S.; Zhao, M.H.; Lai, C.; Wei, Z.; Huang, C.; Xie, G.X.; et al. Use of iron oxide nanomaterials in wastewater treatment: A review. *Sci. Total Environ.* **2012**, *424*, 1–10. [CrossRef] [PubMed]
6. Ha, H.; Olson, J.; Bian, L.; Rogerson, P.A. Analysis of heavy metal sources in soil using kriging interpolation on principal components. *Environ. Sci. Technol.* **2014**, *48*, 4999–5007. [CrossRef]
7. Chen, M.; Xu, P.; Zeng, G.M.; Yang, C.P.; Huang, D.L.; Zhang, J.C. Bioremediation of soils contaminated with polycyclic aromatic hydrocarbons, petroleum, pesticides, chlorophenols and heavy metals by composting: Applications, microbes and future research needs. *Biotechnol. Adv.* **2015**, *33*, 745–755. [CrossRef]
8. Udeigwe, T.K.; Eze, P.N.; Teboh, J.M.; Stietiya, M.H. Application, chemistry, and environmental implications of contaminant-immobilization amendments on agricultural soil and water quality. *Environ. Int.* **2011**, *37*, 258–267. [CrossRef]
9. Hu, G.; Li, J.; Zeng, G. Recent development in the treatment of oily sludge from petroleum industry: A review. *J. Hazard. Mater.* **2013**, *261*, 470–490. [CrossRef]
10. Zeng, G.M.; Chen, M.; Zeng, Z.T. Shale gas: Surface water also at risk. *Nature* **2013**, *499*, 154. [CrossRef]
11. Tang, W.W.; Zeng, G.M.; Gong, J.L.; Liang, J.; Xu, P.; Zhang, C.; Huang, B.-B. Impact of humic/fulvic acid on the removal of heavy metals from aqueous solutions using nanomaterials: A review. *Sci. Total Environ.* **2014**, *468*, 1014–1027. [CrossRef] [PubMed]
12. Zhu, Y.; Yang, L.; Yuan, Q.; Yan, C.; Dong, C.; Meng, C.; Sui, X.; Yao, L.; Yang, F.; Lu, Y.; et al. Airborne particulate polycyclic aromatic hydrocarbon (PAH) pollution in a background site in the North China Plain: Concentration, size distribution, toxicity and sources. *Sci. Total Environ.* **2015**, *466–467*, 357–368. [CrossRef] [PubMed]
13. Yam, R.; Leung, W. Emissions trading in Hong Kong and the Pearl River Delta region -a modeling approach to trade decisions in Hong Kong's electricity industry. *Environ. Sci. Pol.* **2013**, *31*, 1–12. [CrossRef]

14. Witter, A.; Nguyen, M.; Baidar, S.; Sak, P. Coal-tar-based sealcoated pavement: A major PAH source to urban stream sediments. *Environ. Pollut.* **2014**, *185*, 59–68. [CrossRef]

15. Bacosa, H.P.; Inoue, C. Polycyclic aromatic hydrocarbons (PAHs) biodegradation potential and diversity of microbial consortia enriched from tsunami sediments in Miyagi, Japan. *J. Hazard. Mater.* **2015**, *283*, 689–697. [CrossRef]

16. Wan, J.; Zeng, G.M.; Huang, D.L.; Huang, C.; Lai, C.; Li, N.J.; Wei, Z.; Xu, P.; He, X.; Lai, M.Y.; et al. The oxidative stress of phanerochaete chrysosporium against lead toxicity. *Appl. Biochem. Biotechnol.* **2015**, *175*, 1981–1991. [CrossRef]

17. Evans, M.; Davies, M.; Janzen, K.; Muir, D.; Hazewinkel, R.; Kirk, J.; de Boer, D. PAH distributions in sediments in the oil sands monitoring area and western Lake Athabasca: Concentration, composition and diagnostic ratios. *Environ. Pollut.* **2016**, *213*, 671–687. [CrossRef]

18. Maliszewska-Kordybach, B.; Smreczak, B.; Klimkowicz-Pawlas, A. The levels and composition of persistent organic pollutants in alluvial agriculture soils affected by flooding. *Environ. Monit. Assess.* **2013**, *185*, 9935–9948. [CrossRef]

19. Tsibart, A.; Gennadiev, A. Polycyclic aromatic hydrocarbons in soils: Sources, behavior, and indication significance. *Eurasian Soil Sci.* **2013**, *46*, 728–741. [CrossRef]

20. Chen, B.; Yuan, M. Enhanced sorption of polycyclic aromatic hydrocarbons by soil amended with biochar. *J. Soils Sediment.* **2011**, *11*, 62–71. [CrossRef]

21. Koshlaf, E.; Shahsavari, E.; Aburto-Medina, A.; Taha, M.; Haleyur, N.; Makadia, T.H.; Morrison, P.D.; Ball, A.S. Bioremediation potential of diesel-contaminated Libyan soil. *Ecotoxicol. Environ. Saf.* **2016**, *133*, 297–305. [CrossRef] [PubMed]

22. Usman, M.M.; Dadrasnia, A.; Lim, K.T.; Mahmud, A.F.; Ismail, S. Application of Biosurfactants in Environmental Biotechnology; Remediation of Oil and Heavy Metal. *AIMS Bioeng.* **2016**, *3*, 289–304. [CrossRef]

23. Antizar-Ladislao, B.; Lopez-Real, J.; Beck, A. Degradation of polycyclic aromatic hydrocarbons (PAHs) in an aged coal tar contaminated soil under in-vessel composting conditions. *Environ. Pollut.* **2006**, *141*, 459–468. [CrossRef] [PubMed]

24. Puglisi, E.; Cappa, F.; Fragoulis, G.; Trevisan, M.; Del Re, A.A.M. Bioavailability and degradation of phenanthrene in compost amended soils. *Chemosphere* **2007**, *67*, 548–556. [CrossRef]

25. Sayara, T.; Borràs, E.; Caminal, G.; Sarrà, M.; Sánchez, A. Bioremediation of PAHs-contaminated soil through composting: Influence of bioaugmentation and biostimulation on contaminant biodegradation. *Int. Biodeter. Biodegr.* **2011**, *65*, 859–865. [CrossRef]

26. Ortega-Calvo, J.; Tejeda-Agredano, M.; Jimenez-Sanchez, C.; Congiu, E.; Sungthong, R.; Niqui-Arroyo, J.; Cantos, M. Is it possible to increase bioavailability but not environmental risk of PAHs in bioremediation? *J. Hazard. Mater.* **2013**, *261*, 733–745. [CrossRef]

27. Lukić, B.; Panico, A.; Huguenot, D.; Fabbricino, M.; van Hullebusch, E.D.; Esposito, G. Evaluation of PAH removal efficiency in an artificial soil amended with different types of organic wastes. *Euro-Mediterr. J. Environ. Integr.* **2016**, *1*, 5. [CrossRef]

28. Miller, M.; Stratton, G.; Murray, G. Effects of nutrient amendments and temperature on the biodegradation of pentachlorophenol contaminated soil. *Water Air Soil Pollut.* **2004**, *151*, 87–101. [CrossRef]

29. Joo, H.S.; Ndegwa, P.M.; Shoda, M.; Phae, C.G. Bioremediation of oilcontaminated soil using Candida catenulata and food waste. *Environ. Pollut.* **2008**, *156*, 891–896. [CrossRef]

30. Li, J.L.; Chen, B.H. Effect of nonionic surfactants on biodegradation of phenanthrene by a marine bacteria of Neptunomonas naphthovorans. *J. Hazard. Mater.* **2009**, *162*, 66–73. [CrossRef]

31. Gomez, S.M. Optimization of field scale biopiles for bioremediation of petroleum hydrocarbon contaminated soil at low temperature conditions by response surface methodology (RSM). *Int. Biodeter. Biodegr.* **2014**, *89*, 103–109. [CrossRef]

32. Huang, C.; Zeng, G.; Huang, D.; Lai, C.; Xu, P.; Zhang, C.; Cheng, M.; Wan, J.; Hu, L.; Zhang, Y. Effect of Phanerochaete chrysosporium inoculation on bacterial community and metal stabilization in lead-contaminated agricultural waste composting. *Bioresour. Technol.* **2017**, *243*, 294–303. [CrossRef] [PubMed]

33. Lamichhane, S.; Krishna, K.C.B.; Sarukkalige, R. Surfactant-enhanced remediation of polycyclic aromatic hydrocarbons: A review. *J. Environ. Manag.* **2017**, *199*, 46–61. [CrossRef]

34. Liu, Y.; Zeng, G.; Zhong, H.; Wang, Z.; Liu, Z.; Cheng, M.; Liu, G.; Yang, X.; Liu, S. Effect of rhamnolipid solubilization on hexadecane bioavailability: Enhancement or reduction? *J. Hazard. Mater.* **2017**, *322*, 394–401. [CrossRef] [PubMed]

35. Samanta, S.K.; Singh, O.V.; Jain, R.K. Polycyclic aromatic hydrocarbons: Environmental pollution and bioremediation. *Trends Biotechnol.* **2002**, *20*, 243–248. [CrossRef]

36. Watanabe, K.; Futamata, H.; Harayama, S. Understanding the diversity in catabolic potential of microorganisms for the development of bioremediation strategies. *Anton. Van Leeuw.* **2002**, *81*, 655–663. [CrossRef] [PubMed]

37. Yu, Z.; Zeng, G.M.; Chen, Y.N.; Zhang, J.C.; Yu, Y.; Li, H.; Liu, Z.-F.; Tang, L. Effects of inoculation with Phanerochaete chrysosporium on remediation of pentachlorophenol-contaminated soil waste by composting. *Process. Biochem.* **2011**, *46*, 1285–1291. [CrossRef]

38. Gemmell, B.J.; Bacosa, H.P.; Liu, Z.; Buskey, E.J. Can gelatinous zooplankton influence the fate of crude oil in marine environments? *Mar. Pollut. Bull.* **2016**, *113*, 483–487. [CrossRef]

39. Sosa, B.S.; Porta, A.; Lerner, J.E.C.; Noriega, R.B.; Massolo, L. Human health risk due to variations in PM10-PM2.5 and associated PAHs levels. *Atmos. Environ.* **2017**, *160*, 27–35. [CrossRef]

40. EC (European Commission). Communication from the Commission to the Council, the European Parliament, the European Economic and Social Committee and the Committee of the Regions—Thematic Strategy for Soil Protection. COM, 2006; 231final. 2006. Available online: http://ec.europa.eu/environment/soil/pdf/com_2006_0231_en.pdf (accessed on 15 January 2020).

41. EEA (European Environment Agency). Progress in Management of Contaminated Sites (CSI 015) 2007. Available online: http://themes.eea.europa.eu/IMS/ISpecs/ISpecification20041007131746/IAssessment1152619898983/view_content (accessed on 15 January 2020).

42. Fu, P.P.; Xia, Q.; Sun, X.; Yu, H. Phototoxicity and environmental transformation of polycyclic aromatic hydrocarbons (PAHs)-light-induced reactive oxygen species, lipid peroxidation, and DNA damage. *J. Environ. Sci. Health C* **2012**, *30*, 1–41. [CrossRef]

43. Niepceron, M.; Martin-Laurent, F.; Crampon, M.; Portet-Koltalo, F.; Akpa-Vinceslas, M.; Legras, M.; Bru, D.; Bureau, F.; Bodilis, J. Gamma Proteobacteria as a potential bioindicator of a multiple contamination by polycyclic aromatic hydrocarbons (PAHs) in agricultural soils. *Environ. Pollut.* **2013**, *180*, 199–205. [CrossRef] [PubMed]

44. Wenzl, T.; Simon, R.; Anklam, E.; Kleiner, J. Analytical methods for polycyclic aromatic hydrocarbons (PAHs) in food and the environment needed for new food legislation in the European Union. *Trends Anal. Chem.* **2006**, *25*, 716–725. [CrossRef]

45. Gabov, D.N.; Beznosikov, V.A.; Kondratenok, B.M.; Yakovleva, E.V. Polycyclic aromatic hydrocarbons in the soils of technogenic landscapes. *Geochem. Int.* **2010**, *48*, 569–579. [CrossRef]

46. Gennadiev, A.; Tsibart, A. Pyrogenic polycyclic aromatic hydrocarbons in soils of reserved and anthropogenically modified areas: Factors and features of accumulation. *Eurasian Soil Sci.* **2013**, *46*, 28–36. [CrossRef]

47. Luo, L.; Lin, S.; Huang, H.L.; Zhang, S.Z. Relationships between aging of PAHs and soil properties. *Environ. Pollut.* **2012**, *170*, 177–182. [CrossRef]

48. Pereira, T.; Laiana, S.; Rocha, J.; Broto, F.; Comellas, L.; Salvadori, D.; Vargas, V. Toxicogenetic monitoring in urban cities exposed to different airborne contaminants. *Ecotoxicol. Environ. Saf.* **2013**, *90*, 174–182. [CrossRef]

49. Mohan, S.V.; Prasanna, D.; Purushotham, R.B.; Sarma, P.N. Ex situ bioremediation of pyrene contaminated soil in bio-slurry phase reactor operated in periodic discontinuous batch mode: Influence of bioaugmentation. *Int. Biodeter. Biodegr.* **2008**, *62*, 162–169. [CrossRef]

50. Hamdi, H.; Benzarti, S.; Manusadžianas, L.; Aoyama, I.; Jedidi, N. Bioaugmentation and biostimulation effects on PAH dissipation and soil ecotoxicity under controlled conditions. *Soil Biol. Bioch.* **2007**, *39*, 1926–1935. [CrossRef]

51. Mrozik, A.; Piotrowska-Seget, Z. Bioaugmentation as a strategy for cleaning up of soils contaminated with aromatic compounds. *Microbiol. Res.* **2010**, *165*, 363–375. [CrossRef]

52. Sayler, G.S.; Ripp, S. Field applications of genetically engineered microorganisms for bioremediation processes. *Curr. Opin. Biotechnol.* **2000**, *11*, 286–289. [CrossRef]

53. Haug, R.T. *The Practical Handbook of Compost Engineering*; Lewis Publishers: Boca Raton, FL, USA, 1993.

54. Hsu, J.H.; Lo, S.L. Chemical and spectroscopic analysis of organic matter transformations during composting of pig manure. *Environ. Pollut.* **1999**, *104*, 189–196. [CrossRef]

55. Lemmon, C.R.; Pylypiw, H.M. Degradation of diazanon, chlorpyrifos, isofenphos and pendimethalin in grass and compost. *Bull. Environ. Contam. Toxicol.* **1992**, *48*, 409–415. [CrossRef] [PubMed]

56. Breitung, J.; Bruns-Nagel, D.; Steinbach, K.; Kaminski, L.; Gemsa, D.; Von LoÈ w, E. Bioremediation of 2, 4, 6-trinitrotoluene-contaminated soils by two different aerated compost systems. *Appl. Microbiol. Biotechnol.* **1996**, *44*, 795–800. [CrossRef] [PubMed]

57. Semple, K.T.; Reid, B.J.; Fermor, T.R. Impact of composting strategies on the treatment of soils contaminated with organic pollutants. *Environ. Pollut.* **2001**, *112*, 269–283. [CrossRef]

58. Loick, N.; Hobbs, P.J.; Hale, M.D.C.; Jones, D.L. Bioremediation of Poly-Aromatic Hydrocarbon (PAH)-Contaminated Soil by Composting. *Crit. Rev. Environ. Sci. Technol.* **2009**, *39*, 271–332. [CrossRef]

59. Sayara, T.; Sarrà, M.; Sánchez, A. Preliminary screening of co-substrates for bioremediation of pyrene-contaminated soil through composting. *J. Hazard. Mater.* **2009**, *172*, 1695–1698. [CrossRef]

60. Gandolfi, I.; Sicolo, M.; Franzetti, A.; Fontanarosa, E.; Santagostino, A.; Bestetti, G. Influence of compost amendment on microbial community and ecotoxicity of hydrocarbon-contaminated soils. *Bioresour. Technol.* **2010**, *101*, 568–575. [CrossRef]

61. Ryckeboer, J.; Mergaert, J.; Vaes, K.; Klammer, S.; De Clercq, D.; Coosemans, J.; Insam, H.; Swings, J. A survey of bacteria and fungi occurring during composting and self-heating processes. *Ann. Microbiol.* **2003**, *53*, 349–410.

62. Tejada, M.; González, J.L.; Hernández, M.T.; García, C. Application of different organic amendments in a gasoline contaminated soil: Effect on soil microbial properties. *Bioresour. Technol.* **2008**, *99*, 2872–2880. [CrossRef]

63. Tejada, M.; Hernandez, M.T.; Garcia, C. Soil restoration using composted plant residues: Effects on soil properties. *Soil Tillage Res.* **2009**, *102*, 109–117. [CrossRef]

64. Hu, Z.H.; Liu, Y.L.; Chen, G.W.; Gui, X.Y.; Chen, T.H.; Zhan, X.M. Characterization of organic matter degradation during composting of manure straw mixtures spiked with tetracyclines. *Bioresour. Technol.* **2011**, *102*, 7329–7334. [CrossRef] [PubMed]

65. Duong, T.T.T.; Penfold, C.; Marschner, P. Differential effects of composts on properties of soils with different textures. *Biol. Fert. Soils* **2012**, *48*, 699–707. [CrossRef]

66. Wu, G.; Kechavarzi, C.; Li, X.; Sui, H.; Pollard, S.J.T.; Coulon, F. Influence of mature compost amendment on total and bioavailable polycyclic aromatic hydrocarbons in contaminated soils. *Chemosphere* **2013**, *90*, 2240–2246. [CrossRef] [PubMed]

67. Bastida, F.; Jehmlichc, N.; Lima, K.; Morris, B.E.L.; Richnow, H.H.; Hernández, T.; von Bergen, M.; García, C. The ecological and physiological responses of the microbial community from a semiarid soil to hydrocarbon contamination and its bioremediation using compost amendment. *J. Proteomics* **2016**, *135*, 162–169. [CrossRef]

68. Latawiec, A.E.; Swindell, A.L.; Simmons, P.; Reid, B.J. Bringing bioavailability into contaminated land decision making: The way forward? *Crit. Rev. Environ. Sci. Technol.* **2011**, *41*, 52–77. [CrossRef]

69. Amir, S.; Hafidi, M.; Merlina, G.; Hamdi, H.; Revel, J.C. Fate of polycyclic aromatic hydrocarbons during composting of lagooning sewage sludge. *Chemosphere* **2005**, *58*, 449–458. [CrossRef]

70. Yang, Y.; Tao, S.; Zhang, N.; Zhang, D.Y.; Li, X.Q. The effect of soil organic matter on fate of polycyclic aromatic hydrocarbons in soil: A microcosm study. *Environ. Pollut.* **2010**, *158*, 1768–1774. [CrossRef]

71. Tang, J.; Lu, X.; Sun, Q.; Zhu, W. Aging effect of petroleum hydrocarbons in soil under different attenuation conditions. *Agric. Ecosyst. Environ.* **2012**, *149*, 109–117. [CrossRef]

72. Cébron, A.; Faure, P.; Lorgeoux, C.; Ouvrard, S.; Leyval, C. Experimental increase in availability of a PAH complex organic contamination from an aged contaminated soil: Consequences on biodegradation. *Environ. Pollut.* **2013**, *177*, 98–105. [CrossRef]

73. Han, X.; Hu, H.; Shi, X.; Zhang, L.; He, J. Effects of different agricultural wastes on the dissipation of PAHs and the PAH-degrading genes in a PAH-contaminated soil. *Chemosphere* **2017**, *172*, 286–293. [CrossRef]

74. Lukić, B.; Huguenot, D.; Panico, A.; Fabbricino, M.; van Hullebusch, E.D.; Esposito, G. Importance of organic amendment characteristics on bioremediation of PAH-contaminated soil. *Environ. Sci. Pollut. Res. Int.* **2016**, *23*, 15041–15052. [CrossRef] [PubMed]

75. Namkoong, W.; Hwang, E.Y.; Park, J.S.; Choi, J.Y. Bioremediation of diesel-contaminated soil with composting. *Environ. Pollut.* **2002**, *119*, 23–31. [CrossRef]

76. Piskonen, R.; Itävaara, M. Evaluation of chemical pretreatment of contaminated soil for improved PAH bioremediation. *Appl. Microbiol. Biotechnol.* **2004**, *65*, 627–634. [CrossRef] [PubMed]
77. Juhasz, A.L.; Naidu, R. Bioremediation of high molecular weight polycyclic aromatic hydrocarbons: A review of the microbial degradation of benzo[a]pyrene. *Int. Biodeterior. Biodegrad.* **2000**, *45*, 57–88. [CrossRef]
78. Sayara, T.; Sarrà, M.; Sánchez, A. Effects of compost stability and contaminant concentration on the bioremediation of PAHs contaminated soil through composting. *J. Hazard. Mater.* **2010**, *179*, 999–1006. [CrossRef]
79. Sayara, T.; Sarrà, M.; Sánchez, A. Optimization and enhancement of soil bioremediation by composting using the experimental design technique. *Biodegradation* **2010**, *21*, 345–356. [CrossRef]
80. Zappi, M.E.; Rogers, B.A.; Teeter, C.L.; Gunnison, D.; Bajpai, R. Bioslurry treatment of a soil contaminated with low concentrations of total petroleum hydrocarbons. *J. Hazard. Mater.* **1996**, *46*, 1–12. [CrossRef]
81. Jorgensen, K.S.; Puustinen, J.; Suortti, A.M. Bioremediation of petroleum hydrocarbon-contaminated soil by composting in biopiles. *Environ. Pollut.* **2000**, *107*, 245–254. [CrossRef]
82. Viamajala, S.; Peyton, B.M.; Richards, L.A.; Petersen, J.N. Solubilization, solution equilibria, and biodegradation of PAHs under thermophilic conditions. *Chemosphere* **2007**, *66*, 1094–1106. [CrossRef]
83. Bartha, R.; Bossert, I. *The Treatment and Disposal of Petroleum Wastes in Petroleum Microbiology*; Macmillan: New York, NY, USA, 1984; pp. 553–578.
84. Cooney, J.J. The Fate of Petroleum Pollutants in Fresh Water Ecosystems. In *Petroleum Microbiology*; Atlas, R.M., Ed.; Macmillan: New York, NY, USA, 1984; pp. 399–434.
85. Purnomo, A.S.; Koyama, F.; Mori, T.; Kondo, R. DDT degradation potential of cattle manure compost. *Chemosphere* **2010**, *80*, 619–624. [CrossRef]
86. Houot, S.; Verge-Leviel, C.; Poitrenaud, M. Potential mineralization of various organic pollutants during composting. *Pedosphere* **2012**, *22*, 536–543. [CrossRef]
87. Peng, J.J.; Zhang, Y.; Su, J.Q.; Qiu, Q.F.; Jia, Z.J.; Zhu, Y.G. Bacterial communities predominant in the degradation of 13C4-4,5,9,10-pyrene during composting. *Bioresour. Technol.* **2013**, *143*, 608–614. [CrossRef] [PubMed]
88. Antizar-Ladislao, B.; Lopez-Real, J.; Beck, A.J. In-vessel composting-bioremediation of aged coal tar soil: Effect of temperature and soil/green waste amendment ratio. *Environ. Int.* **2005**, *31*, 173–178. [CrossRef] [PubMed]
89. Antizar-Ladislao, B.; Spanova, K.; Beck, A.J.; Russell, N.J. Microbial community structure changes during bioremediation of PAHs in an aged coal-tar contaminated soil by in-vessel composting. *Int. Biodeterior. Biodegrad.* **2008**, *61*, 357–364. [CrossRef]
90. Kriipsalu, M.; Marques, M.; Hogland, W.; Nammari, D.R. Fate of polycyclic aromatic hydrocarbons during composting of oil sludge. *Environ. Technol.* **2008**, *29*, 43–53. [CrossRef]
91. Alves, D.; Villar, I.; Mato, S. Thermophilic composting of hydrocarbon residue with sewage sludge and fish sludge as cosubstrates: Microbial changes and TPH reduction. *J. Environ. Manag.* **2019**, *239*, 30–37. [CrossRef]
92. Zhu, F.; Storey, S.; Ashaari, M.M.; Clipson, N.; Doyle, E. Benzo (a) pyrene degradation and microbial community responses in composted soil. *Environ. Sci. Pollut. R.* **2017**, *24*, 5404–5414. [CrossRef]
93. Ren, X.; Zeng, G.; Tang, L.; Wang, J.; Wan, J.; Wang, J.; Deng, Y.; Liu, Y.; Peng, B. The potential impact on the biodegradation of organic pollutants from composting technology for soil remediation. *Waste Manag.* **2018**, *72*, 138–149. [CrossRef]
94. Mattei, P.; Cincinelli, A.; Martellini, T.; Natalini, R.; Pascale, E.; Renella, G. Reclamation of river dredged sediments polluted by PAHs by co-composting with green waste. *Sci. Total. Environ.* **2016**, *566–567*, 567–574. [CrossRef]
95. Hesnawi, R.M.; McCartney, D. Impact of compost amendments and operating temperature on diesel fuel bioremediation. *Environ. Eng. Sci.* **2006**, *5*, 37–45. [CrossRef]
96. Li, J.Y.; Ye, Q.F.; Gan, J. Influence of organic amendment on fate of acetaminophen and sulfamethoxazole in soil. *Environ. Pollut.* **2015**, *206*, 543–550. [CrossRef] [PubMed]
97. Plaza, C.; Xing, B.S.; Fernández, J.M.; Senesi, N.; Polo, A. Binding of polycyclic aromatic hydrocarbons by humic acids formed during composting. *Environ. Pollut.* **2009**, *157*, 257–263. [CrossRef] [PubMed]
98. Zmora-Nahum, S.; Markovitch, O.; Tarchitzky, J.; Chen, Y. Dissolved organic carbon (DOC) as a parameter of compost maturity. *Soil Biol. Biochem.* **2005**, *37*, 2109–2116. [CrossRef]

99. Adam, I.K.U.; Miltner, A.; Kästner, M. Degradation of 13C-labeled pyrene in soil-compost mixtures and fertilized soil. *Appl. Microbiol. Biotechnol.* **2015**, *99*, 9813–9824. [CrossRef]

100. Senesi, N.; Plaza, C. Role of humification processes in recycling organic wastes of various nature and sources as soil amendments. *Clean-Soil Air Water.* **2007**, *35*, 26–41. [CrossRef]

101. He, X.S.; Xi, B.D.; Cui, D.Y.; Liu, Y.; Tan, W.B.; Pan, H.W.; Li, D. Influence of chemical and structural evolution of dissolved organic matter on electron transfer capacity during composting. *J. Hazard. Mater.* **2014**, *268*, 256–263. [CrossRef]

102. Doublet, J.; Francou, C.; Pétraud, J.P.; Dignac, M.F.; Poitrenaud, M.; Houot, S. Distribution of C and N mineralization of a sludge compost within particle-size fractions. *Bioresour. Technol.* **2010**, *101*, 1254–1262. [CrossRef]

103. Verma, S.L.; Marschner, P. Compost effects on microbial biomass and soil P pools as affected by particle size and soil properties. *J. Soil Sci. Plant. Nut.* **2013**, *13*, 313–328.

104. Jednak, T.; Avdalović, J.; Miletić, S.; Slavković-Beškoski, L.; Stanković, D.; Milić, J.; Llic, M.; Beškoski, V.; Gojgić-Cvijović, G.; Vrvić, M.M. Transformation and synthesis of humic substances during bioremediation of petroleum hydrocarbons. *Int. Biodeterior. Biodegrad.* **2017**, *122*, 47–52. [CrossRef]

105. Ortega-Calvo, J.J.; Saiz-Jimenez, C. Effect of humic fractions and clay on biodegradation of phenanthrene by a Pseudomonas fluorescens strain isolated from soil. *Appl. Environ. Microbiol.* **1998**, *64*, 3123–3126. [CrossRef]

106. Kobayashi, T.; Murai, Y.; Tatsumi, K.; Iimura, Y. Biodegradation of polycyclic aromatic hydrocarbons by Sphingomonas sp. enhanced by water-extractable organic matter from manure compost. *Sci. Total Environ.* **2009**, *407*, 5805–5810. [CrossRef] [PubMed]

107. Nwankwegu, A.S.; Orji, M.U.; Onwosi, C.O. Studies on organic and in-organic biostimulants in bioremediation of diesel-contaminated arable soil. *Chemosphere* **2016**, *162*, 148–156. [CrossRef] [PubMed]

108. Chen, F.; Li, X.; Zhu, Q.; Ma, J.; Hou, H.; Zhang, S. Bioremediation of petroleum-contaminated soil enhanced by aged refuse. *Chemosphere* **2019**, *222*, 98–105. [CrossRef]

109. Wang, Z.Y.; Xu, Y.; Zhao, J.; Li, F.M.; Gao, D.M.; Xing, B.S. Remediation of petroleum contaminated soils through composting and rhizosphere degradation. *J. Hazard. Mater.* **2011**, *190*, 677–685. [CrossRef]

110. Hickman, Z.A.; Reid, B.J. The co-application of earthworms (Dendrobaena veneta) and compost to increase hydrocarbon losses from diesel contaminated soils. *Environ. Int.* **2008**, *34*, 1016–1022. [CrossRef]

111. Bao, H.; Wang, J.; Li, J.; Zhang, H.; Wu, F. Effects of corn straw on dissipation of polycyclic aromatic hydrocarbons and potential application of backpropagation artificial neural network prediction model for PAHs bioremediation. *Ecotoxicol. Environ. Saf.* **2019**, *186*, 109745. [CrossRef]

112. Huesemann, M.H.; Hausmann, T.S.; Fortman, T.J. Does bioavailability limit biodegrada- tion? A comparison of hydrocarbon biodegradation and desorption rates in aged soils. *Biodegradation* **2004**, *15*, 261–274. [CrossRef]

113. Hwang, S.; Cutright, T.J. Biodegradability of aged pyrene and phenanthrene in a natural soil. *Chemosphere* **2002**, *47*, 891–899. [CrossRef]

114. Kennedy, T.A.; Naeem, S.; Howe, K.M.; Knops, J.M.H.; Tilman, D.; Reich, P. Biodiversity as a barrier to ecological invasion. *Nature* **2002**, *417*, 636–638. [CrossRef] [PubMed]

115. Zhang, Y.; Zhu, Y.G.; Houot, S.; Qiao, M.; Nunan, N.; Garnier, P. Remediation of polycyclic aromatic hydrocarbon (PAH) contaminated soil through composting with fresh organic wastes. *Environ. Sci. Pollut. Res. Int.* **2011**, *18*, 1574–1584. [CrossRef] [PubMed]

116. Winquist, E.; Bjorklof, K.; Schultz, E.; Räsänen, M.; Salonen, K.; Anasonye, F.; Cajthaml, T.; Steffen, K.T.; Jørgensen, K.S.; Tuomela, M. Bioremediation of PAH-contaminated soil with fungi- From laboratory to field scale. *Int. Biodeter. Biodegr.* **2014**, *86*, 238–247. [CrossRef]

117. Wu, M.L.; Dick, W.A.; Li, W.; Wang, X.C.; Yang, Q.; Wang, T.T.; Xu, L.M.; Zhang, M.H.; Chen, L.M. Bioaugmentation and biostimulation of hydrocarbon degradation and the microbial community in a petroleum-contaminated soil. *Int. Biodeter. Biodegr.* **2016**, *107*, 158–164. [CrossRef]

118. Canet, R.; Birnstingl, J.G.; Malcolm, D.G.; Lopez-Real, J.M.; Beck, A.J. Biodegradation of polycyclic aromatic hydrocarbons (PAHs) by native microflora and combinations of white-rot fungi in a coal-tar contaminated soil. *Bioresour. Technol.* **2001**, *76*, 113–117. [CrossRef]

119. McFarland, M.J.; Qiu, X.J. Removal of benzo(a)pyrene in soil composting systems amended with the white rot fungus Phanerochaete chrysosporium. *J. Hazard. Mater.* **1995**, *42*, 61–70. [CrossRef]

120. Lebeau, T. Bioaugmentation for in Situ Soil Remediation: How to Ensure the Success of Such a Process. In *Bio- Augmentation, Biostimulation and Biocontrol, Soil Biology*; Singh, A., Parmar, N., Kuhad, R.C., Eds.; Springer: Berlin/Heidelberg, Germany, 2011; Volume 28, pp. 129–186.

121. Tyagi, M.; da Fonseca, M.M.R.; de Carvalho, C.C.C.R. Bioaugmentation and biostimulation strategies to improve the effectiveness of bioremediation processes. *Biodegradation* **2011**, *22*, 231–241. [CrossRef] [PubMed]

122. Zafra, G.; Cortés-Espinosa, D.V. Biodegradation of polycyclic aromatic hydrocarbons by Trichoderma species: A mini review. *Environ. Sci. Pollut. Res.* **2015**, *22*, 19426–19433. [CrossRef] [PubMed]

123. Mueller, J.G.; Cerniglia, C.E.; Pritchard, P.H. Bioremediation of Environments Contaminated by Polycyclic Aromatic Hydrocarbons. In *Bioremediation: Principles and Applications*; Crawford, R.L., Crawford, D.L., Eds.; Cambridge University Press: Cambridge, UK, 1996; pp. 1215–1294.

124. Semple, K.T.; Doick, K.J.; Jones, K.C.; Burauel, P.; Craven, A.; Harms, H. Defining bioavailability and bioaccessibility of contaminated soil and sediment is complicated. *Environ. Sci. Technol.* **2004**, *38*, 228A–231A. [CrossRef] [PubMed]

125. Semple, K.T.; Riding, M.J.; McAllister, L.E.; Sopena-Vazquez, F.; Bending, G.D. Impact of black carbon on the bioaccessibility of organic contaminants in soil. *J. Hazard. Mater.* **2013**, *261*, 808–816. [CrossRef] [PubMed]

126. Stokes, J.D.; Paton, G.; Semple, K.T. Behaviour and assessment of bioavailability of organic contaminants in soil: Relevance for risk assessment and remediation. *Soil Use Manag.* **2006**, *21*, 475–486. [CrossRef]

127. Loeher, R.C.; Mc Millen, S.J.; Webster, M.T. Predictions of biotreatability and actual results: Soils with petroleum hydrocarbons. *Pract. Period. Hazard. Toxic Radioact. Waste Manag.* **2001**, *5*, 78–87. [CrossRef]

128. Trinidade, P.V.O.; Sobral, A.C.L.; Rizzo, S.G.F.; Leite Soriano, A.U. Bioremediation of weathered and recently oil-contaminated soils from Brazil: A compression study. *Chemosphere* **2005**, *58*, 515–522. [CrossRef] [PubMed]

129. Allard, A.S.; Remberger, M.; Neilson, A.H. The negative impact of aging on the loss of PAH components in a creosote-contaminated soil. *Int. Biodeter. Biodegr.* **2000**, *46*, 43–49. [CrossRef]

130. Maletic, S.; Dalmacija, B.; Roncevic, S.; Agbaba, J.; Ugarcina, P.S. Impact of hydrocarbon type, concentration and weathering on its biodegradability in soil. *J. Environ. Sci. Heal. A* **2011**, *46*, 1042–1049. [CrossRef] [PubMed]

131. Delgado-Balbuena, L.; Romero-Tepal, E.M.; Luna-Guido, M.L.; Marsch, R.; Dendooven, L. Removal of anthracene from recently contaminated and aged soils. *Water Air Soil Pollut.* **2013**, *224*, 1420. [CrossRef]

132. Margesin, R.; Labbe, D.; Schninner, F.; Greer, C.W.; Whyte, L.G. Characterization of Hydrocarbon-Degrading Microbial Populations in Contaminated and Pristine Contaminated Soils. *Appl. Environ. Microbiol.* **2003**, *69*, 3085–3092. [CrossRef]

133. Pawar, R.M. The Effect of Soil pH on Bioremediation of Polycyclic Aromatic Hydrocarbons (PAHS). *J. Bioremed. Biodeg.* **2015**, *6*, 291. [CrossRef]

134. Couling, N.R.; Towell, M.G.; Semple, K.T. Biodegradation of PAHs in soil: Influence of chemical structure, concentration and multiple amendment. *Environ. Pollut.* **2010**, *158*, 3411–3420. [CrossRef]

135. Trellu, C.; Mousset, E.; Pechaud, Y.; Huguenot, D.; Van Hullebusch, E.D.; Esposito, G.; Oturan, M.A. Removal of hydrophobic organic pollutants from soil washing/flushing solutions: A critical review. *J. Hazard. Mater.* **2016**, *306*, 149–174. [CrossRef]

136. Cheng, K.Y.; Lai, K.M.; Wong, J.W.C. Effects of pig manure compost and nonionicsurfactant Tween 80 on phenanthrene and pyrene removal from soil vegetated with Agropyron elongatum. *Chemosphere* **2008**, *73*, 791–797. [CrossRef]

137. Adrion, A.C.; Nakamura, J.; Shea, D.; Aitken, M.D. Screening nonionic surfactants for enhanced biodegradation of polycyclic aromatic hydrocarbons remaining in soil after conventional biological treatment. *Environ. Sci. Technol.* **2016**, *50*, 3838–3845. [CrossRef]

138. Shivlata, L.; Satyanarayana, T. Thermophilic and alkaliphilic Actinobacteria: Biology and potential applications. *Front. Microbiol.* **2015**, *6*, 1014. [CrossRef] [PubMed]

139. Shi, Z.; Wang, C.; Zhao, Y. Effects of surfactants on the fractionation, vermiaccumulation, and removal of fluoranthene by earthworms in soil. *Chemosphere* **2020**, *250*, 126332. [CrossRef] [PubMed]

140. Wolf, D.C.; Gan, J. Influence of rhamnolipid biosurfactant and Brij-35 synthetic surfactant on 14C-Pyrene mineralization in soil. *Environ. Pollut.* **2018**, *243*, 1846–1853. [CrossRef] [PubMed]

141. Ambrosoli, R.; Petruzzelli, L.; Minati, J.L.; Marsan, F.A. Anaerobic PAH degradation in soil by a mixed bacterial consortium under denitrifying conditions. *Chemosphere* **2005**, *60*, 1231–1236. [CrossRef] [PubMed]

142. Peng, R.H.; Xiong, A.S.; Xue, Y.; Fu, X.Y.; Gao, F.; Zhao, W.; Tian, Y.-S.; Yao, Q.-H. Microbial biodegradation of polyaromatic hydrocarbons. *FEMS Microbiol. Rev.* **2008**, *32*, 927–955. [CrossRef] [PubMed]

143. Seo, J.S.; Keum, Y.S.; Li, Q.X. Bacterial degradation of aromatic compounds. *Int. J. Environ. Res. Public Health* **2009**, *6*, 278–309. [CrossRef] [PubMed]

144. Mallick, S.; Chakraborty, J.; Dutta, T.K. Role of oxygenases in guiding diverse metabolic pathways in the bacterial degradation of low-molecular-weight polycyclic aromatic hydrocarbons: A review. *Crit. Rev. Microbiol.* **2011**, *37*, 64–90. [CrossRef]

145. Ghosal, D.; Dutta, A.; Chakraborty, J.; Basu, S.; Dutta, T.K. Characterization of the metabolic pathway involved in assimilation of acenaphthene in *Acinetobacter* sp. strain AGAT-W. *Res. Microbiol.* **2013**, *164*, 155–163. [CrossRef]

146. Hammel, K.E. Mechanisms for polycyclic aromatic hydrocarbon degradation by ligninolytic fungi. *Environ. Health Perspect.* **1995**, *103*, 41.

147. Haritash, A.K.; Kaushik, C.P. Biodegradation aspects of polycyclic aromatic hydrocarbons (PAHs): A review. *J. Hazard. Mater.* **2009**, *169*, 1–15. [CrossRef]

148. Cerniglia, C.E. Biodegradation of polycyclic aromatic hydrocarbons. *Biodegradation* **1992**, *3*, 351–368. [CrossRef]

149. Eaton, R.W.; Chapman, P.J. Bacterial metabolism of naphthalene: Construction and use of recombinant bacteria to study ring cleavage of 1,2-dihydroxynaphthalene and subsequent reactions. *J. Bacteriol.* **1992**, *174*, 7542–7554. [CrossRef]

150. van Herwijnen, R.; Wattiau, P.; Bastiaens, L.; Daal, L.; Jonker, L.; Springael, D.; Govers, H.A.J.; Parsons, J.R. Elucidation of the metabolic pathway of fluorene and cometabolic pathways of phenanthrene, fluoranthene, anthracene and dibenzothiophene by *Sphingomonas* sp. LB126. *Res. Microbiol.* **2003**, *154*, 199–206. [CrossRef]

151. Kweon, O.; Kim, S.-J.; Jones, R.C.; Freeman, J.P.; Adjei, M.D.; Edmondson, R.D.; Cerniglia, C.E. A polyomic approach to elucidate the fluoranthene-degradative pathway in Mycobacterium vanbaalenii PYR-1. *J. Bacteriol.* **2007**, *189*, 4635–4647. [CrossRef]

152. Kim, S.J.; Kweon, O.; Jones, R.C.; Freeman, J.P.; Edmondson, R.D.; Cerniglia, C.E. Complete and integrated pyrene degradation pathway in Mycobacterium vanbaalenii PYR-1 based on systems biology. *J. Bacteriol.* **2007**, *189*, 464–472. [CrossRef]

153. Sutherland, J.B.; Rafii, F.; khan, A.A.; Cerniglia, C.E. Mechanisms of Polycyclic Aromatic Hydrocarbon Degradation. In *Microbial Transformation and Degradation of Toxic Organic Chemicals*; Young, L.Y., Cerniglia, C.E., Eds.; Wiley-Liss: New York, NY, USA, 1995.

154. Bezalel, L.; Hadar, Y.; Cerniglia, C.E. Enzymatic mechanisms involved in phenanthrene degradation by the white rot fungus Pleurotus ostreatus. *Appl. Environ. Microbiol.* **1997**, *63*, 2495–2501. [CrossRef] [PubMed]

155. Cerniglia, C.E.; Sutherland, J.B. Degradation of polycyclic aromatic hydrocarbons by fungi. In *Handbook of Hydrocarbon and Lipid Microbiology*; Timmis, K.N., McGenity, T.J., van der Meer, J.R., de Lorenzo, V., Eds.; Springer: Berlin/Heidelberg, Germany, 2010; pp. 2080–2110.

156. Jerina, D.M. The 1982 Bernard, B. brodie Award Lecture. Metabolism of Aromatic hydrocarbons by the cytochrome P-450 system and epoxide hydrolase. *Drug Metab. Dispos.* **1983**, *11*, 1–4.

157. Chang, B V; Chang, W.; Yuan, S.Y. Anaerobic degradation of polycyclic aromatic hydrocarbons in sludge. *Adv. Environ. Res.* **2003**, *7*, 623–628. [CrossRef]

158. Tortella, G.R.; Diez, M.C.; Duran, N. Fungal diversity and use in decomposition of environmental pollutants. *Crit. Rev. Microbiol.* **2005**, *31*, 197–212. [CrossRef] [PubMed]

159. Aydin, S.; Karaçay, H.A.; Shahi, A.; Gökçe, S.; Ince, B.; Ince, O. Aerobic and anaerobic fungal metabolism and Omics insights for increasing polycyclic aromatic hydrocarbons biodegradation. *Fungal Biol. Rev.* **2017**, *31*, 61–72. [CrossRef]

applied sciences

Article

Chemical and Spectroscopic Investigation of Different Soil Fractions as Affected by Soil Management

Francesco De Mastro, Claudio Cocozza, Gennaro Brunetti * and Andreina Traversa

Dipartimento di Scienze del Suolo, della Pianta e degli Alimenti, University of Bari, Via Amendola 165/A, 70126 Bari, Italy; francesco.demastro@uniba.it (F.D.M.); claudio.cocozza@uniba.it (C.C.); andreina.traversa@uniba.it (A.T.)

* Correspondence: gennaro.brunetti@uniba.it

Received: 14 March 2020; Accepted: 5 April 2020; Published: 9 April 2020

Abstract: The interaction of organic matter with the finest soil fractions (<20 µm) represents a good way for its stabilization. This study investigates the effects of conventional (CT), minimum (MT), and no (NT) tillage, fertilization, and non-fertilization, and soil depth (0–30, 30–60, and 60–90 cm) on the amount of organic carbon (OC) in four soil fractions. Diffuse reflectance infrared Fourier transform spectroscopy (DRIFT) was performed to obtain information about the OC quality and the mineralogical composition of these fractions. The CT shows the highest amount of the finest fraction while the fertilization enhances the microbial community with the increase of soil micro-aggregates (250–53 µm). The coarse fraction (>250 µm) is highest in the upper soil layer, while the finest fraction is in the deepest one. The greatest OC content is observed in the topsoil layer and in the finest soil fraction. DRIFT analysis suggests that organic components are more present in the finest fraction, calcite is mainly localized in the coarse fraction, quartz is in micro-aggregates and 53–20 µm fraction, and clay minerals are in the finest fraction.

Keywords: tillage; fertilization; soil depth; organic carbon; clay minerals; diffuse reflectance; infrared Fourier transform spectroscopy

---

1. Introduction

Previous studies have used the particle size fractionation for obtaining information about the influence of land use and depth on the distribution of soil organic carbon (SOC) [1,2]. The various soil fractions can differently immobilize organic carbon (OC) through the formation of organo-mineral complexes [3,4]. In particular, quartz particles exhibit only weak bonding affinities to organic matter (OM), while clay size particles (i.e., sesquioxides and phyllosilicates) have a large surface area and numerous sorption sites [3,4]. The physical protection of OM through its occlusion within clay minerals limits its microbial decomposition, which reduces the C mineralization [5]. Therefore, the sand related OM represents the active pool of soil organic matter (SOM), the OM linked to the silt size fraction is the intermediate pool, and the clay related OM represents the passive and the older SOM pool [4]. The microaggregates, composed mainly of clay minerals, represents the most efficient way to stabilize the SOM [6,7] by forming bridges between the exchangeable cations of layer silicates and functional groups of organic compounds [8]. The formation of macroaggregates is favored by the decomposition of fresh plant residues and fungal hyphae [9]. The SOM in the macroaggregates is available for microbial utilization while the protected microaggregates form a long-term reserve of mineral-associated C that is not "humified" and can be attacked by microorganisms once exposed [9].

It is known that chemical fertilizers can modify soil physical, chemical, and biological properties with clear consequences on soil aggregates [10]. A different development of the root system, stimulated by fertilization, influences the production and release of root exudates, directly involved in the formation of aggregates [11]. In addition, a greater growth of the root system following the fertilization increases the SOM [11]. Several previous studies have demonstrated that tillage practices influence the content and the dynamic of SOM [7,12–14]. Soil tillage increases the turnover of macroaggregates by inhibiting the formation of microaggregates within macroaggregates and, thus, reducing the sites where the OM is stabilized [5].

In order to identify the minerals and their changes among the different soil fractions, the diffuse reflectance infrared Fourier transform (DRIFT) spectroscopy has been utilized. This technique can be considered rapid, inexpensive, and precise, and can be applied to estimate the water-bearing minerals, such as clay minerals together to other sheet silicates as muscovite, illite, smectite, kaolinite, and chlorite [15], and the presence of organic matter [16]. DRIFT spectroscopy allows us to analyze the matrices without pressing them by avoiding the error due to scattering [17], and to have a band intensity four times greater than that of IR spectroscopy due to the non-mixing of the soil sample with KBr [18].

The objective of this study was to investigate the effects of different soil managements on the quantity of SOC associated with several soil-size fractions. In addition, with DRIFT analysis, we tried to better understand the OM interaction with the mineral parts of the different soil fractions.

2. Materials and Methods

2.1. Study Area and Experimental Design

The trial was conducted in the experimental station of the University of Bari (Italy) located at Policoro (40°10′20″ N; 16°39′04″ E; altitude: 15 m above sea level). The soil texture was classified as silt loam (sand 8%, silt 68%, clay 24%), according to the USDA [19]. Since 2005, a two-year rotation of durum wheat with faba bean in a split-block design with three field replications has been introduced. Treatments were as follows: i) no tillage and no fertilization (NT), ii) NT and crop fertilization (30 kg P_2O_5 ha^{-1}: NTF), iii) minimum tillage (20 cm deep subsoiling in late August and 15 cm deep disk harrowing in November) and no fertilization (MT), iv) MT and crop fertilization (MTF), v) conventional tillage (35 cm deep moldboard plowing in late August and 15 cm deep disk harrowing in November) and no fertilization (CT), and vi) CT and crop fertilization (CTF). More details about treatments and soil properties are reported elsewhere [20,21].

After more than a decade of trial, each faba bean plot (30 × 30 m) was sampled in July 2017 at three different depths (0–30, 30–60, and 60–90 cm) using an auger, after the harvest, and the removal of aboveground crop residues. Due to the soil homogeneity, nine sub-samples have been collected from each plot using a grid sampling scheme.

2.2. Particle-Size Physical Fractionation and Determination of Organic Carbon in Fractions

Fraction-size separation was obtained by ultrasonic dispersion, according to Amelung and Zech [22], and wet sieving, according to Bornemann et al. [23] (Figure 1).

About 150 mL of milli-Q® ultrapure water were added to 30 g of air-dried soil and the suspension was gently sonicated by placing the probe tip 15 mm below the water surface and using a probe-type sonicator Sigma Aldrich, 500-Watt model (60 J mL^{-1}). This weak sonication was used for preserving micro-aggregates from disruption [24]. The first fraction (macro-aggregates fraction, A: 2000–250 μm) was separated from the suspension by wet sieving (250 μm), and the filtered remnant was sonicated a second time at 440 J mL^{-1} and separated by wet sieving using sieves with different meshes (53 μm and 20 μm). After this step, the obtained fractions were: fraction B (microaggregates fraction, 250–53 μm), fraction C (coarse silt-sized fraction 53–20 μm), and fraction D (free fine silt plus clay fraction, <20 μm). All fractions were dried at 35 °C before elemental analysis. The water-soluble organic fraction was isolated and discharged from each D fraction, according to Zsolnay [25], to avoid any interference. Briefly, an aliquot of each air-dried D fraction was suspended in water (1:10, w/v), and mechanically shaken for 15 min. The suspensions were then centrifuged at 6000 rpm for 15 min and the supernatant was removed.

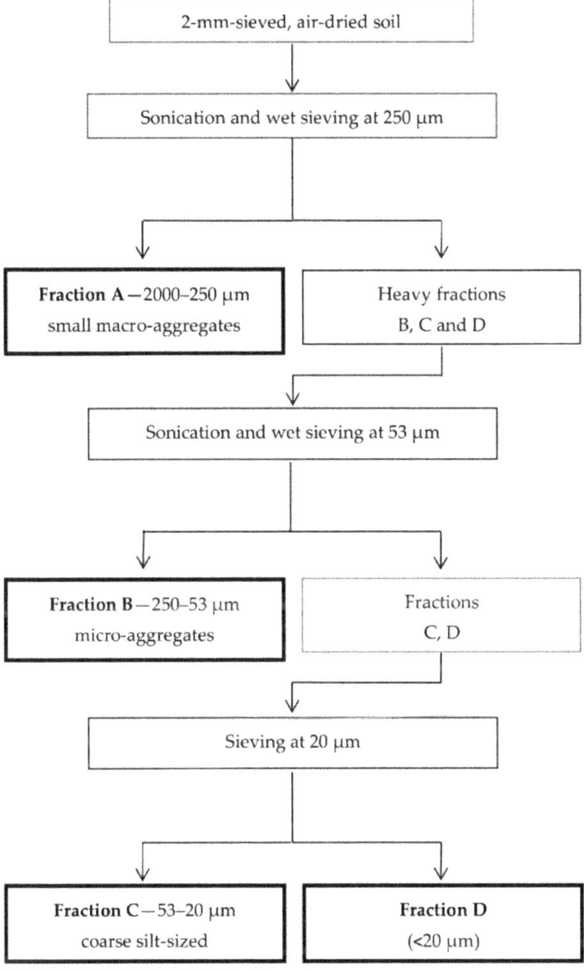

Figure 1. Fractionation scheme of soil.

The OC content of all soil fractions was determined in triplicate using a Flash 2000 CIINS-O Elemental Analyser (Thermo Scientific) calibrated by an organic analytical standard consisting of a low organic content soil with 1.55% (w/w) of carbon. About 6–7 mg of each soil fraction were dried at 40 °C and pre-treated with hydrochloric acid (HCl 1%) to dissolve carbonates. The OC stocks were calculated by multiplying the C concentrations and the corresponding particle-size weights.

2.3. Spectroscopic Analysis of Particle-Size Fractions

Diffuse Reflectance Fourier Transform (DRIFT) spectra were recorded for each fraction in triplicate and in transmittance mode using a Thermo Nicolet Nexus FT-IR spectrophotometer, which was equipped with a Nicolet Omnic 6.0 software. Before DRIFT analysis, air-dried samples were thoroughly mixed to obtain a representative sample and then finely ground in a mill. About 200 mg of the mixture was filled in a cup and the surface was smoothed with a plastic slide. Spectra were recorded in the range of 4000 to 400 cm^{-1}, with 4 cm^{-1} resolution and 16 scans min^{-1} for each acquisition.

2.4. Statistical Analysis

All analyses performed on soil fractions were conducted in triplicate. The analysis of variance (four-way ANOVA) and the Tukey's test (R software, version 3.2.3) were used to measure the effect of fertilization, tillage, and depth on the OC content for each soil fraction.

3. Results and Discussion

3.1. Effects of Treatments on the Amount of Each Soil Fraction and on their Organic Carbon Content

The physical fractionation recovered 98% of the mass and 99% of the OC from all samples. Such percentages were comparable to those previously reported by other authors [26–28], which indicates that the loss of material was very low and confirms the efficiency of the fractionation method adopted.

Table 1 shows the amounts of soil dry matter obtained from each fraction (g kg^{-1} soil), while Table 2 reports the corresponding statistical analyses, as influenced by treatments (soil depth, tillage, and fertilization).

Table 1. Amount of soil dry matter in the size fractions (g kg^{-1} soil) (mean ± standard deviation).

Sample	Dry Matter (g kg^{-1})			
CT	A (>250 μm)	B (250–53 μm)	C (53–20 μm)	D (<20 μm)
0–30 cm	6.8 ± 0.2	89.5 ± 12.0	220.2 ± 26.2	689.0 ± 0.9
30–60 cm	3.8 ± 0.2	46.0 ± 0.9	199.0 ± 3.8	763.7 ± 0.9
60–90 cm	2.0 ± 0.0	15.2 ± 0.7	134.7 ± 3.8	867.0 ± 1.9
CTF				
0–30 cm	8.3 ± 0.5	174.7 ± 17.9	202.5 ± 11.1	604.5 ± 12.0
30–60 cm	4.8 ± 0.2	174.3± 21.2	203.7 ± 17.9	617.8 ± 6.8
60–90 cm	1.8 ± 0.7	39.7 ± 2.4	135.8 ± 4.0	835.2 ± 3.5
MT				
0–30 cm	10.3 ± 2.8	123.2 ± 28.5	311.7 ± 36.3	557.3 ± 43.4
30–60 cm	5.3 ± 0.0	95.2 ± 10.1	197.5 ± 2.6	722.5 ± 7.3
60–90 cm	1.3 ± 0.0	28.8 ± 0.2	179.2 ± 9.7	826.8 ± 10.6
MTF				
0–30 cm	17.7 ± 2.4	285.5 ± 14.4	203.3 ± 1.4	500.7 ± 9.4
30–60 cm	9.0 ± 2.8	251.5 ± 7.8	223.7 ± 9.9	534.0 ± 3.8
60–90 cm	3.8 ± 0.7	153.5 ± 2.1	190.0 ± 0.9	716.3 ± 1.4
NT				
0–30 cm	5.3 ± 0.0	70.7 ± 3.3	232.7 ± 0.5	695.3 ± 4.2
30–60 cm	4.2 ± 0.2	45.7 ± 2.4	199.2 ± 0.2	764.7 ± 10.4
60–90 cm	1.7 ± 0.0	19.0 ± 3.8	172.0 ± 10.8	826.0 ± 7.1
NTF				
0–30 cm	8.7 ± 0.9	192.3 ± 2.8	239.0 ± 2.4	563.5 ± 4.5
30–60 cm	4.2 ± 0.2	171.8 ± 8.2	260.0 ± 2.4	559.3 ± 21.7
60–90 cm	6.7 ± 0.5	269.5 ± 4.0	174.0 ± 4.7	575.3 ± 10.4

CT: Conventional tillage. CTF: Conventional tillage fertilized. MT: Minimum tillage. MTF: Minimum tillage fertilized. NT: No tillage. NTF: No tillage fertilized.

On average, 70% to 75% of the soil fractions consisted of fine silt and clay (<20 μm). The A, B, and C fractions decreased with depth (Table 1), while the D fraction showed an inverse trend with the only exception being the deepest layer of NTF treatment. The quantity of the smallest fraction (<20 μm) increased with depth and ranged from 500 to 867 mg kg^{-1}.

The fertilized plots resulted in the highest amount of B fraction and the lowest amount of D fraction with respect to the unfertilized ones. Since microaggregates are the result of microbial decomposition of SOM from the macroaggregates [9,29], the highest amount of B fraction in fertilized soils could derive from the higher microbial activity promoted by the same fertilization. For example, Liao et al. [30] found the higher fungal abundance in microaggregates (250–53 µm) regardless of the type of fertilization.

Table 2. Analysis of variance and mean values of the amount of each soil fraction subdivided by soil depth, tillage, and fertilization treatment. The standard deviation is in parentheses.

Size Fractions	Dry Matter (g kg^{-1})			
	A (>250 um)	B (250–53 um)	C (53–20 um)	D (<20 um)
Depth	***	**	**	***
Tillage	**	**	n.s.	**
Fertilization	**	***	n.s.	***
Depth				
0–30	0.28 b (0.12)	4.67 b (2.30)	7.04 b (2.01)	18.05 a (2.54)
30–60	0.15 a (0.06)	3.92 ab (2.37)	6.41 b (0.80)	19.81 b (2.96)
60–90	0.08 a (0.05)	2.62 a (2.94)	4.92 a (0.68)	23.23 c (3.15)
Tillage				
NT	0.15 a (0.06)	3.84 ab (2.80)	6.38 a (1.05)	19.92 a (3.03)
MT	0.23 b (0.17)	4.68 b (2.79)	6.52 a (2.12)	19.28 a (3.93)
CT	0.13 a (0.07)	2.69 a (2.03)	5.47 a (1.20)	21.88 b (3.16)
Fertilization				
No	0.13 a (0.08)	1.77 a (1.14)	6.15 a (1.93)	22.37 b (2.90)
Yes	0.21 b (0.14)	5.70 b (2.18)	6.10 a (1.12)	18.35 a (3.02)

CT: Conventional tillage. MT: Minimum tillage. NT: No tillage. The values in each column followed by a different letter are significantly different according to Tukey's test. n.s.: not significant. *** significant at the $P \leq 0.001$.

With regard to the tillage, the highest amount of D fraction and the lowest amount of B fractions were obtained from CT soils, which is likely due to the major physical disturbance and microbiological activity induced by the conventional tillage that increased macro-aggregates and micro-aggregates turnover [5].

Table 3 shows the analysis of variance and mean values of OC content of soil fractions, as affected by soil depth, tillage, fertilization, and size fraction. The interactions among these parameters were not significant (data not shown) except the one between the OC content of each soil fraction and soil depth ($P \leq 0.001$) since, as expected, the OC content of all fractions decreased with depth.

The significant decrease of the OC content of all fractions from the upper layer (on average, 125.9 mg OC kg^{-1} fraction) to the deepest one (54.5 mg OC kg^{-1} fraction) resembles the common stratification of SOC along the profiles. The highest value of OC was found in the D fraction due to the entrapment of organic components in the finest fractions of soil, as reported by Gregorich et al. [31]. This highlighted the role of clay particles in the OM stabilization due to their high specific surface area and charge. In fact, clay minerals are considered the most active constituents in the formation of organo-mineral complexes [32] and are responsible for long-term preservation of soil OM, even over millennia [33].

No kind of tillage significantly affected the OC content of each fraction likely due to the balance in the soil achieved because of the long-term experiment, as reported by Rita et al. [34] who found no significant difference in OC fraction content among several 30-year-old land-use trials. In addition, the main OM input, which is the above-ground crop residues, has been removed from the field at the end of the crop cycles in all treatments. With regard to the latter topic, it has been demonstrated that the aboveground crop residues are important for building up soil fertility not only as input of OM, but also because they cover the soil during the hot weather, which conserves SOM [35].

Table 3. Analysis of variance and mean values of the OC in soil fractions, subdivided by soil depth, tillage, fertilization treatment, and size of fractions. The standard deviation is in parentheses.

	Organic Carbon (mg kg^{-1})
Depth	***
Tillage	n.s.
Fertilization	n.s.
Size	***
Depth (cm)	
0–30	125.9 c (10.9)
30–60	89.8 b (7.8)
60–90	54.5 a (4.7)
Tillage	
NT	94.9 a (8.2)
MT	81.8 a (7.1)
CT	93.5 a (8.1)
Fertilization	
No	92.8 a (6.6)
Yes	87.3 a (6.2)
Size	
A	7.7 a (0.8)
B	22.0 ab (2.2)
C	54.8 b (5.5)
D	275.7 c (27.6)

CT: Conventional tillage. MT: Minimum tillage. NT: No tillage. A (>250 μm). B (250–53 μm). C (53–20 μm). D (<20 μm). The values in each column followed by a different letter are significantly different according to Tukey's test. n.s.: not significant. *** Significant at the $P \leq 0.001$.

Lastly, the fractional OC content did not differ between fertilized and not fertilized plots since the adopted fertilization was only inorganic and rather low.

3.2. Effects of Treatments on the Spectroscopic Properties of Each Soil Fraction

Figure 2 shows the DRIFT spectra of different soil fractions under various tillage.

The peak at about 3623 cm^{-1} can be ascribed to the O-H vibration in the octahedral layers of 2:1 and/or 1:1 silicates [36]. The same peak can be highlighted by removing the SOM through an appropriate procedure [37]. This peak was observed in B, C, and D fractions regardless of the type of tillage, with a slight increase of the relative intensity as the fraction size decreased. This was in line with other papers [38–41] showing that phyllosilicates (kaolinite, chlorite, smectite, illite) are the main components of the clay fraction of soils. The peak at 2927 cm^{-1}, ascribed to the stretching of the aliphatic C-H group, was evident only in the A fraction, and could be due to the signal of organic matter consisting of labile plant residues [42]. Additionally, in this case, this peak can be highlighted by removing minerals through HF treatment [43]. The peak at about 2517 cm^{-1} can be attributed to the CO$_3$ stretching and calcite bending, as indicated by peaks at about 1450, 867, and 698 cm^{-1} [15,44]. These peaks, especially those at 1450 and 867 cm^{-1}, were more pronounced in the spectrum of the A fraction. High percentage of calcite in the sand-sized fraction of a Mediterranean soil was also found in a previous work [3] and was ascribed to its lithogenic origin. The peaks at about 1991, 1868, 1793, and 698 cm^{-1} were related to Si-O bending of quartz minerals [15,43,44] as well as the peaks at about 791 cm^{-1} related to the Si-O stretching of quartz minerals [15,43]. Overall, these peaks were slightly more pronounced in all fractions B and C possibly because of the physical breakup of sand size quartz into silt dimension. In contrast, the reduced intensity of the same peaks in D fractions suggested a limited physical alteration of quartz minerals in the finest particles and, therefore, an intermediate stage of soil evolution [45]. The peak at about 1630 cm^{-1} was ascribed to aromatic C=C skeletal vibrations, C=O stretching of quinone and amide groups, C=O of H-bonded conjugated ketones, and it was typical of the organic components [46]. As expected, this peak was absent in the A fraction, appeared in the B fraction, and became more pronounced from fraction C to D.

The presence of the peak linked to aromatic structures in the fraction D could be due to the presence of organic matter involved in the formation of organo-mineral associations and in the coating of the mineral surface by sorption or precipitation processes [47–49]. The broad band at about 1030 cm^{-1} can be ascribed to the stretching of the carbohydrate and polysaccharides-like substances [50]. It was found mainly in the silt-clay and free fine silt plus clay fractions (fraction C and D). Many previous studies have reported high proportions of these compounds in the mineral-associated organic matter fraction [5,51–54]. Polysaccharides of microbial origin mainly bind clay particles by promoting the formation of microaggregates of <50 μm [55]. Glicoproteins of fungal origin, such as glomalin, contain about 85% of sugars and are decomposed very slowly in soil [56]. In contrast, the lower relative intensity of the previously mentioned peak in fraction A suggested the presence of polysaccharides of plant origin responsible for the formation of easily degradable macro-aggregates [56]. The bands at about 529 and 478 cm^{-1} can be ascribed to Si-O-Al and Si-O-Si vibrations, respectively, and are distinctive of phyllosilicates [41]. The relative intensity of the second band decreased in proportion to the size of the fractions. Ndzana et al. [41] reported similar results suggesting that the crystalline structure of phyllosilicates weakened in the finest soil fraction.

DRIFT spectra recorded from NT, MT, and CT samples were similar. Therefore, Figure 3 shows only the DRIFT spectra of each soil-size fraction isolated from the NT treatment along the soil profile. The only slight difference evident in all the soil-size fractions was the peak among 1493 and 1450 cm^{-1}, which increased its relative intensity with depth. This suggests a slight increase of calcite along the soil profile due to dissolution/precipitation phenomena typical of aridic climates and high soil pH [45]. In addition, a slight reduction of the relative intensity of the peak at 1628 cm^{-1} was observed only in the D fraction, according to the reduction of organic matter content with depth.

Figure 4 reports the DRIFT spectra of each soil fraction isolated from the 0–30 layer of the NT treatment fertilized and unfertilized. The spectra of fractions A, B, and C were very similar between the two levels of fertilization. The fraction D of NTF treatment showed a greater relative intensity of the peak at 1027 cm^{-1} compared to NT treatment, possibly due to the greater quantity of polysaccharides coming from microorganisms whose activity is certainly favored by fertilization, as reported by De Mastro et al. a [20].

Figure 5 shows the DRIFT spectra of each soil fraction isolated from the 0–30 layer under NT, MT, and CT practices. Even in this case, the spectra of fractions A, B, and C were more similar regardless of the kind of tillage. The peak at 1027 cm^{-1} showed a greater relative intensity in fraction D of MT and CT treatments compared to NT since the higher aeration of the formers enhanced the microbial activity.

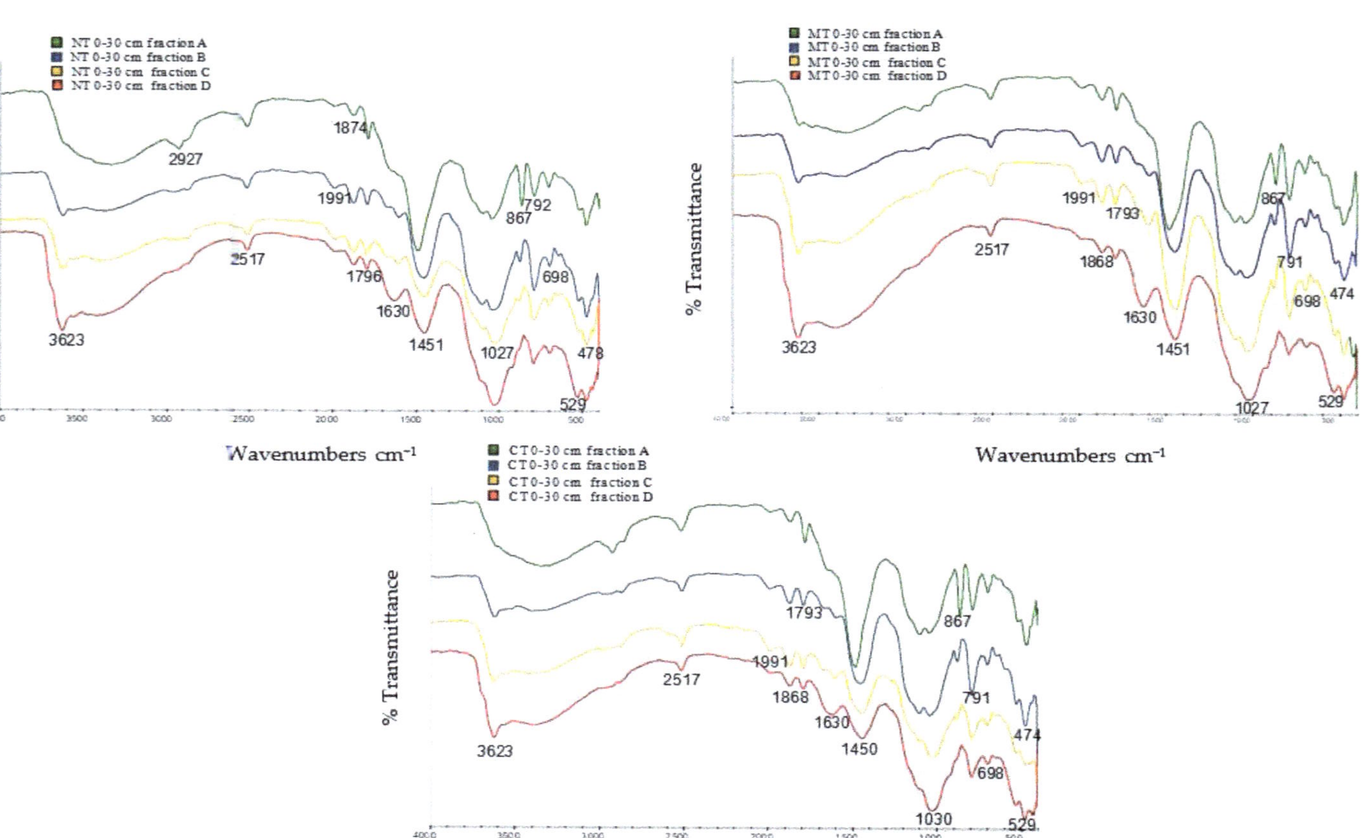

Figure 2. Diffuse reflectance infrared Fourier transform spectra of all soil fractions isolated from different tillage treatments at 0–30 cm of depth.

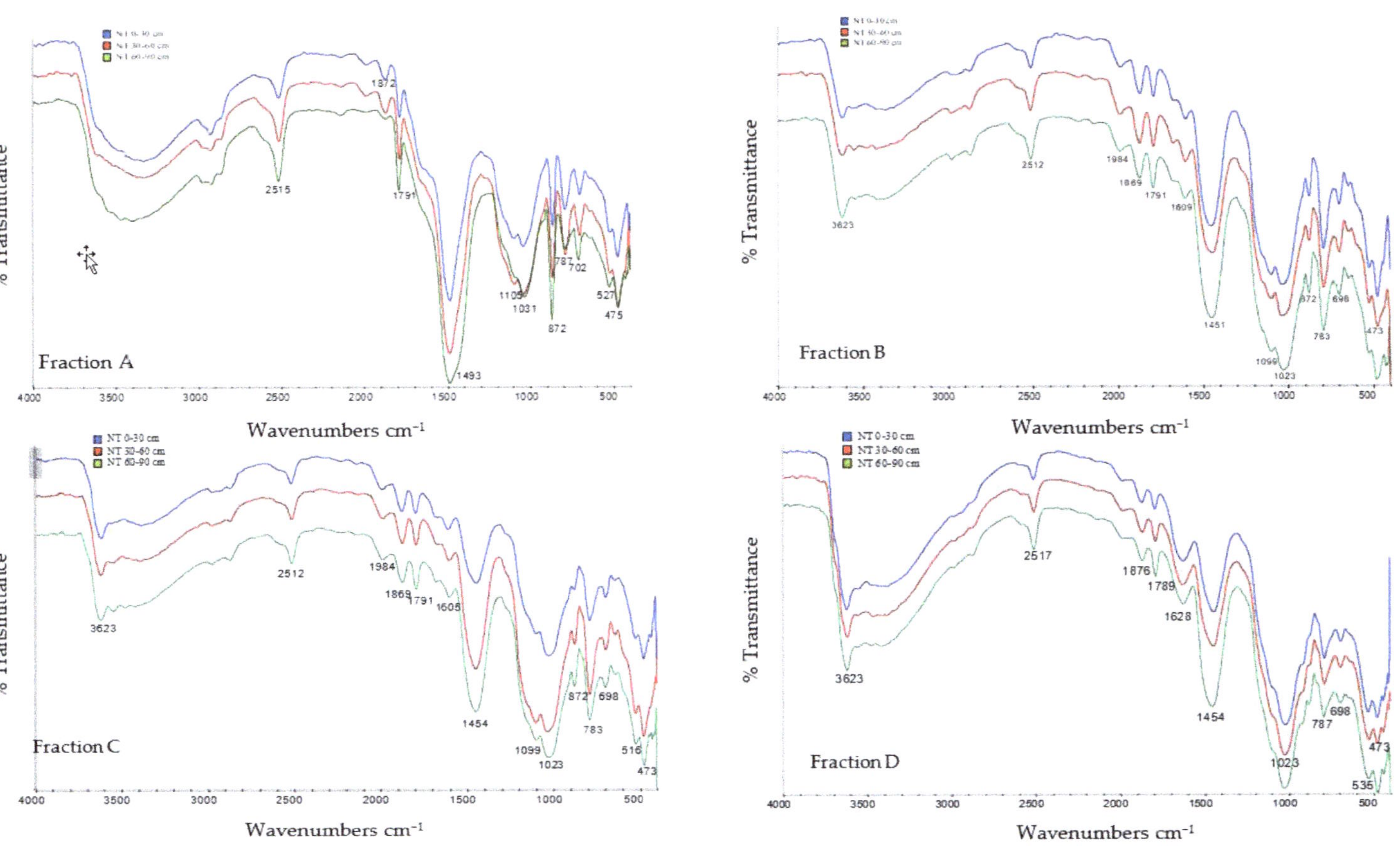

Figure 3. Diffuse reflectance infrared Fourier transform spectra of all soil fractions (A, B, C, and D) isolated from no tillage (NT) treatment at different soil depths.

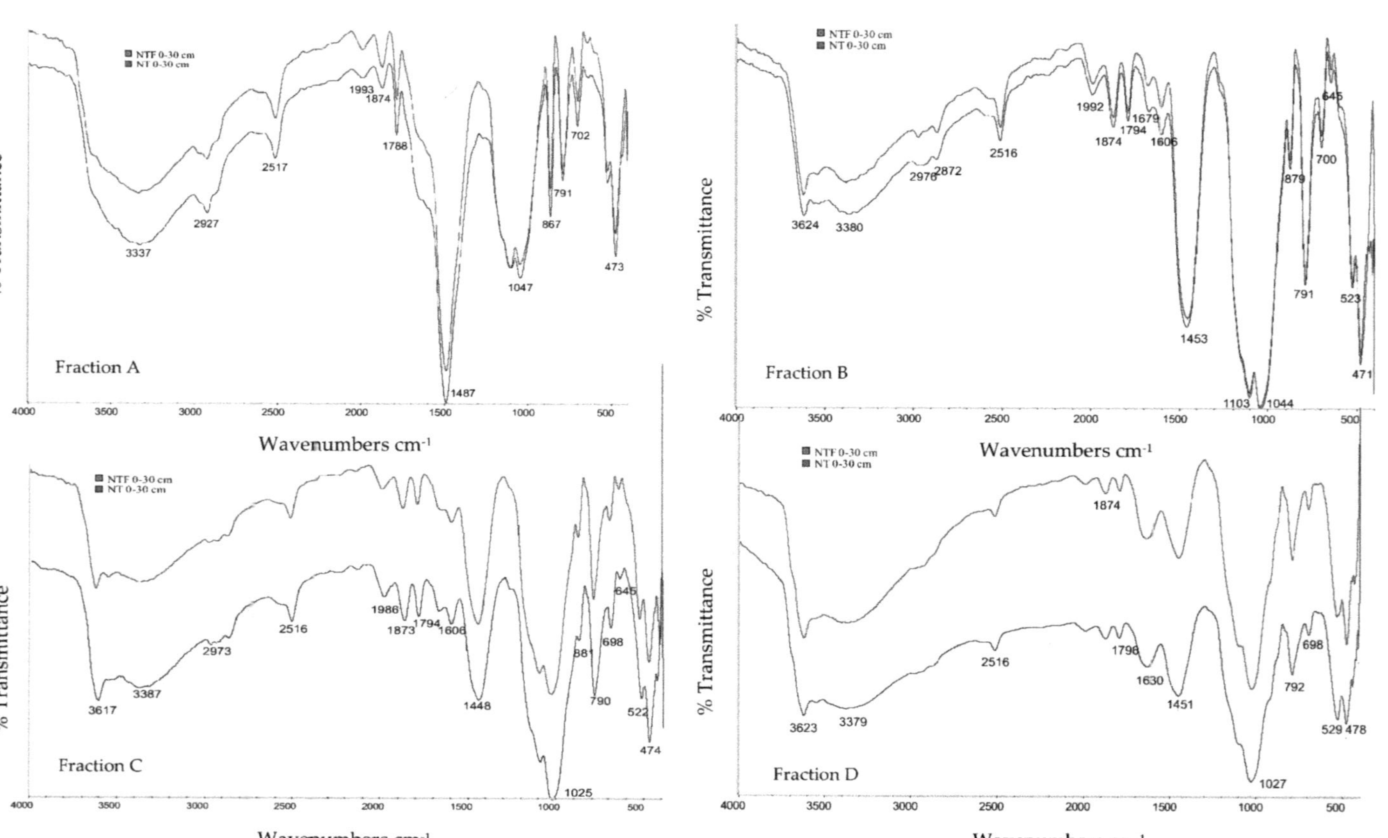

Figure 4. Diffuse reflectance infrared Fourier transform spectra of the different soil fractions (A, B, C, and D) isolated from no tillage (NT) treatment (0–30 cm) fertilized and not fertilized.

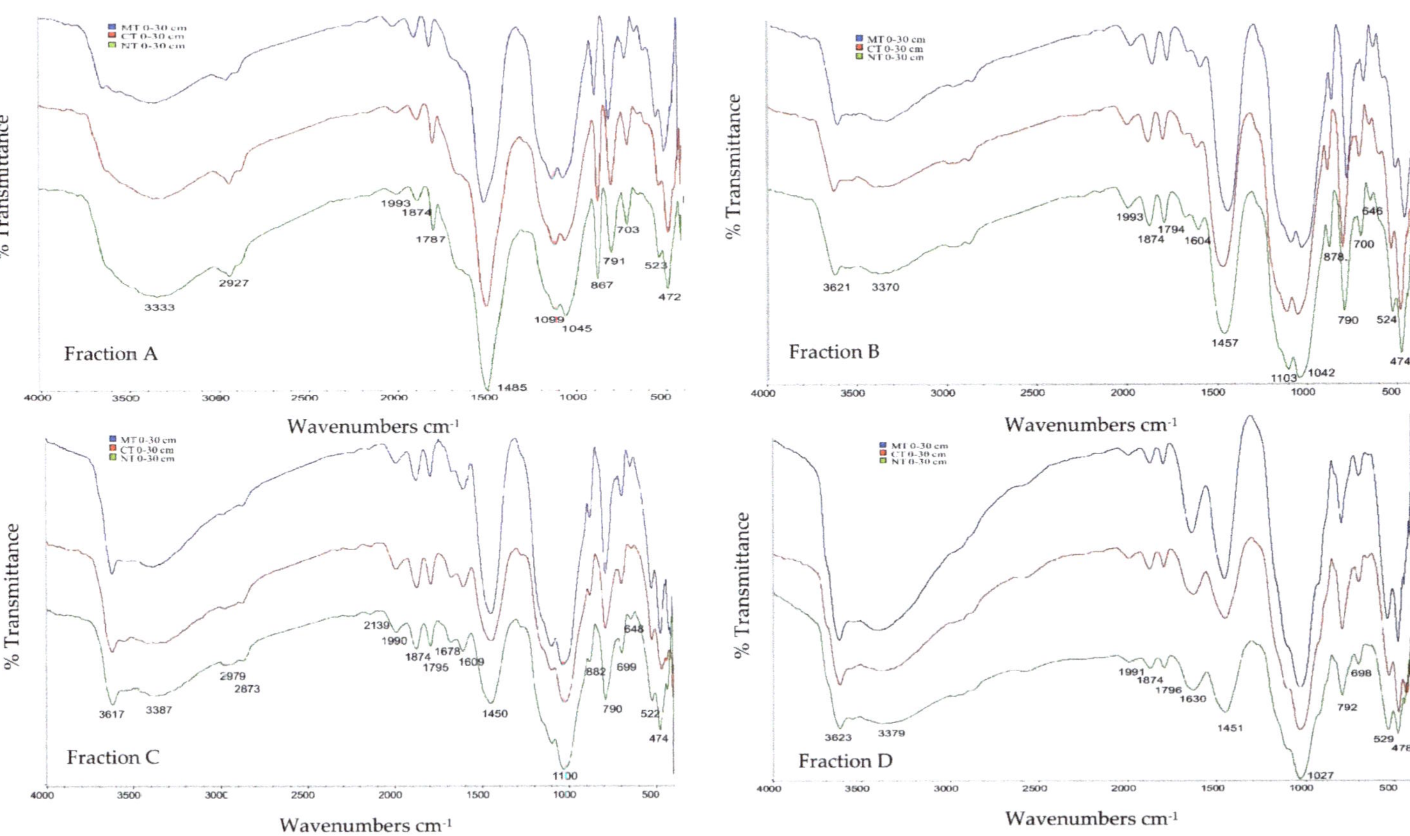

Figure 5. Diffuse reflectance infrared Fourier transform spectra of all soil fractions (A, B, C, and D) isolated from different tillage treatments at 0–30 cm of depth.

4. Conclusions

The soil management influenced the quantity of soil fractions since the CT enhanced the finest ones, whereas the fertilization increased the B fractions, which possibly fuels the development of a microbial community that fosters microaggregate formation. In general, soil depth influenced the amount of each fraction, with higher amounts of fraction A in the upper soil layer and higher amounts of the finest one is more evident in the deepest. The OC content was primarily influenced by fraction size and soil depth since higher OC content was found in the topsoil layer (0–30 cm) and in the finest soil fraction (fraction D), as confirmed by the DRIFT analysis. The different tillage may increase the mass of the soil fractions but not their OC content. However, MT and CT affected positively the quality of the OM stimulating a relative major production of polysaccharides of microbial origin that possibly stabilize the finest size fractions due to their major recalcitrance. The same result is obtained with fertilization. The DRIFT analysis can provide information about the quality of the main minerals present in the different soil-size fractions, since fraction A appeared mostly rich in calcite, fractions B and C appeared mostly rich in quartz, and the finest fraction showed the highest content of phyllosilicates. The different quality of minerals among the soil fractions suggested an early-intermediate stage of weathering and did not change with tillage and fertilization, except for calcite whose relative intensity increased with depth.

Author Contributions: Conceptualization, G.B., F.D.M., A.T., and C.C. Methodology, G.B., F.D.M., A.T., and C.C. Software, G.B., F.D.M., A.T., and C.C. Validation, G.B., F.D.M., A.T., and C.C. Formalanalysis, F.D.M. Investigation, F.D.M. Resources, G.B. and C.C. Data curation, G.B. and A.T. Writing—original draft preparation, A.T. Writing—review and editing, G.B., F.D.M., A.T., and C.C. Visualization, G.B. and C.C. Supervision, G.B. Project administration, G.B. Funding acquisition, G.B. All authors have read and agreed to the published version of the manuscript.

Funding: This research received no external funding.

Conflicts of Interest: The authors declare no conflict of interest.

References

1. Salvo, L.; Hernández, J.; Ernst, O. Distribution of soil organic carbon in different size fractions, under pasture and crop rotations with conventional tillage and no-till systems. *Soil Till. Res.* **2010**, *109*, 116–122. [CrossRef]

2. Von Lützow, M.; Kögel-Knabner, I.; Ekschmitt, K.; Flessa, H.; Guggenberger, G.; Matzner, E.; Marschner, B. SOM fractionation methods: Relevance to functional pools and to stabilization mechanisms. *Soil Biol. Biochem.* **2007**, *39*, 2183–2207. [CrossRef]

3. Brunetti, G.; Mezzapesa, G.N.; Traversa, A.; Bonifacio, E.; Farrag, K.; Senesi, N.; D'Orazio, V. Characterization of clay- and silt-sized fractions and corresponding humic acids along a Terra Rossa soil profile. *Clean* **2016**, *44*, 1–10. [CrossRef]

4. Von Lützow, M.; Kögel-Knabner, I.; Ekschmitt, K.; Matzner, E.; Guggenberger, G.; Marschner, B.; Flessa, H. Stabilization of organic matter in temperate soils: Mechanisms and their relevance under different soil conditions—A review. *Eur. J. Soil Sci.* **2006**, *57*, 426–445. [CrossRef]

5. Plaza, C.; Courtier-Murias, D.; Fernández, J.M.; Polo, A.; Simpson, A.J. Physical, chemical, and biochemical mechanisms of soil organic matter stabilization under conservation tillage systems: A central role for microbes and microbial by-products in C sequestration. *Soil Biol. Biochem.* **2013**, *57*, 124–134. [CrossRef]

6. Jones, E.; Singh, B. Organo-mineral interactions in contrasting soils under natural vegetation. *Front. Environ. Sci.* **2014**, *2*, 1–15. [CrossRef]

7. Poirier, V.; Basile-Doelsch, I.; Balesdent, J.; Borschneck, D.; Whalen, J.K.; Angers, D.A. Organo-mineral interactions are more important for organic matter retention in subsoil than topsoil. *Soil Syst.* **2020**, *4*, 4. [CrossRef]

8. Mikutta, R.; Mikutta, C.; Kalbitz, K.; Scheel, T.; Kaiser, K.; Jahn, R. Biodegradation of forest floor organic matter bound to minerals via different binding mechanisms. *Geochim. Cosmochim. Acta* **2007**, *71*, 2569–2590. [CrossRef]

9. Neuman, J. Soil organic matter maintenance in no-till and crop rotation management systems. In *Reference Module in Earth Systems and Environmental Sciences*; Elsevier: Amsterdam, The Netherlands, 2017. [CrossRef]

10. Wen, Y.; Liu, W.; Deng, W.; He, X.; Yu, G. Impact of agricultural fertilization practices on organo-mineral associations in four long-term field experiments: Implications for soil C sequestration. *Sci. Total Environ.* **2019**, *651*, 591–600. [CrossRef]

11. Keiluweit, M.; Bougoure, J.J.; Nico, P.S.; Pett-Ridge, J.; Weber, P.K.; Kleber, M. Mineral protection of soil carbon counteracted by root exudates. *Nat. Clim. Chang.* **2015**, *5*, 588–595. [CrossRef]

12. Liang, A.; Chen, S.; Zhang, X.; Chen, X. Short-term effects of tillage practices on soil organic carbon turnover assessed by δ^{13} C abundance in particle-size fractions of black soils from Northeast China. *Sci. World J.* **2014**, *2014*, 514183. [CrossRef] [PubMed]

13. Page, K.L.; Dang, Y.P.; Dalal, R.C. The ability of conservation agriculture to conserve soil organic carbon and the subsequent impact on soil physical, chemical, and biological properties and yield. *Front. Sustain. Food Syst.* **2020**, *4*, 1–17. [CrossRef]

14. Lal, R. Soil carbon sequestration impacts on global climate change and food security. *Science* **2004**, *304*, 1623–1627. [CrossRef] [PubMed]

15. Udvardi, B.; Kovács, I.J.; Kónya, P.; Földvári, M.; Füri, J.; Budai, F.; Falus, G.; Fancsik, T.; Szabó, C.; Szalai, Z.; et al. Application of attenuated total reflectance Fourier transform infrared spectroscopy in the mineralogical study of a landslide area, Hungary. *Sedimen. Geol.* **2014**, *313*, 1–14. [CrossRef]

16. Li, Y.; Cai, J.; Song, G.; Ji, J. DRIFT spectroscopic study of diagenetic organic–clay interactions in argillaceous source rocks. *Spectrochim. Acta A* **2015**, *148*, 138–145. [CrossRef]

17. Renner, G.; Schmidt, T.C.; Schram, J. Characterization and quantification of microplastics by infrared spectroscopy. In *Comprehensive Analytical Chemistry*; Rocha-Santos, T.A.P., Duarte, A.C., Eds.; Elsevier: Amsterdam, The Netherlands, 2017; Volume 75, pp. 67–118.

18. Piccolo, A.; Conte, P. Advances in nuclear magnetic resonance and infrared spectroscopies of soil organic particles. In *Structure and Surface Reactions of Soil Particles. Analytical and Physical Chemistry of Soil*; Huang, P.M., Senesi, N., Buffle, J., Eds.; Wiley: New York, NY, USA, 1998; Volume 4, pp. 183–250.

19. *Soil Survey Staff, Keys to Soil Taxonomy*, 12th ed.; USDA-Natural Resources Conservation Service: Washington, DC, USA, 2014.

20. De Mastro, F.; Cocozza, C.; Traversa, A.; Savy, D.; Abdelrahman, H.M.; Brunetti, G. Influence of crop rotation, tillage and fertilization on chimical and spectroscopic characteristics of humic acids. *PLoS ONE* **2019**, *14*, e0219099. [CrossRef]

21. De Mastro, F.; Brunetti, G.; Traversa, A.; Cocozza, C. Effect of crop rotation, fertilization and tillage on main soil properties and its water extractable organic matter. *Soil Res.* **2019**, *57*, 365–373. [CrossRef]

22. Amelung, W.; Zech, W. Minimisation of organic matter disruption during particle-size fractionation of grassland epipedons. *Geoderma* **1999**, *92*, 73–85. [CrossRef]

23. Bornemann, L.; Welp, G.; Amelung, W. Particulate organic matter at the field scale: Rapid acquisition using MID-Infrared spectroscopy. *Soil Sci. Am. J.* **2010**, *74*, 1147–1156. [CrossRef]

24. Kölbl, A.; Leifeld, J.; Kögel-Knabner, I.A. Comparison of two methods for the isolation of free and occluded particulate organic matter. *J. Plant Nutr. Soil Sci.* **2005**, *168*, 660–667. [CrossRef]

25. Zsolnay, A. Dissolved humus in soil waters. In *Humic Substances in Terrestrial Ecosystems*; Piccolo, A., Ed.; Elsevier: Amsterdam, The Netherlands, 1996; pp. 171–223.

26. Plante, A.F.; Conant, R.T.; Stewart, C.E.; Paustian, K.; Six, J. Impact of soil texture on the distribution of soil organic matter in physical and chemical fractions. *Soil Sci. Soc. Am. J.* **2006**, *70*, 287–296. [CrossRef]

27. Moni, C.; Rumpel, C.; Virto, I.; Chabbi, A.; Chenu, C. Relative importance of sorption versus aggregation for organic matter storage in subsoil horizons of two contrasting soils. *Eur. J. Soil Sci.* **2010**, *61*, 958–969. [CrossRef]

28. O'Brien, S.L.; Jastrow, J.D. Physical and chemical protection in hierarchical soil aggregates regulates soil carbon and nitrogen recovery in restored perennial grasslands. *Soil Biol. Biochem.* **2013**, *61*, 1–13. [CrossRef]

29. Watteau, F.; Villemin, G. Soil microstructures examined through transmission electron microscopy reveal soil-microorganisms interactions. *Front. Environ. Sci.* **2018**, *6*, 1–10. [CrossRef]

30. Liao, H.; Zhang, Y.; Zuo, Q.; Du, B.; Chen, W.; Wei, D.; Huang, Q. Contrasting response of bacterial and fungal communities to aggregate-size fractions and long-term fertilizations in soils of northeastern China. *Sci. Total Environ.* **2018**, *635*, 784–792. [CrossRef]

31. Gregorich, E.G.; Liang, B.C.; Drury, C.F.; Mackenzie, A.F.; McGill, W.B. Elucidation of the source and turnover of water soluble and microbial biomass carbon in agricultural soils. *Soil Biol. Biochem.* **2000**, *32*, 581–587. [CrossRef]

32. Chenu, C.; Plante, A.F.; Puget, P. Organo-mineral relationships. In *Encyclopedia of Soil Science*; Lal, R., Ed.; CRC Press: Boca Raton, FL, USA, 2006; pp. 1227–1230.

33. Zhou, Z.G.; Chen, N.; Cao, X.Y.; Chua, T.; Mao, J.D.; Mandel, R.D.; Bettis, E.A.; Thompson, M.L. Composition of clay-fraction organic matter in Holocene paleosols revealed by advanced solid-state NMR spectroscopy. *Geoderma* **2014**, *223*, 54–61. [CrossRef]

34. Rita, J.C.; Gama-Rodrigues, E.F.; Gama Rodrigues, A.C.; Polidoro, J.C.; Machado, R.C.; Baligar, V.C. C and N content in density fractions of whole soil and soil size fraction under cacao agroforestry systems and natural forest in Bahia, Brazil. *Environ. Manag.* **2011**, *48*, 134–141. [CrossRef]

35. Abdelrahman, H.; Cocozza, C.; Olk, D.C.; Ventrella, D.; Montemurro, F.; Miano, T. Changes in labile fractions of soil organic matter during the conversion to organic farming. *J. Soil Sci. Plant Nutr.* **2020**. [CrossRef]

36. Terra, F.S.; Demattê, J.A.; Rossel, R.A.V. Spectral libraries for quantitative analyses of tropical Brazilian soils: Comparing vis–NIR and mid-IR reflectance data. *Geoderma* **2015**, *255*, 81–93. [CrossRef]

37. Margenot, A.J.; Calderón, F.J.; Magrini, K.A.; Evans, R.J. Application of DRIFTs, 13 C NMR, and PY-MBMS to characterize the effects of soil science oxidation assays on soil organic matter composition in a Mollic Xerofluvent. *Appl. Spectr.* **2017**, *71*, 1506–1518. [CrossRef] [PubMed]

38. Drewnik, M.; Skiba, M.; Szymański, W.; Żyła, M. Mineral composition vs. soil forming processes in loess soils—A case study from Krakow (Southern Poland). *Catena* **2014**, *119*, 166–173. [CrossRef]

39. Viennet, J.C.; Hubert, F.; Ferrage, E.; Tertre, E.; Legout, A.; Turpault, M.P. Investigation of clay mineralogy in a temperate acidic soil of a forest using X-ray diffraction profile modeling: Beyond the HIS and HIV description. *Geoderma* **2015**, *241*, 75–86. [CrossRef]

40. Liu, Y.L.; Yao, S.H.; Han, X.Z.; Zhang, B.; Banwart, S. Chapter six-soil mineralogy changes with different agricultural practices during 8-year soil development from the parent material of a Mollisol. *Adv. Agron.* **2017**, *142*, 143–179.

41. Ndzana, G.M.; Huang, L.; Zhang, Z.; Zhu, J.; Liu, F.; Bhattacharyya, R. The transformation of clay minerals in the particle size fractions of two soils from different latitude in China. *Catena* **2019**, *175*, 317–328. [CrossRef]

42. Jafarzadeh-Haghighi, A.H.; Shamshuddin, J.; Hamdan, J.; Zainuddin, N. Structural composition of organic matter in particle-size fractions of soils along a climo-biosequence in the Main Range of Peninsular Malaysia. *Open Geosci.* **2016**, *8*, 503–513. [CrossRef]

43. Margenot, A.J.; Calderón, F.J.; Parikh, S.J. Limitations and potential of spectral subtractions in Fourier Transform Infrared Spectroscopy of soil samples. *Soil Sci. Soc. Am. J.* **2016**, *80*, 10–26. [CrossRef]

44. Nguyen, T.T.; Janik, L.J.; Raupach, M. Diffuse reflectance infrared Fourier transform (DRIFT) spectroscopy in soil studies. *Aust. J. Soil Res.* **1991**, *29*, 49–67. [CrossRef]

45. Sposito, G. *The Chemistry of Soils*; Oxford University Press, Inc.: New York, NY, USA, 2008.

46. Senesi, N.; D'Orazio, V.; Ricca, G. Humic acids in the first generation of EUROSOILS. *Geoderma* **2003**, *116*, 325–344. [CrossRef]

47. Kaiser, K.; Zech, W. Dissolved organic matter sorption by mineral constituents of subsoil clay fractions. *J. Plant Nutr. Soil Sci.* **2000**, *163*, 531–535. [CrossRef]

48. Eusterhues, K.; Rennert, T.; Knicker, H.; Kögel-Knabner, I.; Totsche, K.U.; Schwertmann, U. Fractionation of organic matter due to reaction with ferrihydrite: Co precipitation versus adsorption. *Environ. Sci. Technol.* **2011**, *45*, 527–533. [CrossRef] [PubMed]

49. Kleber, M.; Eusterhues, K.; Keiluweit, M.; Mikutta, C.; Mikutta, R.; Nico, P.S. Mineral–organic associations: Formation, properties, and relevance in soil environments. *Adv. Agron.* **2015**, *130*, 1–140.

50. Fu, H.; Quan, X. Complexes of fulvic acid on the surface of hematite, goethite, and akaganeite: FTIR observation. *Chemosphere* **2006**, *63*, 403–410. [CrossRef] [PubMed]

51. Kleber, M.; Nico, P.S.; Plante, A.; Filley, T.; Kramer, M.; Swanston, C.; Sollins, P. Old and stable soil organic matter is not necessarily chemically recalcitrant: Implication for modeling concepts and temperature sensitivity. *Glob. Chang. Biol.* **2011**, *17*, 1097–1107. [CrossRef]

52. Hatton, P.J.; Kleber, M.; Zeller, B.; Moni, C.; Plante, A.F.; Townsend, K.; Gelhaye, L.; Lajtha, K.; Derrien, D. Transfer of litter-derived N to soil mineral–organic associations: Evidence from decadal 15N tracer experiments. *Org. Geochem.* **2012**, *42*, 1489–1501. [CrossRef]

53. Keiluweit, M.; Bougoure, J.J.; Zeglin, L.; Myrold, D.D.; Weber, P.K.; Pett-Ridge, J.; Kleber, M.; Nico, P.S. Nano-scale investigation of the association of microbial nitrogen residues with iron (hydr)oxides in a forest soil O-horizon. *Geochim. Cosmochim. Acta* **2012**, *95*, 213–226. [CrossRef]
54. Zaccone, C.; Beneduce, L.; Lotti, C.; Martino, G.; Plaza, C. DNA occurrence in organic matter fractions isolated from amended, agricultural soils. *Appl. Soil Ecol.* **2018**, *130*, 134–142. [CrossRef]
55. Puget, P.; Angers, D.A.; Chenu, C. Nature of carbohydrates associated with water-stable aggregates of two cultivated soils. *Soil Biol. Biochem.* **1999**, *31*, 55–63. [CrossRef]
56. Gunina, A.; Kuzyakov, Y. Sugars in soil and sweets for microorganisms: Review of origin, content, composition and fate. *Soil Biol. Biochem.* **2015**, *90*, 87–100. [CrossRef]

applied sciences

Article

Assessing Water Infiltration and Soil Water Repellency in Brazilian Atlantic Forest Soils

Sergio Esteban Lozano-Baez [1,*], Miguel Cooper [2], Silvio Frosini de Barros Ferraz [3], Ricardo Ribeiro Rodrigues [1], Laurent Lassabatere [4], Mirko Castellini [5] and Simone Di Prima [4,6]

1 Laboratory of Ecology and Forest Restoration (LERF), Department of Biological Sciences, "Luiz de Queiroz" College of Agriculture, University of São Paulo, Av. Pádua Dias 11, Piracicaba, SP 13418-900 Brazil; rrresalq@usp.br
2 Department of Soil Science, "Luiz de Queiroz" College of Agriculture, University of São Paulo, Av. Pádua Dias 11, Piracicaba, SP 13418-900; Brazil; mcooper@usp.br
3 Forest Hydrology Laboratory, "Luiz de Queiroz" College of Agriculture, University of São Paulo, Av. Pádua Dias 11, Piracicaba, SP 13418-900, Brazil; silvio.ferraz@usp.br
4 Université de Lyon; UMR5023 Ecologie des Hydrosystèmes Naturels et Anthropisés, CNRS, ENTPE, Université Lyon 1, 3 rue Maurice Audin, 69518 Vaulx-en-Velin, France; laurent.lassabatere@entpe.fr (L.L.); simone.diprima@entpe.fr (S.D.P.)
5 Council for Agricultural Research and Economics, Research Centre for Agriculture and Environment (CREA-AA), Via C. Ulpiani 5, 70125 Bari, Italy; mirko.castellini@crea.gov.it
6 Department of Agricultural Sciences, University of Sassari, Viale Italia, 39, 07100 Sassari, Italy
* Correspondence: sergio.lozano@usp.br; Tel.: +55-19-3429-4100

Received: 22 February 2020; Accepted: 7 March 2020; Published: 12 March 2020

Abstract: This study presents the results of the soil hydraulic characterization performed under three land covers, namely pasture, 9-year-old restored forest, and remnant forest, in the Brazilian Atlantic Forest. Two types of infiltration tests were performed, namely tension (Mini-Disk Infiltrometer, MDI) and ponding (Beerkan) tests. MDI and Beerkan tests provided complementary information, highlighting a clear increase of the hydraulic conductivity, especially at the remnant forest plots, when moving from near-saturated to saturated conditions. In addition, measuring the unsaturated soil hydraulic conductivity with different water pressure heads allowed the estimation of the macroscopic capillary length in the field. This approach, in conjunction with Beerkan measurements, allowed the design better estimates of the saturated soil hydraulic conductivity under challenging field conditions, such as soil water repellency (SWR). This research also reports, for the first time, evidence of SWR in the Atlantic Forest, which affected the early stage of the infiltration process with more frequency in the remnant forest.

Keywords: Beerkan method; infiltration; forest restoration; soil water repellency

---

1. Introduction

The United Nations has declared the period 2021–2030 the decade of restoration to scale up existing initiatives, such as the Bonn Challenge, to restore degraded ecosystems [1]. It is expected that restoration will not only help to slow climate change through carbon sequestration, provide food, and increase biodiversity [2], but will also have hydrological benefits because of the perceived association between forest cover and soil hydrological ecosystem services [3,4]. In this context, it is necessary to better understand the consequences of forest regrowth on soil hydrological processes, such as water infiltration, which is fundamental to maintain productive soil-water-plant interactions, and also to control soil erosion and runoff, soil moisture content, and groundwater recharge in the ecosystems [5–7].

Estimating saturated and unsaturated soil hydraulic conductivities is crucial for interpreting and modeling soil hydrological processes. In addition, knowledge of these properties may provide information on the impact of land use on soils characteristics [8], which are rarely considered in studies of forest restoration [9]. During the last years, many infiltration methods and devices have been developed to determine soil hydraulic properties [10]. Among them, the Beerkan method [11] is becoming very popular in soil science because it constitutes a simple and an inexpensive way to determine the saturated soil hydraulic conductivity, K_s, in the field [12,13]. On the other hand, the mini-disk infiltrometer (MDI) is a routinely used method for measuring infiltration rates under negative pressure head in the field. The MDI is easily transportable and easy to use on hillslopes, thus, it substantially facilitates the replicability of the measurements [10].

Our previous study in the Atlantic Forest of Brazil [14] used the Beerkan protocol at three land covers, namely pasture, 9-year-old restored forest, and remnant forest. Our results showed that water repellency impacted water infiltration, yielding convex shaped cumulative infiltration curves. However, this observation was not carefully assessed. Similarly, many studies on tropical soils have reported some indirect effects of water repellency on water infiltration, such phenomenon is still poorly documented, especially in comparison with temperate regions [5,15–17]. Soil water repellency (SWR) or hydrophobicity is a transient soil property with which soils increase the resistance to wetting and infiltration. It is spatially and temporally very variable [18,19]. This is caused mainly by amphiphilic molecules produced by plants and organism, and generally occurs after forest fires or dry periods. Other factors that can be related to water repellency are the soil texture, soil temperature, pH, water content, soil organic carbon, land use, and plant cover [18–21]. In addition, recent research highlighted that climate change could increase the water repellency of soils, due to the increasing occurrence of extreme events such as droughts, which create the soil conditions (i.e., high temperatures and low soil water content) that promote the water repellency [22].

Currently SWR is receiving increased attention in the scientific literature, due to the important hydrological effects. For example, SWR reduces infiltration capacity, increases runoff rates as well as leaching of agrochemicals and soil erosion, also it can affect negatively the crop production, nutrients, and plant-available water [17,23,24]. On the other hand, SWR has positive impacts on soil aggregate stability and organic carbon sequestration [22]. Müller and Deurer [17] reported the benefit of SWR for the arid and semi-arid climates, considering that this soil property reduces the loss of soil water by evaporation and allows the rainwater to reach deeper depths. Despite these efforts, our understanding of SWR is still limited [18], especially when subcritical phenomena occur.

This investigation aims to broaden our previous work [14], using the same location in the Brazilian Atlantic Forest. In particular, the specific objective was to compare both unsaturated and saturated soil hydraulic conductivity determined with simple and low-cost field infiltration methods (MDI and Beerkan), for three land covers, namely pasture, 9-year-old restored forest, and remnant forest. This paper includes the first measurements of SWR in the Brazilian Atlantic Forest and its relevance regarding soil hydraulic properties, which had never been investigated so far, to the best of our knowledge.

2. Materials and Methods

2.1. Field Sites and Soil Sampling

The study area (22°53′ S, 46°54′ W) is located in the county of Campinas, São Paulo State, Southeast Brazil. The area is located inside the sub-basin of Atibaia River (2800 km^2), which belongs to the Piracicaba River basin. The vegetation is classified as seasonal semideciduous forest. The zone is characterized by a complex geology located at the transition between the Atlantic Plateau and the Peripheral Depression geomorphological provinces, with Ultisols and Entisols as main soils [25].

The elevation varies from 600–900 m a.s.l. The climate is classified as Cwa according to the Köppen classification, with annual rainfalls of 1700 mm and mean annual temperature of 20 °C [26].

The three investigated land covers (pasture, P, restored forest, R, and remnant forest, F) correspond with those of Lozano-Baez et al. [14], with the use of the same 18 plots (7 × 7 m in size). These plots represent two pasture (P1 and P2), two restored forest (R3 and R4), and two remnant forest (F5 and F6). Each of these sites is further divided into three blocks (i.e., upslope, U, midslope, M, and downslope, D). For a detailed description of the field sites, the reader may refer to our previous work [13]. In brief, for a given plot, three undisturbed soil cores (5 cm in height and 5 cm in diameter) were collected at the 0–5 cm depth. With these samples we determined in the laboratory the initial volumetric soil water content, θ_i (cm^3 cm^{-3}), and the soil bulk density, ρ_b (g cm^{-3}). Three disturbed soil samples (0–10 cm depth) were also collected to determine the soil texture and the soil organic carbon content (OC). The soil texture was determined by the hydrometer method [27] and the OC was determined by the Walkley-Black method.

2.2. Unsaturated and Saturated Soil Hydraulic Conductivity Measurements

Unsaturated soil hydraulic conductivity was measured using the Mini-Disc Infiltrometer (MDI) [28]. A total of 108 MDI experiments were carried out in the study sites. At each plot, we randomly selected six points with a minimum distance between measurements of 2 m. At the same sampling point, we used three different water pressure head values, h, in the ascending sequences −20, −5, and 0 mm, in order to sample several subdomains of the pore size distribution. Unsaturated soil hydraulic conductivity was calculated according to the method proposed by Zhang [29]. Before the MDI tests started, we removed the litter and leaves, and the sampled soil surface was gently levelled and smoothed. We used a thin layer of fine sand to ensure the contact between the infiltrometer and the soil (Figure 1). The thickness of the layer of sand was negligible and did not modify the imposed pressure head of the MDI. Visual readings of the water level were taken every 30 s until steady infiltration was nearly reached. For further descriptions of the MDI, details of measurements and calculations of hydraulic conductivity see Decagon Devices Inc. (Washington, DC, USA) [28].

Figure 1. Picture in the forest site showing the mini-disk infiltrometer and the steel ring used for the Beerkan infiltration test.

The saturated soil hydraulic conductivity, K_s (mm h^{-1}), was measured with ponding infiltration experiments of the Beerkan type [13]. At each plot, we performed seven Beerkan tests, for a total of 126 experiments. We used a steel ring with an inner diameter of 16 cm inserted to a depth of about 1 cm into the soil surface (Figure 1). In each infiltration point, a known volume of water (150 mL) was repeatedly poured into the cylinder at a small height above soil surface (i.e., a few cm) and the energy of the water was dissipated with the hand fingers to minimize the soil disturbance. Then, the time needed for each poured volume to complete infiltration was logged. This procedure was repeated until the difference in infiltration time between three consecutives trials became negligible.

The equilibration time, t_s (s), namely the duration of the transient phase of the infiltration process, was estimated according to the suggested criterion by Bagarello et al. [30] for analyzing cumulative infiltration data. More specifically, the t_s value was determined as the first value for which:

$$\hat{E} = \left| \frac{I(t) - I_{reg}(t)}{I(t)} \right| \leq E \tag{1}$$

where $I_{reg}(t)$ is estimated from regression analysis considering the last points, and E defines a given threshold to check linearity. Equation (1) is applied from the end of the experiment until finding the first data point that fits the condition $\hat{E} \leq E$ [31,32]. An illustrative example of t_s estimation using the commonly used value of $E = 2\%$ [30] is shown in Figure 2a. Transient infiltration conditions therefore occur from time 0 until time t_s (i.e., when $\hat{E} > 2$), while steady-state conditions establishes for all data points measured after time t_s (i.e., when $\hat{E} \leq 2$).

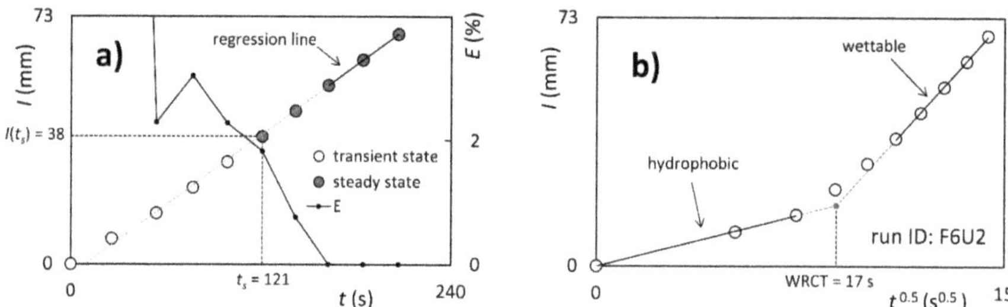

Figure 2. (a) Example of estimation of the equilibration time, t_s (s), and infiltrated depth at the equilibration time, $I(t_s)$ (mm) from cumulative infiltration and (b) water repellency cessation time, WRCT (s), as the intersection point of two straight lines, representing the initial (hydrophobic) and the late (wettable) stages of the I vs. $t^{0.5}$ plot of a Beerkan infiltration run affected by soil water repellency (SWR).

At the end of each infiltration test, we collected a disturbed soil sample within the infiltration surface to determine the saturated gravimetric water content, and thus the saturated volumetric water content, θ_s (cm^3 cm^{-3}), considering the values of dry bulk density, ρ_b, previously determined.

2.3. Estimating the Saturated Soil Hydraulic Conductivity, K_s

We estimated K_s by the Simplified method based on the near Steady-state phase of a Beerkan Infiltration run (SSBI), recently proposed by Bagarello et al. [32]. This method estimates K_s through an infiltration experiment of the Beerkan type [13] and an estimate of the macroscopic capillary length, λ_c (mm), expressing the relative importance of the capillary over gravity forces during water movement in unsaturated soil [33–35]. Firstly, the experimental steady-state infiltration rate, i_s (mm h^{-1}), is estimated by linear regression analysis of the last data points of the cumulative infiltration, I (mm), versus time, t (h), plot, describing the near steady-state condition. Then, SSBI estimates the saturated soil hydraulic conductivity, K_{sS} (mm h^{-1}) (the subscript S is used to indicate SSBI), as follows [33]:

$$K_{sS} = \frac{i_s}{\frac{\gamma \gamma_w \lambda_c}{r_d} + 1} \tag{2}$$

where γ and γ_w are dimensionless constants [36,37] related to the infiltration front shape, that are commonly set at 0.75 and 1.818, and r_d (mm) is the radius of the containment ring. Two different scenarios were considered to apply the SSBI method. The first scenario considered the MDI experiments, carried out with pressure heads of $h_{-20} = -20$ mm and $h_0 = 0$, to estimate λ_c by the following equation [38]:

$$\lambda_c = \frac{h_{-20} - h_0}{\ln\left(Q_{s,h_{-20}} / Q_{s,h_0}\right)} \tag{3}$$

where $Q_{s,h_{-20}}$ and Q_{s,h_0} (mm^3 h^{-1}) are the steady flow rates corresponding to h_{-20} and h_0, respectively, and they were estimated as follows:

$$Q_s = i_s \pi r_d^2 \tag{4}$$

For this scenario, we firstly averaged for each plot the individual i_s values, then plot-dependent λ_c values were estimated by Equation (3) (Table S1).

The second K_{sS} dataset was obtained considering $\lambda_c = 83$ mm, since it represents the suggested first approximation value for most soils types [37,39].

The Beerkan Estimation of Soil Transfer parameters (BEST) method [11] was also applied to estimate the saturated soil hydraulic conductivity, K_{sB} (mm h^{-1}) (the subscript B is used to indicate BEST). More specifically, among the three existing BEST algorithms, we used the BEST-steady algorithm [40], that estimates K_{sB}, by the following equation [41]:

$$K_{sB} = \frac{C\, i_s}{A\, b_s + C} \tag{5}$$

where b_s (mm) is the intercept of the regression line fitted to the last data points of the I versus t plot. The A (mm^{-1}) and C constants are defined for the specific case of the Brooks and Corey [42] relation and taking into account soil moisture initial conditions as follows [36]:

$$A = \frac{\gamma}{r_d(\theta_s - \theta_i)} \tag{6}$$

$$C = \frac{1}{2\left[1 - \left(\frac{\theta_i}{\theta_s}\right)^{\eta}\right](1 - \beta)} \ln\left(\frac{1}{\beta}\right) \tag{7}$$

where β is a coefficient commonly set at 0.6, and η is a shape parameter that is estimated from the analysis of the particle size data with the pedotransfer function included in the BEST procedure [11].

Following Bagarello et al. [32], the BEST-steady algorithm was chosen to check the SSBI method, compering K_{sS} and K_{sB} in terms of factors of difference, FoD, calculated as the highest value between K_{sB} and K_{sS} divided by the lowest value between K_{sB} and K_{sS}. Differences between K_{sS} and K_{sB} not exceeding a factor of two were considered indicative of satisfactory K_s predictions [33].

2.4. Soil Water Repellency Carachterization

Some of the Beerkan runs provided cumulative infiltration curves with convex shapes, signaling the occurrence of SWR phenomena [43,44].Then, the water repellency secession time, WRCT (s), was estimated from the intersection point of two straight lines, representing the initial and the late stages of I vs. $t^{0.5}$ relationship [45,46] (Figure 2b). The persistence of water repellency was measured using the water drop penetration time (WDPT) test. This test is widely used to determine the persistence of water repellency, it is easy to perform in field and presents the hydrological implications of hydrophobicity, because the amount of surface runoff is affected by the time required for the infiltration of droplets [47]. At each plot, we selected five sampling points. The WDPT was carry out by placing 10 drops (0.05 mL) of distilled water on to the soil surface and recording the time for their complete infiltration. Following other investigations [48,49] the infiltration recording was stopped after 3600 s. Moreover, if the drop did not infiltrate after this time interval, the value of 3600 s was assigned for the WDPT [47].

2.5. Data Analysis

Following similar investigations [10,12,14], unique values of clay, silt, sand, OC, ρ_b, θ_i, and θ_s were determined for each plot by averaging the measured values. For these soil parameters, we assumed a normal distribution, thus no transformation was performed on these data before statistical analysis. In addition, the K_{sB}, K_{sS}, K_{-20}, K_{-5}, K_0, and WDPT data were assumed to be log-normally distributed since the statistical distribution of these data is generally log-normal [50]. Statistical comparison was conducted using two-tailed *t*-tests, whereas the Tukey Honestly Significant Difference test was applied to compare our data set. The ln-transformed K_{sS}, K_{sB}, K_{-20}, K_{-5}, K_0 and WDPT data were used for the statistical treatment. A probability level, $\alpha = 0.05$, was used for all statistical analyses. It is reasonable to presume that infiltrometer data can also vary depending on the initial soil moisture and its effect on SWR [31], therefore the Spearman's rank correlation coefficients (*r*) were used to evaluate the relative influence of the soil properties on the infiltration process. For all the statistical analyses the Minitab© computer program (Minitab Inc., State College, PA, USA) was used.

3. Results and Discussion

3.1. Soil Properties

The 18 plots showed appreciable differences in soil texture. Sandy loam (i.e., P1M, P1D, R3U, R3M, R3D, and F5U) and sandy clay loam (i.e., P1U, R4S, R4M, R4D, F5M, and F5D) were the dominant soil textures among the plots, followed by clay loam (i.e., P2M, F6U, and F6M) and loam (i.e., P2U, P2D, and F6D). The soil texture of the plots is presented in the USDA textural triangle (Figure 3).

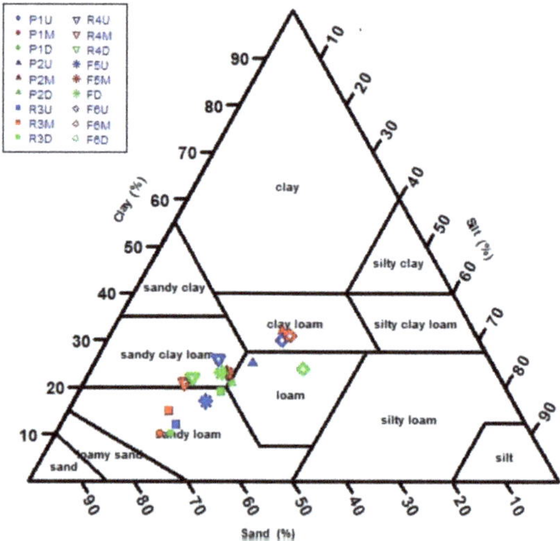

Figure 3. Textural distribution of the 18 plots in the USDA textural triangle.

As pointed out by our previous study [14], the OC at the pasture sites in the soil depth 0–0.10 m was similar to remnant forest, while restored forest sites presented the lowest OC values. The highest ρ_b values were observed in the restored forest R4, where the exposure of the soil and trampling pressure during the land-use history was greater in comparison with restored forest R3. Forest soils were characterized by the lowest ρ_b values, which can be related to the heterogeneous soil structure and higher soil macroporosity in this cover [51,52]. At the time of sampling, the θ_i ranged from 0.12 to 0.32 cm^3 cm^{-3} and the soil was significantly wetter in plots P1U, P2M, R4S, R4M, and F5I (Figure S1).

3.2. Assessing SSBI Estimates

Both SSBI scenarios always yielded physically plausible estimates (i.e., positive K_s values). For the first scenario (i.e., λ_c estimated from multi tension experiments), the K_{sS} values ranged between 5.9 and 1486.8 mm h^{-1}. The mean FoD was equal to 1.36 (maximum value = 2.74) and the individual values were less than 2 and 1.5 for 89% and 78% of the cases, respectively (Figure 4). For the second scenario (i.e., λ_c = 83 mm), K_{sS} data ranged between 3.7 and 934.5 mm h^{-1}. The mean FoD was equal to 1.51 (maximum value = 2.37) and the individual values were less than 2 and 1.5 in the 90% and 53% of the cases, respectively. Therefore, using the estimated λ_c values resulted in a slightly better estimation of K_{sS}, yielding a lower mean FoD value, thus, only the first scenario was considered in the subsequent analysis.

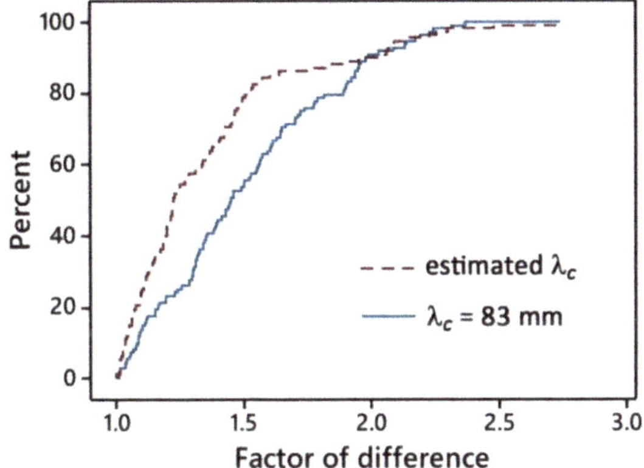

Figure 4. Empirical cumulative distribution function plot of the factors of difference between the saturated soil hydraulic conductivity values estimated by the BEST -steady (K_{sB}) and SSBI methods (K_{sS}). K_{sS} data were estimated considering λ_c = 83 mm (blue solid line) and the mean λ_c values estimated for each sampled plot from the MDI experiments carried out with a suction of 0 and −20 mm (red dashed line).

3.3. Comparing BEST versus SSBI Estimates Under Soil Water Repellency Conditions

BEST-steady failed to estimate K_s in case of convex-shaped cumulative infiltration curves, which led to negative b_s values and consequently to null K_{sB}. The K_{sB} data ranged between 4.5 and 1394 mm h^{-1} (almost three orders of magnitude). The BEST-steady algorithm yielded physically plausible estimates (i.e., positive K_s values) for 108 of 126 infiltration runs (i.e., 85.7% of cases). The percentage of successful runs was of 95.2% both for the pasture and restored forest (40 of 42 runs). With reference to the remnant forest, BEST led to a failure rate value of 33.3%, leading to lacks of estimates for 14 of 42 infiltration runs. In these cases, cumulative infiltration curves had convex shapes, which are typical for hydrophobia i.e., [42,45,46]. Such hydrophobia may result from significant amounts of organic matter content i.e., [52–54], originating from fauna and flora activities [55]. Soil texture also plays a major role on SWR, in particular, SWR is expected to increase for decreasing clay content. In this sense, our plots (i.e., F5U, R3U, R3D, P1M) with more sand content exhibited higher WDPT values. On the other hand, for the forest plots (i.e., F6U, F6M, F6D) the significant amounts of organic matter had a main role in generating relevant WDPT values also on finer textured soils [23,56].

The BEST-steady algorithm was unable to provide positive K_s values, showing that BEST can only be used when the soil does not exhibit hydrophobic effect, as suggested by Lassabatere et al. [57]. As shown in Figure 5, at increasing failure rates of the BEST method corresponded higher WDPT values, suggesting that where hydrophobic condition occurred, mainly in the remnant forest plots, it was the main cause of failure of BEST-steady. More specifically, BEST-steady requires both the slope and the intercept of regression line fitted to the last data points on the *I* vs. *t* plot. The magnitude of b_s depends on the entire cumulative infiltration curve (including the transient phase) [58,59], therefore that term is sensitive to SWR that impedes the early wetting phase of the infiltration process. When soil hydrophobicity occurred, the *I* vs. $t^{0.5}$ plot exhibited the characteristic "hockey-stick-like" shape [46], hiding the estimation of K_s trough BEST-steady [42]. On the other hand, SSBI differs by the term expressing steady-state condition, considering exclusively the final infiltration rate [13]. The exclusive use of this term allowed to consider only the final stage of the infiltration process, i.e., when the hydrophobicity effect on infiltration was diminished. In this investigation eighteen Beerkan infiltration tests exhibited a clear hockey-stick-like shape, mainly at the remnant forest plots, that allowed calculation of WRCT as the intersection point of two straight lines, representing the initial and the late stages of *I* vs. $t^{0.5}$ relationships [45] (Figure 2b).

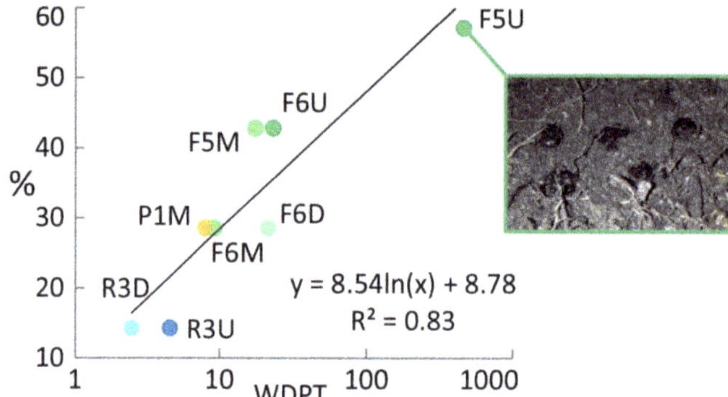

Figure 5. Comparison between the water drop penetration time, WDPT (s), and the failure rate of the BEST-steady algorithm (%). The picture represents water droplets resisting infiltration into forest soil due to the water repellency.

Table 1 shows the results of the WRCT and the equilibration time calculations. Water repellency always affected the very early stage of the infiltration process since the WRCT values raged between 14 and 93 s, and they were always lower than the t_s values. Therefore, for all the experiments steady-state infiltration rates (i_s) were always reached before the end of the runs and after that the influence of hydrophobicity had ceased, so the K_s values estimated by the use of the SSBI method could be always properly estimated considering the last data points of the infiltration curves. Limiting the hydraulic characterization to the stabilized phase avoided the uncertainties due to specific shape of the cumulative infiltration and a no clear distinction between the early- and late-time infiltration process because soil hydrophobic phenomena [58]. In other words, the results presented in this study suggest that if hydrophobicity affects the first stage of a Beerkan infiltration test, the SSBI estimates should characterize the hydraulic property of the soil properly. We believe that this result has practical importance because the use the SSBI method allowed us to maintain the integrity of the dataset, and to compare the hydraulic behavior of different sites with different land uses, where soil hydrophobicity only occurs in some circumstance.

Moreover, maintaining a small water head on the soil surface may be useful to study the infiltration process in macroporous repellent soils [48]. The SSBI method, covering the soil surface with a practically null depth of water, hence, lower than the commonly water-entry values for repellent soils [59], could allow the operator to characterize water infiltration occurring through either structural or other gaps in the water repellent layer or as fingered flow through zones of hydrophilic or less water repellent soil [60]. On the contrary, establishing several cm of ponded head of water on the infiltration surface is expected to overwhelm SWR [61]. In this investigation, the detection of hockey-stick-like shapes suggested that maintaining a small water head on the soil surface helped to prevent excessive positive pressure from overcoming SWR [62], and allowed the detection of water repellency.

Table 1. Values of the intercept, b_s (mm) of regression line fitted to the last data points describing the steady-state conditions on the I vs. t plot, total duration, t_{end} (s), total infiltrated depth, I_{end} (mm), infiltrated depth at the equilibration time, $I(t_s)$ (mm), equilibration time, t_s (s), and water repellency cessation time, WRCT (s), for the eighteen Beerkan infiltration runs affected by hydrophobicity.

ID	b_s (mm)	t_{end} (s)	I_{end} (mm)	$I(t_s)$ (mm)	t_s (s)	WRCT (s)
P1M6	−13.2	5053	52.7	30.1	3290	93
P1M7	−8.1	4601	45.2	22.5	2678	89
R3U2	−4.7	457	52.7	30.1	277	25
R3D7	−2.1	327	52.7	22.5	145	20
F5U1	−8.4	363	82.9	45.2	210	23
F5U2	−2.7	329	82.9	7.5	39	21
F5U3	−6.4	325	75.3	22.5	115	22
F5U4	−10.4	682	75.3	45.2	439	33
F5M2	−3.3	219	75.3	22.5	72	17
F5M3	−3.3	160	60.2	37.6	103	15
F5M4	−10.7	245	97.9	22.5	74	19
F6U1	−2.3	474	67.8	22.5	166	26
F6U2	−3.2	207	67.8	37.6	121	17
F6U5	−13.3	188	75.3	52.7	140	18
F6M5	−10.4	495	82.9	37.6	253	27
F6M6	−4.3	208	82.9	22.5	64	17
F6D2	−10.1	148	82.9	60.2	112	14
F6D4	−1.3	325	75.3	52.7	229	20

Lower SWR was detected in wetter soils. The correlation between θ_i and ln(WDPT) was significant ($r = -0.67$, $p = 0.002$) (Table 2). This result was in line with the reasoning that the soil water content governs the interaction between soil particles and amphiphilic organic molecules, resulting from degradation of tree tissues, that coat soil particles and may be responsible for SWR [45]. The transition from wettable to hydrophobic status (and vice versa) is generally associated to a critical range of soil moisture [63]. The lower water content of this range defines the condition below which the medium is water repellent, the higher identifies the condition above which the medium is wettable. K_s data were positively correlated to ln(WDPT). This is logical, since both macropore flow (which affects the magnitude of K_s) and water repellency phenomena were relevant at the remnant forest plots. In brief, the correlation between these two variables is not the result of a causal connection but the concomitancy of two processes: hydrophobia and macropore flow, which also lead to mainly subcritical water repellency. In addition, we conclude that hydrophobia had no effect on the estimation of the saturated hydraulic conductivity. Indeed, in opposite case, K_{sS} and ln(WDPT) would have a negative correlation. Consequently, we assumed that the SSBI method proved efficient for detecting SWR and estimating properly the soil saturated hydraulic conductivity, at the same time. Lastly, K_{sB} and K_{sS} had a positive correlation with a value close to unity. The two estimators provide close estimates, as discussed above with the FoD. We then can conclude that soil hydrophobicity only affected the failure rate of the BEST-steady algorithm (Figure 5), without affecting the quality of its estimate when the method worked.

Table 2. Spearman's rank correlation coefficients of the hydraulic measured properties.

Variables	θ_i	ln(WDPT)	ln(K_{sS})
ln(WDPT)	−0.67		
p-Value	0.002		
ln(K_{sS})	−0.59	0.74	
p-Value	0.009	<0.001	
ln(K_{sB})	−0.61	0.73	0.97
p-Value	0.007	0.001	<0.001

3.4. Unsaturated versus Saturated Soil Hydraulic Conductivity

The two types of infiltration tests, i.e., tension and ponding experiments, highlighted a clear increase of the hydraulic conductivity, especially at the remnant forest plots, when moving from near-saturated to saturated conditions (Figure 6). It is important to underscore that saturated and unsaturated conditions could be affected by the different soil texture. Our previous study [14] shows that soils with higher clay content (i.e., P2M, F6M, F6D) evidenced greater variations, by contrast, sandy soils (i.e., P1U, P1M, F5D) had lower variation. The mean values of the ratios between saturated, K_{sS}, and near-saturated soil hydraulic conductivity, K_{-20}, were 10.7, 21.5, and 118.3, for the pasture, restored forest and remnant forest, respectively. A similar trend was also detected when the K_{-5} values were considered, with the mean values of the ratios equal to 2.2, 5.6, and 23.7. Similar results were also obtained when K_{sB} values were considered, with the values of the ratios equal to 10.6, 17.5, 92.0, and 2.2, 4.6, 17.4, for the K_{-20} and K_5 data, respectively. We also noticed a discrepancy between K_{sS} and K_0 data, especially at the Forest site, because only under ponded conditions at the surface the macropores are activated [63]. The increase of the difference between saturated and unsaturated conditions can be explained by the activation of macroporosity at the forest plots [64]. Overall, the soil in the remnant forest is heterogeneous and characterized by a dominance of complex macropores. For example, a higher soil macroporosity and total porosity have been reported in the same forest soil by our previous work [14]. This soil macroporosity resulted from the better soil structure, which is caused by the high amount of biopores, roots, soil fauna activity and greater inputs of organic matter [52,63,64]. Moreover, soil variability at the scale of a few meters could have been less represented by the MDI, due to the small diameter of the infiltrometer (i.e., 4.5 cm).

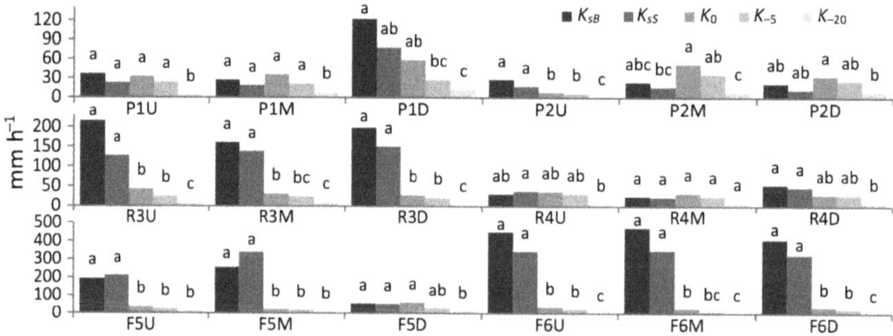

Figure 6. Comparison of the mean saturated soil hydraulic conductivity values estimated with BEST-steady, K_{sB} (mm h^{-1}), and the SSBI method, K_{sS} (mm h^{-1}), and hydraulic conductivity, K_0, K_{-5}, and K_{-20} (mm h^{-1}), values measured with the minidisk infiltrometer under a tension of 0, −5, and −20 mm. For a given plot, means that do not share a letter are significantly different according to the Tukey honestly significant difference test ($p < 0.05$).

4. Conclusions

To improve the soil hydraulic characterization of the soils under different land uses in the Atlantic Forest of Brazil, we measured and compared the unsaturated and saturated soil hydraulic conductivity, using the MDI and Beerkan method for three land covers, namely pasture, 9-year-old restored forest, and remnant forest. This research reports, for the first time, provide evidence of SWR in the Atlantic Forest, especially in the remnant forest. Our measurements demonstrated that SWR affected the early stage of the infiltration process. The comparison between alternative methods to estimate K_s allowed to account for the effect of SWR on water infiltration measurements. In particular, when there are evidences of SWR, our results suggest using the SSBI method instead of BEST-steady to avoid the failure of the analysis in case of string SWR. Indeed, the SSBI method allowed to maintain the integrity of the infiltration dataset, facilitating the hydraulic comparison between different land uses. Tension

(MDI) and ponding (Beerkan) infiltration tests provided a complementary information, highlighting a clear increase of the hydraulic conductivity, especially at the remnant forest plots, when moving from near-saturated to saturated conditions. This information is relevant to assess the infiltration recovery after forest restoration, as it signals soil structure heterogeneity and higher soil macroporosity. In addition, measuring the unsaturated soil hydraulic conductivity with different water pressure heads also allowed to estimate λ_c in the field. This approach, in conjunction with Beerkan measurements, allowed to generate better K_s estimates based on field measurements, and also avoided any subjectivity caused by assuming a constant λ_c value, which is often selected based on general descriptions of soil textural and structural characteristics when estimating K_s from ponding infiltration experiments [34]. Nonetheless, developing alternative methods for estimating λ_c is desirable for alleviating the amount of work necessary to accurately estimate K_s.

In this investigation we used the water repellency secession time (WRCT) and water drop penetration time (WDPT) to assess SWR. The SWR was observed in pasture and forest soils with higher sand content. As expected, SWR phenomena were less severe for increasing soil moisture conditions and more common on remnant forest soils. SWR has important hydrological effects, including water supply in forest ecosystems. Thus, SWR cannot be neglected in forest soil hydraulic studies and it must be accounted when developing hydrological models. Future research will focus on understanding the interactions between vegetation, soil biology and soil properties (i.e., including physical, chemical, and mineralogical properties) that are promoting SWR in Atlantic forest soils. Details on the effects of forest restoration on water repellency are severely lacking. It is also important to consider the temporal and spatial dynamic of the soil infiltration and water repellency. For example, future studies could quantify the spatial extent at the larger scale, upscaling the measurement from point to catchment scale.

Supplementary Materials: The following are available online at http://www.mdpi.com/2076-3417/10/6/1950/s1, Figure S1: Comparison between the mean soil organic carbon content (OC in g KgPress release^{-1}), dry soil bulk density (ρ_b in g cm^{-3}), initial volumetric soil water content (θ_i in cm^3 cm^{-3}), and saturated volumetric soil water content (θ_s in cm^3 cm^{-3}), values for the 18 sampled plots. Bars indicate standard deviation. For a given variable and plot, means that do not share a letter are significantly different according to the Tukey honestly significant difference test ($P < 0.05$). The subscript letter refers to the landscape position (Upslope, Middleslope and Downslope) in each site, Table S1: Mean values for each sampled plot (P1, P2, R3, R4, F5 and F6) of the steady-state infiltration rates, i_{s,h_0} and $i_{s,h_{-20}}$ (mm h^{-1}), flow rates, Q_{s,h_0} and $Q_{s,h_{-20}}$ (mm^3 h^{-1}), obtained from the MDI experiments carried out with a pressure head $h_0 = 0$ and $h_{-20} = -20$ mm, and macroscopic capillary length, λ_c (mm), estimated by Equation (3).

Author Contributions: S.E.L.-B. carried out the data collection and wrote the initial draft; M.C., S.F.d.B.F. and R.R.R. conceived and designed the experiment; L.L., M.C. and S.D.P. revised, discussed, modified, and supplemented the ideas for the final draft. All authors have read and agreed to the published version of the manuscript.

Funding: This research was supported by the Fundação de Amparo à Pesquisa do Estado de São Paulo (BIOTA/FAPESP Program: 2013/50718-5 and 1999/09635-0), and Conselho Nacional de Desenvolvimento Científico e Tecnológico (CNPq 561897/2010-7 and Miguel Cooper's, Ricardo Ribeiro Rodrigues's and Silvio Frosini de Barros Ferraz's scientific productivity fellowships). This work was also supported through the "Programma Operativo Nazionale (PON) Ricerca e Innovazione 2014—2020 (Linea 1—Mobilità dei ricercatori, AIM1853149, CUP: J54I18000120001) funded by the European Regional Development Fund (ERDF) and the Italian Ministry of Education, University and Research (MIUR).

Acknowledgments: We are grateful to Luiz Felippe Salemi who gave their valuable suggestions. S.E.L.-B. would like to thank Julia Gardies, Daigard Ricardo Ortega, Monica Borda and Miller Ruiz for their support during field work.

Conflicts of Interest: The authors declare no conflict of interest.

References

1. UN Press Release. *New UN Decade on Ecosystem Restoration Offers Unparalleled Opportunity for Job Creation, Food Security and Addressing Climate Change*; FAO: New York, NY, USA, 2019.

2. Giannini, V.; Bertacchi, A.; Bonari, E.; Silvestri, N. Rewetting in Mediterranean reclaimed peaty soils and its potential for phyto-treatment use. *J. Environ. Manag.* **2018**, *208*, 92–101. [CrossRef] [PubMed]

3. Brancalion, P.H.S.; Niamir, A.; Broadbent, E.; Crouzeilles, R.; Barros, F.S.M.; Almeyda Zambrano, A.M.; Baccini, A.; Aronson, J.; Goetz, S.; Reid, J.L.; et al. Global restoration opportunities in tropical rainforest landscapes. *Sci. Adv.* **2019**, *5*, 1–11. [CrossRef] [PubMed]

4. Chazdon, R.L. Beyond Deforestation: Restoring Forests and Ecosystem Services on Degraded Lands. *Science* **2008**, *320*, 1458–1460. [CrossRef] [PubMed]

5. Ziegler, A.D.; Giambelluca, T.W.; Tran, L.T.; Vana, T.T.; Nullet, M.A.; Fox, J.; Vien, T.D.; Pinthong, J.; Maxwell, J.F.; Evett, S. Hydrological consequences of landscape fragmentation in mountainous northern Vietnam: Evidence of accelerated overland flow generation. *J. Hydrol.* **2004**, *287*, 124–146. [CrossRef]

6. Elsenbeer, H. Hydrologic flowpaths in tropical rainforest soilscapes? a review. *Hydrol. Process.* **2001**, *15*, 1751–1759. [CrossRef]

7. Hassler, S.K.; Zimmermann, B.; van Breugel, M.; Hall, J.S.; Elsenbeer, H. Recovery of saturated hydraulic conductivity under secondary succession on former pasture in the humid tropics. *For. Ecol. Manag.* **2011**, *261*, 1634–1642. [CrossRef]

8. Gonzalez-Sosa, E.; Braud, I.; Dehotin, J.; Lassabatere, L.; Angulo-Jaramillo, R.; Lagouy, M.; Branger, F.; Jacqueminet, C.; Kermadi, S.; Michel, K. Impact of land use on the hydraulic properties of the topsoil in a small French catchment. *Hydrol. Process.* **2010**, *24*, 2382–2399. [CrossRef]

9. Zwartendijk, B.W.; van Meerveld, H.J.; Ghimire, C.P.; Bruijnzeel, L.A.; Ravelona, M.; Jones, J.P.G. Rebuilding soil hydrological functioning after swidden agriculture in eastern Madagascar. *Agric. Ecosyst. Environ.* **2017**, *239*, 101–111. [CrossRef]

10. Alagna, V.; Bagarello, V.; Di Prima, S.; Iovino, M. Determining hydraulic properties of a loam soil by alternative infiltrometer techniques: Hydraulic Properties of a Loam Soil by Infiltrometer Techniques. *Hydrol. Process.* **2016**, *30*, 263–275. [CrossRef]

11. Lassabatere, L.; Angulo-Jaramillo, R.; Soria Ugalde, J.M.; Cuenca, R.; Braud, I.; Haverkamp, R. Beerkan estimation of soil transfer parameters through infiltration experiments—BEST. *Soil Sci. Soc. Am. J.* **2006**, *70*, 521. [CrossRef]

12. Castellini, M.; Di Prima, S.; Iovino, M. An assessment of the BEST procedure to estimate the soil water retention curve: A comparison with the evaporation method. *Geoderma* **2018**, *320*, 82–94. [CrossRef]

13. Angulo-Jaramillo, R.; Bagarello, V.; Di Prima, S.; Gosset, A.; Iovino, M.; Lassabatere, L. Beerkan Estimation of Soil Transfer parameters (BEST) across soils and scales. *J. Hydrol.* **2019**, *576*, 239–261. [CrossRef]

14. Lozano-Baez, S.; Cooper, M.; Ferraz, S.; Ribeiro Rodrigues, R.; Pirastru, M.; Di Prima, S. Previous Land Use Affects the Recovery of Soil Hydraulic Properties after Forest Restoration. *Water* **2018**, *10*, 453. [CrossRef]

15. Elsenbeer, H.; Newton, B.E.; Dunne, T.; de Moraes, J.M. Soil hydraulic conductivities of latosols under pasture, forest and teak in Rondonia, Brazil. *Hydrol. Process.* **1999**, *13*, 7. [CrossRef]

16. Vogelmann, E.S.; Reichert, J.M.; Reinert, D.J.; Mentges, M.I.; Vieira, D.A.; de Barros, C.A.P.; Fasinmirin, J.T. Water repellency in soils of humid subtropical climate of Rio Grande do Sul, Brazil. *Soil Tillage Res.* **2010**, *110*, 126–133. [CrossRef]

17. Müller, K.; Deurer, M. Review of the remediation strategies for soil water repellency. *Agric. Ecosyst. Environ.* **2011**, *144*, 208–221. [CrossRef]

18. Doerr, S.H.; Shakesby, R.A.; Walsh, R. Soil water repellency: Its causes, characteristics and hydro-geomorphological significance. *Earth-Sci. Rev.* **2000**, *51*, 33–65. [CrossRef]

19. DeBano, L.F. *Water Repellent Soils: A State-of-the-Art*; U.S. Department of Agriculture, Forest Service, Pacific Southwest Forest and Range Experiment Station: Berkeley, CA, USA, 1981; pp. 1–21.

20. Jordán, A.; Zavala, L.M.; Mataix-Solera, J.; Nava, A.L.; Alanís, N. Effect of fire severity on water repellency and aggregate stability on Mexican volcanic soils. *CATENA* **2011**, *84*, 136–147. [CrossRef]

21. Lichner, L.; Hallett, P.D.; Drongová, Z.; Czachor, H.; Kovacik, L.; Mataix-Solera, J.; Homolák, M. Algae influence the hydrophysical parameters of a sandy soil. *CATENA* **2013**, *108*, 58–68. [CrossRef]

22. Goebel, M.O.; Bachmann, J.; Reichstein, M.; Janssens, I.A.; Guggenberger, G. Soil water repellency and its implications for organic matter decomposition—Is there a link to extreme climatic events? *Glob. Chang. Biol.* **2011**, *17*, 2640–2656. [CrossRef]

23. Mao, J.; Nierop, K.G.J.; Dekker, S.C.; Dekker, L.W.; Chen, B. Understanding the mechanisms of soil water repellency from nanoscale to ecosystem scale: A review. *J. Soils Sediments* **2019**, *19*, 171–185. [CrossRef]

24. de Oliveira, L.H.d.S.; Valladares, G.S.; Coelho, R.M.; Criscuolo, C. Soil vulnerability to degradation at Campinas municipality, SP. *Geografia (Londrina)* **2014**, *22*, 65–79.

25. Mello, M.H.; Pedro Junior, M.J.; Ortolani, A.A.; Alfonsi, R.R. *Chuva e Temperatura: Cem anos de Observações em Campinas*; Boletim Tecnico; IAC: Campinas, Brazil, 1994.

26. Gee, G.; Or, D. Particle-size analysis. In *Methods of Soil Analysis: Physical Methods*; Dane, J.H., Topp, C., Eds.; Soil Science Society of America: Madison, WI, USA, 2002; pp. 255–293. ISBN 978-0-89118-841-4.

27. Decagon Devices, Inc. *Minidisk Infiltrometer User's Manual*; Decagon Devices, Inc.: Pullman, WA, USA, 2014.

28. Zhang, R. Determination of Soil Sorptivity and Hydraulic Conductivity from the Disk Infiltrometer. *Soil Sci. Soc. Am. J.* **1997**, *61*, 1024–1030. [CrossRef]

29. Bagarello, V.; Iovino, M.; Reynolds, W. Measuring hydraulic conductivity in a cracking clay soil using the Guelph permeameter. *Trans. ASAE* **1999**, *42*, 957–964. [CrossRef]

30. Angulo-Jaramillo, R.; Bagarello, V.; Iovino, M.; Lassabatere, L. Saturated Soil Hydraulic Conductivity. In *Infiltration Measurements for Soil Hydraulic Characterization*; Springer International Publishing: Berlin/Heidelberg, Germany, 2016; pp. 43–180. ISBN 978-3-319-31786-1.

31. Di Prima, S.; Castellini, M.; Najm, M.R.A.; Stewart, R.D.; Angulo-Jaramillo, R.; Winiarski, T.; Lassabatere, L. Experimental assessment of a new comprehensive model for single ring infiltration data. *J. Hydrol.* **2019**, *573*, 937–951. [CrossRef]

32. Bagarello, V.; Di Prima, S.; Iovino, M. Estimating saturated soil hydraulic conductivity by the near steady-state phase of a Beerkan infiltration test. *Geoderma* **2017**, *303*, 70–77. [CrossRef]

33. Elrick, D.E.; Reynolds, W.D. Methods for analyzing constant-head well permeameter data. *Soil Sci. Soc. Am. J.* **1992**, *56*, 320. [CrossRef]

34. Reynolds, W.D.; Elrick, D.E. Ponded Infiltration from a Single Ring: I. Analysis of Steady Flow. *Soil Sci. Soc. Am. J.* **1990**, *54*, 1233. [CrossRef]

35. Raats, P.A.C. Analytical Solutions of a Simplified Flow Equation. *Trans. ASAE* **1976**, *19*, 0683–0689. [CrossRef]

36. Haverkamp, R.; Ross, P.J.; Smettem, K.R.J.; Parlange, J.Y. Three-dimensional analysis of infiltration from the disc infiltrometer: 2. Physically based infiltration equation. *Water Resour. Res.* **1994**, *30*, 2931–2935. [CrossRef]

37. Reynolds, W.; Elrick, D. 3.4.3.2.b Pressure infiltrometer. In *Methods of Soil Analysis, Part 4, PhysicalMethods*; Dane, J.H., Topp, G.C., Eds.; SSSA Book Series, No. 5. Soil Science Society of America: Madison, WI, USA, 2002; Volume 4, pp. 826–836.

38. Reynolds, W.D.; Elrick, D.E. Determination of hydraulic conductivity using a tension infiltrometer. *Soil Sci. Soc. Am. J.* **1991**, *55*, 633–639. [CrossRef]

39. Reynolds, W.; Elrick, D.; Youngs, E. 3.4.3.2 Ring or cylinder infiltrometers (vadose zone). In *Methods of Soil Analysis, Part 4, PhysicalMethods*; Dane, J.H., Topp, G.C., Eds.; SSSA Book Series, No. 5. Soil Science Society of America: Madison, WI, USA, 2002; pp. 818–820.

40. Bagarello, V.; Di Prima, S.; Iovino, M. Comparing Alternative Algorithms to Analyze the Beerkan Infiltration Experiment. *Soil Sci. Soc. Am. J.* **2014**, *78*, 724. [CrossRef]

41. Di Prima, S.; Lassabatere, L.; Bagarello, V.; Iovino, M.; Angulo-Jaramillo, R. Testing a new automated single ring infiltrometer for Beerkan infiltration experiments. *Geoderma* **2016**, *262*, 20–34. [CrossRef]

42. Brooks, R.H.; Corey, T. *Hydraulic Properties of Porous Media*; Hydrolog. Paper 3; Colorado State University: Fort Collins, CO, USA, 1964.

43. Concialdi, P.; Di Prima, S.; Bhanderi, H.M.; Stewart, R.D.; Abou Najm, M.R.; Lal Gaur, M.; Angulo-Jaramillo, R.; Lassabatere, L. An open-source instrumentation package for intensive soil hydraulic characterization. *J. Hydrol.* **2020**, *582*, 124492. [CrossRef]

44. Alagna, V.; Iovino, M.; Bagarello, V.; Mataix-Solera, J.; Lichner, L'. Application of minidisk infiltrometer to estimate water repellency in Mediterranean pine forest soils. *J. Hydrol. Hydromech.* **2017**, *65*, 254–263. [CrossRef]

45. Lichner, L.; Felde, J.M.N.L.; Büdel, B.; Leue, M.; Gerke, H.H.; Ellerbrock, R.H.; Kollár, J.; Rodný, M.; Šurda, P.; Fodor, N.; et al. Effect of vegetation and its succession on water repellency in sandy soils. *Ecohydrology* **2018**, *11*, e1991. [CrossRef]

46. Wessel, A.T. On using the effective contact angle and the water drop penetration time for classification of water repellency in dune soils. *Earth Surf. Process. Landforms* **1988**, *13*, 555–561. [CrossRef]

47. Di Prima, S.; Bagarello, V.; Angulo-Jaramillo, R.; Bautista, I.; Cerdà, A.; del Campo, A.; González-Sanchis, M.; Iovino, M.; Lassabatere, L.; Maetzke, F. Impacts of thinning of a Mediterranean oak forest on soil properties influencing water infiltration. *Hydrol. Hydromech.* **2017**, *65*, 276–286. [CrossRef]

48. Dekker, L.W.; Doerr, S.H.; Oostindie, K.; Ziogas, A.K.; Ritsema, C.J. Water Repellency and Critical Soil Water Content in a Dune Sand. *Soil Sci. Soc. Am J.* **2001**, *65*, 1667–1674. [CrossRef]

49. Lee, D.M.; Elrick, D.E.; Reynolds, W.D.; Clothier, B.E. A comparison of three field methods for measuring saturated hydraulic conductivity. *Can. J. Soil Sci.* **1985**, *65*, 563–573. [CrossRef]

50. Salemi, L.F.; Groppo, J.D.; Trevisan, R.; de Moraes, J.M.; de Barros Ferraz, S.F.; Villani, J.P.; Duarte-Neto, P.J.; Martinelli, L.A. Land-use change in the Atlantic rainforest region: Consequences for the hydrology of small catchments. *J. Hydrol.* **2013**, *499*, 100–109. [CrossRef]

51. Cooper, M.; Medeiros, J.C.; Rosa, J.D.; Soria, J.E.; Toma, R.S. Soil functioning in a toposequence under rainforest in São Paulo, Brazil. *Revista Brasileira de Ciência do Solo* **2013**, *37*, 392–399. [CrossRef]

52. Goebel, M.-O.; Bachmann, J.; Woche, S.K.; Fischer, W.R. Soil wettability, aggregate stability, and the decomposition of soil organic matter. *Geoderma* **2005**, *128*, 80–93. [CrossRef]

53. Lipsius, K.; Mooney, S.J. Using image analysis of tracer staining to examine the infiltration patterns in a water repellent contaminated sandy soil. *Geoderma* **2006**, *136*, 865–875. [CrossRef]

54. Buczko, U.; Bens, O.; Fischer, H.; Hüttl, R.F. Water repellency in sandy luvisols under different forest transformation stages in northeast Germany. *Geoderma* **2002**, *109*, 1–18. [CrossRef]

55. Lassabatere, L.; Angulo-Jaramillo, R.; Yilmaz, D.; Winiarski, T. BEST method: Characterization of soil unsaturated hydraulic properties. In *Advances in Unsaturated Soils*; CRC Press: London, UK, 2013; pp. 527–532.

56. Di Prima, S.; Concialdi, P.; Lassabatere, L.; Angulo-Jaramillo, R.; Pirastru, M.; Cerda, A.; Keesstra, S. Laboratory testing of Beerkan infiltration experiments for assessing the role of soil sealing on water infiltration. *CATENA* **2018**, *167*, 373–384. [CrossRef]

57. Di Prima, S.; Stewart, R.D.; Mirko, C.; Bagarello, V.; Abou Najm, M.R.; Pirastru, M.; Giadrossich, F.; Iovino, M.; Angulo-Jaramillo, R.; Lassabatere, L. Estimating the macroscopic capillary length and derivatives from Beerkan infiltration experiments. *J. Hydrol..* submitted.

58. Wang, Z.; Feyen, J.; Ritsema, C.J. Susceptibility and predictability of conditions for preferential flow. *Water Resour. Res.* **1998**, *34*, 2169–2182. [CrossRef]

59. Bagarello, V.; Basil, G.; Caltabellota, G.; Giordano, G.; Iovino, M. Testing soil water repellency in a Sicilian area two years after a fire. *J. Agric. Eng.* **2019**, 1–23. [CrossRef]

60. Ebel, B.A.; Moody, J.A. Rethinking infiltration in wildfire-affected soils. *Hydrol. Process.* **2013**, *27*, 1510–1514. [CrossRef]

61. Nyman, P.; Sheridan, G.; Lane, P.N.J. Synergistic effects of water repellency and macropore flow on the hydraulic conductivity of a burned forest soil, south-east Australia. *Hydrol. Process.* **2010**, *24*, 2871–2887. [CrossRef]

62. Lozano-Baez, S.E.; Cooper, M.; Ferraz, S.F.B.; Rodrigues, R.R.; Castellini, M.; Di Prima, S. Recovery of Soil Hydraulic Properties for Assisted Passive and Active Restoration: Assessing Historical Land Use and Forest Structure. *Water* **2019**, *11*, 86. [CrossRef]

63. Lassabatere, L.; Di Prima, S.; Bouarafa, S.; Iovino, M.; Bagarello, V.; Angulo-Jaramillo, R. BEST-2K Method for Characterizing Dual-Permeability Unsaturated Soils with Ponded and Tension Infiltrometers. *Vadose Zone J.* **2019**, *18*, 180124. [CrossRef]

64. Lassabatere, L.; Yilmaz, D.; Peyrard, X.; Peyneau, P.E.; Lenoir, T.; Šimůnek, J.; Angulo-Jaramillo, R. New Analytical Model for Cumulative Infiltration into Dual-Permeability Soils. *Vadose Zone J.* **2014**, *13*, 1–15. [CrossRef]

applied sciences

Article

Evaluation of Spent Grain Biochar Impact on Hop (*Humulus lupulus* L.) Growth by Multivariate Image Analysis

Tiziana Amoriello [1,]*, Simona Fiorentino [2], Valerio Vecchiarelli [2] and Mauro Pagano [3,]*

[1] CREA Research Centre for Food and Nutrition, Via Ardeatina 546, 00178 Roma, Italy
[2] Centro Appenninico del Terminillo Carlo Jucci, Università degli Studi di Perugia, Via Comunali 43, 02100 Rieti, Italy; simona.fiorentino@unipg.it (S.F.); valerio.vecchiarelli@unipg.it (V.V.)
[3] CREA Research Centre for Engineering and Agro-Food Processing, Via della Pascolare 16, 00015 Monterotondo (Roma), Italy
* Correspondence: tiziana.amoriello@crea.gov.it (T.A.); mauro.pagano@crea.gov.it (M.P.); Tel.: +39-06-514941 (T.A.)

Received: 20 November 2019; Accepted: 8 January 2020; Published: 10 January 2020

Abstract: Biochar is generally considered as an effective soil amendment, which can improve soil organic matter and nutrients content and enhance crop productivity. In this study, biochar derived from brewers' spent grain (BSG) was used in a pot and field experiment to assess whether its addition to soil could affect hop plant growth. The experiment was conducted in Central Italy during the period March–August 2017. Three different German cultivars of hop plant (Hallertau Magnum, Perle, Spalter spalt) were considered. Biochar was added to the pot soil at 20% level. Its effect on the roots was evaluated using multivariate image analysis (MIA) and the statistical technique of general linear models (GLM), whereas the shoots, bines length and yield using GLM. Results showed that biochar significantly improved root growth ($p < 0.0001$). Regarding shoots, no variability for the genotypes was observed during the vegetative period, whereas slight differences resulted before plant dormancy, especially for the Hallertau Magnum cultivar. No differences in the number of leaves or bines length were observed between the two treatments for all cultivars. The addition of biochar to the soil significantly improved yield (number of cones). These results highlighted that BSG-derived biochar can be useful to improve hop plant growth and cones production.

Keywords: amendment; biochar; brewers' spent grain; hop; image analysis; plant growth

---

1. Introduction

Beer is one of most consumed beverages in the world, and plays an important role in the global economy. In 2018, the overall beer production amounted to about 1.94 billion hectolitres. However, beer processing produces a huge amount of waste: for every hectolitre of beer produced, about 20 kg of spent grain, 0.2–0.4 kg of spent yeast and 0.3 kg of spent hops/hot trub are generated. In particular, the 85% of total solid waste is represented by brewers' spent grain (BSG), the residue left after barley malting and separation of the wort during the brewing process, and it contains the husk and the outer layer of barley kernel [1]. BSG is mainly composed of protein (more than 20%) and fibers, represented by a lignino-cellulosic material, whose main constituents are hemicellulose (28–35%), cellulose (17–25%) and lignin (7–27%) [1–3].

Appropriate management of these waste streams has become a challenging issue. The solid by-products from the brewing process are mainly disposed of as waste or sold as animal feed, due to their considerable amount of valuable compounds (proteins, lipids, carbohydrates, polyphenols, and minerals) and nitrogen-containing nutrients. However, brewing industries are interested in new solutions to reduce the amount of waste produced and transform by-products in added value products. Therefore, in a perspective of a zero waste approach and in order to reduce waste storage and logistics costs, alternatives to valorize these residues and to recover and reuse them in a sustainable and profitable way are continuously proposed and developed. Moreover, by-products can be transformed in combustible gas, to be reused in the beer production cycle, or in soil amendments for agricultural applications [4,5].

A valid alternative can be the transformation of BSG into biochar, which can be used as an effective soil amendment for the production of vegetables. Biochar is a carbon-rich, fine-grained and porous material with stable physical and chemical properties, produced from waste biomass through pyrolysis, e.g., a thermal decomposition of lignocellulose biomass by heating at elevated temperatures (generally between 350 and 700 °C) under limited oxygen conditions [6–8]. It has stable aromatic C structures, low O and H to C ratios, low bulk density, moderate cation exchange capacity (CEC), and high ash content, pH and surface area [9,10]. Due to these properties, it is recognized as a multifunctional material which can be applied for long-term C sequestration and climate change mitigation [11,12]. It can be used successfully in agriculture to increase the organic C content of soil and reduce the leaching of nutrients; it could have positive effects on soils contaminated with heavy metals and organic pollutants [6,12–15]. Furthermore, biochar can be useful as soil amendment, due to its high porosity and sorption capacity and large surface area, reducing soil bulk density, improving soil structure, and increasing soil water holding capacity [7]. Biochar application can increase plant root growth, root penetration, and nutrient and water uptake [16]. In fact, N concentration in soil is increased by biochar during critical stages of plant growth, and N uptake and fertilizer recovery from roots are mostly guaranteed [7,17]. Biochar amendment also significantly enhances microbial activity and abundance, probably due to a greater adsorption of various nutrients [18,19]. Such positive influence could have effects on soil structure and indirectly on plant growth and rooting patterns [20].

Numerous studies have been conducted in controlled environments to evaluate the impact of biochar on soil properties and plant growth. However, the effects are not always positive: benefits on soil and plant growth strongly depend on the feedstock sources used to produce the biochar, production conditions (pyrolysis temperature, heating rate and residence time), differences in soil properties and specie-specific root growth patterns [15,21–23]. Therefore, this study aims to assess the potential ability of BSG-derived biochar for soil amendment and its influence on hop plant growth.

2. Materials and Methods

2.1. Biochar Production and Characterization

A fifty kilograms sample of wet BSG was chosen randomly from a big container at a craft brewery. Subsequently, three replicates of about 1 kg each were selected randomly from this sample, oven-dried at 105 ± 1 °C for 78 h and successively subjected to elemental analysis for the determination of the CHN content, according to UNI EN 15104: 2011. Ash content was analyzed by heating ground dried BSG samples (maximum grain size 2 mm) at 550 °C ± 1 °C for 8 h in a muffle furnace, according to UNI EN 14775. 2010. Three analytical replicates were performed for all parameters analyzed and for each sample.

Biochar was prepared from BSG using the pyrolytic reactor Elsa D17 (pat. BLUECOMB) and through a thermochemical process that can reach carbonization temperatures of 400–500 °C [24,25].

Biochar was chemically characterized by Fourier transformed infrared spectroscopy analysis (FT-IR), according to the methodology described by Amoriello et al. [26]. Spectra were collected at room temperature with a FT-IR spectrometer (iS 50 FT-IR Nicolet Thermo Fisher Scientific Inc., Waltham, MA, USA) equipped with a single-reflection horizontal ATR cell with a diamond crystal [26].

2.2. Botanical and Agronomic Characteristics of Hop Plant

Hop (*Humulus lupulus* L.) is a dioecious perennial plant, belonging to the *Cannabaceae* family. It reaches maturity after the first three years and remains productive for over 20–25 years [27]. The hop plant consists of a perennial rootstock ("crown") of rhizomes below ground, annual climbing bines above ground, which provide the canopy and photosynthetic capacity to support flowering, and flowers that develop at the terminal buds of lateral branches and are harvested as green cones. The perennial root system can grow more than 4 m deep and up to 5 m laterally. Rhizomes feed the growth of the productive canopy and ensure the survival of the plant from successive seasons. The growing season spans from March to August–September in the northern hemisphere. Hops emerge in early spring and grow up to a height of 5–7 m on poles or under a trellis system. Flowering starts in late June or early July in the northern hemisphere. The cones mature for picking between August and September, depending on climate conditions and genotype. The female strobiles (cones) represent the most interesting parts of the plant from a technological point of view, as they are one of the essential ingredients of the brewing industry, providing aroma, bitterness, flavor, and antimicrobial properties to beer [28].

2.3. Experimental Site and Design

The study was carried out in Rieti, Italy (latitude 42°24′29″52 N, longitude 12°51′36″36 E) during the period March–August 2017. The sampling site is characterized by relatively cold and rainy winters and hot and dry summers. The long-term average annual temperature recorded between 1980 and 2009 was 12.1 °C; the mean temperature during the growing season was 16.0 °C; the average maximum temperature for the summer months was 30.1 °C; the average minimum temperature for the coldest months was −1.6 °C. In summer, maximum temperatures were often over 30 °C. The annual mean precipitation was equal to 1021 mm and it fell mainly from October to December. Summer rainfall was irregularly distributed and the total mean amount over June–August was 136 mm. In 2017, the annual average temperature was 12.7 °C; the average temperature in the growing season was 17.4 °C; the average maximum temperature for the hottest months (July to August) was 33.2 °C. The annual precipitation, precipitation for the growing season, and precipitation over the hottest months was 721 mm, 186 mm, and 27 mm, respectively. The annual relative humidity was 68%.

A completely randomized experimental design was conducted during the hop growing season in a climate uncontrolled greenhouse environment from March to 5 May and outside from 6 May to 28 August. Three different cultivars of German hop plant (Hallertau Magnum, Perle, Spalter spalt) were considered for the study due to their economic importance and their sensitivity to environmental conditions [27]. These three varieties grow well in most climates, but best in warm, dry and sunny regions. Hallertau Magnum produces very good yields (1700–2300 kg ha^{-1}); the cone's structure is very large and longish size. Perle also produces good yields (1600–2100 kg ha^{-1}); cone size is small to medium. Spalter spalt yield amounts 1750–2000 kg ha^{-1}; the cone size is small to medium.

In March, hop rhizomes were placed in plastic pots (15 cm × 14 cm) with soil containing acid peat, expanded perlite and clay. Each pot contained 8 L of soil. The packed pot soil had a pH of 6.0, a dry bulk density of 1.1 g cm^{-3}, an electric conductivity of 0.45 dS m^{-1} and a porosity of 90% *v/v*. Hop rhizomes for each variety were randomly assigned between pots with soil added with 20% of biochar (1.6 L of biochar and 6.4 L of soil containing acid peat, expanded perlite and clay) and pots without biochar (control). Trials with biochar were replicated seven times, whereas controls (without biochar) were run in triplicate. We chose an unbalanced sampling to better evaluate the variability of the biochar effect on hop plants. Young plants were transferred from pots to the field at the beginning of May, corresponding to the moment of bines elongation. No biochar was added in field. Until June, the plants were irrigated with 0.5 L of water every two days; in the two warmest months (July and August), the same amount of water was given every day.

Plant development was monitored through the percentage of roots in pot (as described in the Section 2.4) and the number of leaves at the beginning of May; shoot diameter was measured by caliper at five different times (T1: 30 June, T2: 18 July, T3: 27 July, T4: 3 November, T5: 21 November), i.e., before flowering, near cones maturity, before plant dormancy; measurement of climbing bines length and number of cones at maturity.

2.4. Statistical Analysis and Multivariate Image Analysis

In order to quantify the effect of biochar on root growth for each genotype, a non-destructive technique, e.g., a multivariate image analysis, was carried out according to Fongaro et al. [29]. At the beginning of May, plants were extracted from pots and the images of roots and soil for each plant were acquired twice using a digital camera Nikon D750 at a high resolution and a color depth of 16 bits, saving the captured images in uncompressed RAW format. The images were acquired by photographing the basis and four external opposite sides. To create the final data set, a region of interest (ROI) of 472 × 472 pixels, representative of the whole sample surface, was extracted from each image using the Lightroom Classic CC 2018.

The images were processed by PLS Toolbox 8.5 (Eigenvectors Research, Inc., Manson, WA, USA) for MATLAB R2016b (The MathWorks Inc., Natick, MA, USA). Red (R), green (G), blue (B) values, measures in the RGB space color, were calculated. A principal component analysis (PCA) on the RGB values was carried out, providing the pixel distribution in the score space (score plot). The first principal component (PC1) contained most of the original information, with a decreasing amount in the remaining score images. PC1 score image. PC1 score image was converted from the RGB space color into the HSI (hue, saturation, intensity) space color. Then, the Hue channel was extracted to obtain the hue image. A pixel segmentation in different intensity ranges, corresponding on different material (soil and root), was carried out to select roots. In this way, it was possible to quantify the percentage of roots present in each pot and for each cultivar.

The influence of genotype, treatment and the two-way interaction on roots percentage or number of leaves or bines length or number of cones was tested using the technique of General Linear Models (GLM). The statistical model was:

$$Y_{ij} = \beta_0 + \beta_1 \cdot \tau_i + \beta_2 \cdot \delta_j + \beta_3\, \tau_i \cdot \delta_j + \varepsilon_{ij} \tag{1}$$

where Y_{ij} = roots percentage or number of leaves or bines length or number of cones, β_0 = mean effect common to all observations; $\beta_{1,2,3}$ = unknown regression parameters; τ_i = treatment ($i = 1, 2; 1$ = with biochar, 2 = without biochar); δ_j = cultivar ($j = 1, 2, 3; 1$ = Hallertau Magnum, 2 = Perle, 3 = Spalter spalt); ε_{ij} = error term.

GLM was also applied to monitor the plant development in time. The statistical model was:

$$Y_{ijk} = \beta_0 + \beta_1\, \tau_i + \beta_2\, \delta_j + \beta_3\, \gamma_k + \beta_4\, \tau_i \cdot \delta_j + \beta_5 \tau_i \cdot \gamma_k + \beta_6\, \delta_j \cdot \gamma_k + \beta_7\, \tau_i \cdot \delta_j \cdot \gamma_k + \varepsilon_{ijk} \tag{2}$$

where Y_{ijk} = diameter of shoots, β_0 = mean effect common to all observations; $\beta_{1\text{-}7}$ = unknown regression parameters; τ_i = treatment ($i = 1, 2; 1$ = with biochar, 2 = without biochar); δ_j = cultivar ($j = 1, 2, 3; 1$ = Hallertau Magnum, 2 = Perle, 3 = Spalter spalt); γ_k = time ($k = 1, \ldots ,5; 1$ = 30 June, 2 = 18 July, 3 = 27 July, 4 = 3 November, 5 = 21 November); ε_{ijk} = error term.

Post hoc Tamhane test ($p < 0.05$) was carried out in order to evaluate differences between groups (with or without biochar) for each considered variable. Statistical analysis was performed with SPSS 20.0 software (SPSS, Inc., Chicago, IL, USA).

3. Results

3.1. Brewers' Spent Grain and Biochar Characterization

BSG analysis showed an average moisture content of 8.8%, and a content of ashes, C, H, N contents of 5.3%, 45.7%, 9.0% and 4.2%, respectively, on the dry weight basis.

Brewers' spent grain (BSG) and biochar were characterized by FT-IR spectra (Figure 1).

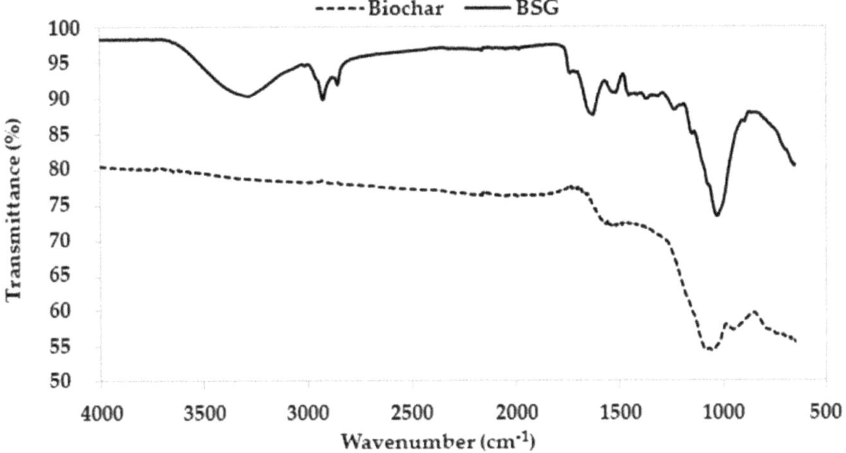

Figure 1. FT-IR spectra of brewers' spent grain (BSG) and biochar.

The two spectra showed different profiles and changes of functional groups, demonstrating the influence of the pyrolytic treatment on the chemical structure of the raw material. In fact, the organic matter pyrolysis caused water loss and variation in concentration of mineral components due to the heat-induced mass loss [30]. BSG profile showed a peak at 3274 cm^{-1} corresponding to a vibrational band consistent with hydroxyl groups (O-H); two peaks at 2926 and 2853 cm^{-1}, corresponding to aliphatic C-H stretch; a peak at 1742 cm^{-1}, associated with the presence of aldehydes, ketones and carboxyl groups (C=O); a band with a peak at 1633 cm^{-1}, assigned to aromatic lignin components (C=C); a peak at 1030 cm^{-1}, corresponding to C-O-C stretch. The thermal decomposition strongly influenced the intensity of biochar bands from 4000 to 1600 cm^{-1} (carboxylic bonds, amides and aliphatic hydrocarbon) [30]. In particular, the disappearance of the peaks at 2926 and 2853 cm^{-1} may be due to the fact that methyl groups, which are the weakest functional groups, as well as the OH groups, break during pyrolysis. The CH peaks shift from being more aliphatic to more aromatic (and eventually disappear altogether) [31]. The cleavage of these groups contributes to mass loss during thermal decomposition and to the production of "non-condensable" gas. Moderate differences between BSG and biochar profiles from 1600 to 1500 cm^{-1} may be attributed to the formation of carbonate-carboxyl group during pyrolysis. At last, biochar showed similar functional groups, between 600 and 1500 cm^{-1}. These results were in accordance with previous studies on BSG-derived biochar [32,33]. The FT-IR analysis highlighted the presence of functional groups such as hydroxyl and carboxyl groups, necessary to consider biochar a soil amendment for improving of the cation exchange capacity and as a potential adsorbent [34].

3.2. Multivariate Image Analysis

Multivariate image analysis was applied to extract roots region and to quantify differences in percentage for the two treatments, with and without biochar. Representative images from roots and soil samples for each cultivar and treatment after RGB processing and PCA score images are shown in Figure 2.

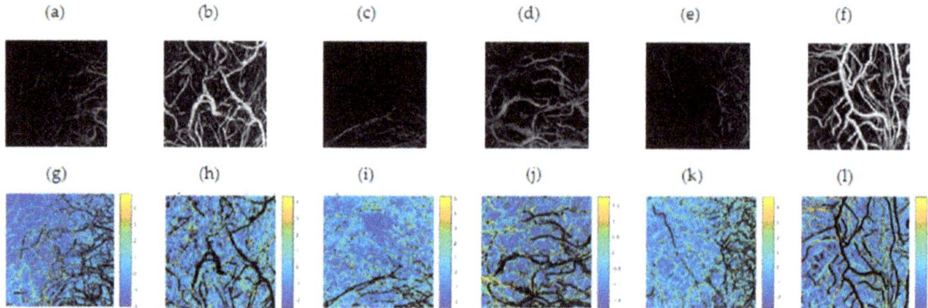

Figure 2. RGB and PCA score images of roots and soil samples representative of different cultivars and treatments, obtained from image analysis process. (**a**) RGB image of Hallertau Magnum without biochar; (**b**) RGB image of Hallertau Magnum with biochar; (**c**) RGB image of Perle without biochar; (**d**) RGB image of Perle with biochar; (**e**) RGB image of Spalter Spalt without biochar; (**f**) RGB image of Spalter Spalt with biochar; (**g**) PCA score image of Hallertau Magnum without biochar; (**h**) PCA score image of Hallertau Magnum with biochar; (**i**) PCA score image of Perle without biochar; (**j**) PCA score image of Perle with biochar; (**k**) PCA score image of Spalter Spalt without biochar; (**l**) PCA score image of Spalter Spalt with biochar.

For all the images, the PC1 score image explained all the variance, being related to pixels of both the root and soil. Comparing the PC1 score images and the relative RGB images, it was possible to distinguish areas with different color intensity, indicated with blue (soil) and yellow (root). Thus, the same false color was assigned to the pixels having the same characteristics, well distinguishing the two regions in the PC1 score images. The pixel segmentation allowed the extraction of the root areas (in black) and the quantification of the roots percentages. Results obtained were plotted in Figure 3. Box plots showed a strong influence on root growth present in the pots with biochar in comparison to those without biochar.

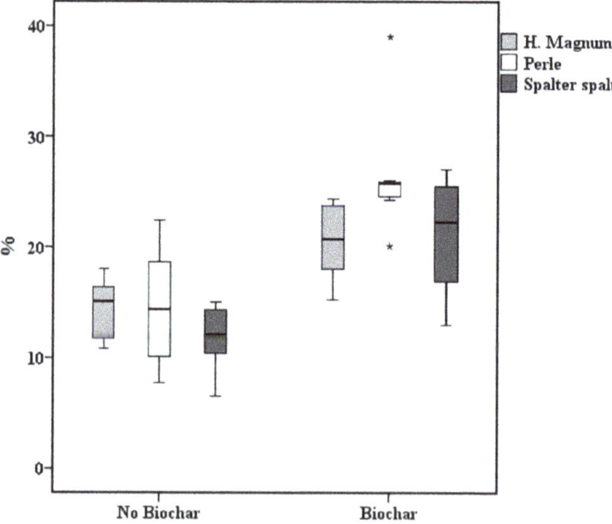

Figure 3. Box plots of root percentages of three different hop cultivars (Hallertau Magnum, Perle and Spalter spalt) and two treatments (presence or absence of biochar). Outliers are displayed as filled stars.

3.3. Assessment of Hop Plant Response to Biochar Amendment

The evaluation of effect of BSG-derived biochar was carried out using the technique of General Linear Models (GLM). The GLM analysis of roots percentage showed that the two single factors (genotype and treatment) were statistically significant, unlike the two-way interactions between cultivar and treatment. The treatment was the main contributing factor to the percentage of roots ($p < 0.0001$), followed by cultivar ($p = 0.0330$). In fact, as clearly shown in Figure 3, the addition of biochar in the soil produces a greater development of the roots. This phenomenon is even more evident for the Perle cultivar.

Regarding shoot development, specifically the shoot diameter, no variability ascribable to the genotypes was observed for the first three times (Table 1).

Table 1. Measurements of diameter of shoots (mean ± standard deviation) at five different times (T1: 30 June, T2: 18 July, T3: 27 July, T4: 3 November, T5: 21 November). Diameters within columns of each cultivar and treatment followed by different letters (a or b) are significantly different.

Time	Hallertau Magnum		Perle		Spalter Spalt	
	No Biochar	Biochar	No Biochar	Biochar	No Biochar	Biochar
T1	7 ± 2 a	8 ± 2 a	8 ± 3 a	6 ± 1 a	8 ± 3 a	5 ± 1 a
T2	8 ± 3 a	9 ± 2 a	9 ± 1 a	7 ± 1 a	8 ± 3 a	7 ± 2 a
T3	9 ± 2 a	10 ± 2 a	10 ± 1 a	8 ± 1 a	9 ± 2 a	8 ± 2 a
T4	15 ± 3 b	15 ± 3 b	11 ± 1 a	12 ± 2 b	11 ± 1 a	12 ± 3 ab
T5	16 ± 2 b	16 ± 4 b	11 ± 1 a	13 ± 4 b	11 ± 1 a	13 ± 3 b

On the contrary, significant but slight differences resulted among diameters measured before plant dormancy, especially for the Hallertau Magnum cultivar. It is interesting to note that the Perle and Spalter Spalt cultivars tended to be more affected by the presence of biochar during the last phase of plant development, although these differences are not statistically significant. This trend was confirmed by the GLM analysis: only the single genotype and time factors were highly significant ($p < 0.0001$), whereas the single treatment factor, all the two-way interactions and the three-way interaction were not statistically significant.

The GLM analysis on number of leaves at the beginning of May and climbing bines length at harvest highlighted no significant differences between cultivar and treatment and their two-way interaction, as shown in Table 2.

Table 2. Measurements of number of leaves (mean ± standard deviation) at the beginning of May, and climbing bines length and number of cones at maturity. Statistical differences between treatments (with biochar or without biochar) for each cultivar and parameter are displayed with different letters (a or b).

	Hallertau Magnum		Perle		Spalter Spalt	
	No Biochar	Biochar	No Biochar	Biochar	No Biochar	Biochar
Number of leaves	20 ± 8 a	18 ± 6 a	19 ± 4 a	15 ± 3 a	26 ± 14 a	25 ± 14 a
Bines length (cm)	547 ± 55 a	570 ± 30 a	552 ± 27 a	532 ± 61 a	555 ± 94 a	550 ± 75 a
Number of cones	97 ± 17 a	144 ± 20 b	84 ± 14 a	123 ± 18 b	75 ± 12 a	106 ± 16 b

As regards the number of harvested cones, statistical differences ($p < 0.0001$) were observed between treatments for each cultivar: the cones of the plants treated with biochar resulted more numerous compared to those of plants without biochar (Table 2). Instead, the genotype factor and the two-way cultivar-treatment interaction resulted not significant by the GLM analysis.

4. Discussion

Despite the high adaptability to different environmental conditions, hop cultivars of European origin are more susceptible to changes in pedo-climatic conditions compared to the American genotypes [27]. Hops require a large amount of water during the growing seasons to optimize yield and quality. Furthermore, soil has to be nutrient-rich, light, well drained, well supplied with moisture, but free from waterlogging, with an optimal pH range between 6.0 and 6.5., although hops also grow on soils with pH from 4.8 to 8.0 [27].

Our study wanted to verify if biochar can positively affect hop growth. As mentioned above, biochar can improve soil water retention, cation exchange capacity, the content of different nutrients, and reduce soil bulk density and N leaching [7]. Moreover, biochar can enhance the uptake of N and other nutrients by improving the root development and the whole plant physiological status [17]. Then, biochar may reduce plant uptake of potentially toxic elements (such as Pb, Cd and As) from soils, due to the adsorption of metals to the negative charges on the surface of biochar particles [35]. However, the efficacy of biochar application depends on various factors concerning both soil and plant. Understanding these factors is essential for maximizing soil productivity and minimizing potential deleterious environmental effects. For example, the positive effect of biochar amendment is more remarkable in a coarse-textured than in a fine-textured soil, and sandy soils are more responsive than clay-rich soils; this positive effect could be also maximized reducing biochar particle sizes in the range of 0.5–1.0 mm [19]. The biomass feedstock could influence the hydraulic properties of biochar: biochar with high porosity and surface area can improve water adsorption [36].

Previous studies have shown discordant experimental evidences about the supposed beneficial effects of biochar on root growth. Numerous papers reported positive responses on root growth parameters [37–41]. On the contrary, other authors [39,42,43] showed negligible benefits on root growth. These discrepancies could depend on different factors: plant species, growing environment, soil properties, feedstock, biochar properties, pyrolysis conditions [15,21–23]. Furthermore, improved concentrations of dissolved organic C after biochar addition and the adsorption of phenolic compounds by biochar may positively affect root growth [38,43]. Moreover, biochar may stimulate a faster root turnover and changes in root morphology [43]. Although there are few detailed studies on the influence of biochar on the root system, it is generally recognized that plants with a longer root length, a larger surface area and more root tips may be able to get more nutrients and grow better [44–46]. In the present study, the addition of biochar amendment positively affected hop root growth, increasing root length and mass density, as well represented in Figure 2. Moreover, a higher presence of roots could be positively correlated with nutrient uptake, and induce potential benefits in plant growth and development, including aboveground biomass. Benefits and limitations of biochar on plant growth could depend on the types of biochar and the incorporation rates [47–49]. In our study, no statistical differences were observed between number of leaves at the moment of bines elongation and climbing bines length at harvest.

Regarding hop yield, biochar seemed to have positively affected cones production. This production varies greatly during the first years of the plant vegetation. In fact, the hop plant reaches productive and qualitative stability only after the third year. Nevertheless, biochar could have helped the plant when the water requirement was greater. In fact, the summer period, characterized by dry weather conditions, caused long periods of water stress, compensated only by emergency irrigation. Hops require a high water availability for successful plant growth and cones production [27], and changes in soil moisture retention may have been one of the most important factor in explaining positive biochar effects on crop yield. Due to its high porosity and specific surface area, biochar increases soil porosity, and, consequently, the soil hydraulic properties, such as the water storage capacity and the absorption capacity [42]. Furthermore, the induced macro-porosity due to the larger particle size of biochar positively affects soil water permeability [50].

Based on our findings, the application of biochar could be a promising strategy to enhance hop plant growth and cones production. However, further investigation on plant physiological mechanisms and hop cones quality is required to identify optimal levels of biochar application and to better understand the effect of biochar on hop plant growth.

5. Conclusions

The present study was conducted to assess the effect of BSG-derived biochar on the hop plant development. The results reinforce the hypothesis that biochar can be a promising and effective soil amendment for hop crops, supplying key nutrients for the plant growth and improving soil properties.

The recovery and reuse of the brewing industry by-products is a good approach of circularity in this industrial sector from the environment protection and waste management standpoint. In fact, the reuse of brewers' spent grain, as feedstock to obtain biochar, will contribute to the adoption of environmentally sustainable practices and a more efficient use of resources, also reducing the use of chemical input. Moreover, it can contribute to decrease the negative economic and environmental impact of the disposal of by-products, decreasing the CO_2 production, and increasing the sustainability of the farming sector.

Author Contributions: All authors conceptualized and designed the experiments. M.P., V.V. and S.F. collected the data. T.A. performed the statistical analysis and image analysis. T.A. wrote the manuscript in consultation with all other authors. M.P., V.V. and S.F. revised and improved the manuscript. All authors have read and agreed to the published version of the manuscript.

Funding: This research received no external funding.

Acknowledgments: The valuable technical assistance of Marco Castellani (Department of Agricultural and Forestry Sciences at the University of Tuscia) is gratefully acknowledge. The authors also thank Paola Manzi (CREA Research Centre for Food and Nutrition) for providing the FT-IR instrument.

Conflicts of Interest: The authors declare no conflict of interest.

References

1. Mussatto, S.I.; Dragone, G.; Roberto, I.C. Brewers'spent grain: Generation, characteristics and potential applications. *J. Cereal Sci.* **2006**, *43*, 1–14. [CrossRef]
2. Lynch, K.M.; Steffen, E.J.; Arendt, E.K. Brewers' spent grain: A review with emphasis on food and health. *J. Inst. Brew.* **2016**, *122*, 553–568. [CrossRef]
3. Kanauchi, O.; Mitsuyama, K.; Araki, Y. Development of a functional germinated barley foodstuff from brewers' spent grain for the treatment of ulcerative colitis. *J. Am. Soc. Brew. Chem.* **2001**, *59*, 59–62.
4. Ferreira, S.; Monteiro, E.; Brito, P.; Castro, C.; Calado, L.; Vilarinho, C. Experimental Analysis of Brewers' Spent Grains Steam Gasification in an Allothermal Batch Reactor. *Energies* **2019**, *12*, 912. [CrossRef]
5. Skendi, A.; Harasym, J.; Galanakis, C.M. Recovery of High Added-Value Compounds from Brewing and Distillate By-Products. In *Sustainable Recovery and Reutilization of Cereal Processing By-Products*; Galanakis, C.M., Ed.; Elsevier: Amsterdam, The Netherlands, 2018; pp. 189–214.
6. Harter, J.; Krause, H.M.; Schuettler, S.; Ruser, R.; Fromme, M.; Scholten, T.; Kappler, A.; Behrens, S. Linking N_2O emissions from biochar-amended soil to the structureand function of the N-cycling microbial community. *ISME J.* **2014**, *8*, 660–674. [CrossRef]
7. Cao, H.; Ning, L.; Xun, M.; Feng, F.; Li, P.; Yue, S.; Song, J.; Zhang, W.; Yang, H. Biochar can increase nitrogen use efficiency of *Malus hupehensis* by modulating nitrate reduction of soil and root. *Appl. Soil Ecol.* **2019**, *135*, 25–32. [CrossRef]
8. Meyer, S.; Glaser, B.; Quicker, P. Technical, economical, and climate-related aspects of biochar production technologies: A literature review. *Environ. Sci. Technol.* **2011**, *45*, 9473–9483. [CrossRef]
9. Lehmann, J. Bio-energy in the black. *Front. Ecol. Environ.* **2007**, *5*, 381–387. [CrossRef]
10. Field, J.L.; Keske, C.M.H.; Birch, G.L.; Defoort, M.W.; Cotrufo, M.F. Distributed biochar and bioenergy coproduction: A regionally specific case study of environmental benefits and economic impacts. *GCB Bioenergy* **2013**, *5*, 177–191. [CrossRef]
11. Atkinsons, C.J.; Fitzgerald, J.D.; Hipps, N.A. Potential mechanism for achieving agricultural benefits from biochar application to temperate soils: A review. *Plant Soil* **2010**, *337*, 1–18. [CrossRef]
12. Cayuela, M.L.; Van Zwieten, L.; Singh, B.P.; Jeffery, S.; Roig, A.; Sanchez-Monedero, M.A. Biochar's role in mitigation soil nitrous oxide emissions: A review and meta-analysis. *Agric. Ecosyst. Environ.* **2014**, *191*, 5–16. [CrossRef]

13. Basso, A.S.; Miguez, F.E.; Laird, D.A.; Horton, R.; Westgate, M. Assessing potential of biochar for increasing water-holding capacity of sandy soils. *GCB Bioenergy* **2013**, *5*, 132–143. [CrossRef]

14. Lehmann, J.; Joseph, S. *Biochar for Environmental Management: Science and Technology*; Earthscan: London, UK, 2009; pp. 13–32.

15. Wu, S.; He, H.; Inthapanya, X.; Yang, C.; Lu, L.; Zeng, G.; Han, Z. Role of biochar on composting of organic wastes and remediation of contaminated soils—A review. *Environ. Sci. Pollut. Res.* **2017**, *24*, 16560–16577. [CrossRef] [PubMed]

16. Sandhu, S.S.; Ussiri, D.A.N.; Kumar, S.; Chintala, R.; Papiernik, S.K.; Malo, D.D.; Schumacher, T.E. Analyzing the impacts of three types of biochar on soil carbon fractions and physiochemical properties in a corn-soybean rotation. *Chemosphere* **2017**, *184*, 473–481. [CrossRef]

17. Backer, R.; Saeed, W.; Seguin, P.; Smith, D.L. Root traits and nitrogen fertilizer recovery efficiency of corn grown in biochar-amended soil under greenhouse conditions. *Plant Soil* **2017**, *415*, 465–477. [CrossRef]

18. Kookana, R.K.; Sarmah, A.; Van Zwieten, L.; Van Krull, E.; Singh, B. Biochar application to soil: Agronomic and environmental benefits and unintended consequences. *Adv. Agron.* **2011**, *112*, 103–143.

19. Kavitha, B.; Reddy, P.V.L.; Kim, B.; Lee, S.S.; Pandey, S.K.; Kim, H.H. Benefits and limitations of biochar amendment in agricultural soils: A review. *J. Environ. Manag.* **2018**, *227*, 146–154. [CrossRef]

20. Ding, Y.; Liu, Y.; Liu, S.; Li, Z.; Tan, X.; Huang, X.; Zeng, G.; Zhou, L.; Zheng, B. Biochar to improve soil fertility. A review. *Agron. Sustain. Dev.* **2016**, *36*, 36. [CrossRef]

21. Ahmad, M.; Rajapaksha, A.U.; Lim, J.E.; Zhang, M.; Bolan, N.; Mohan, D.; Vithanage, M.; Lee, S.S.; Ok, Y.S. Biochar as a sorbent for contaminant management in soil and water: A review. *Chemosphere* **2014**, *99*, 19–33. [CrossRef]

22. O'Connor, D.; Peng, T.; Zhang, J.; Tsang, D.C.W.; Alessi, D.S.; Shen, Z.; Bolan, N.S.; Hou, D. Biochar application for the remediation of heavy metal polluted land: A review of in situ field trials. *Sci. Total Environ.* **2018**, *619–620*, 815–826. [CrossRef]

23. Zhao, L.; Cao, X.; Mašek, O.; Zimmerman, A. Heterogeneity of biochar properties as a function of feedstock sources and production temperatures. *J. Hazard. Mater.* **2013**, *256–257*, 1–9. [CrossRef] [PubMed]

24. Sorgonà, A.; Longo, L.; Proto, A.R.; Cavalletti, P.; Cecchini, M.; Salvati, L.; Gallucci, F.; Colantoni, A. Characterization of biochar and syngas obtained from pellets of grape vine and sun flower husk using a pyrolysis system. *Procedia Soc. Behav. Sci.* **2016**, *223*, 871–878. [CrossRef]

25. Sperandio, G.; Amoriello, T.; Carbone, K.; Fedrizzi, M.; Monteleone, A.; Tarangioli, S.; Pagano, M. Increasing the Value of Spent Grain from Craft Microbreweries for Energy Purposes. *Chem. Eng. Trans.* **2017**, *58*, 487–492.

26. Amoriello, T.; Ciccoritti, R.; Carbone, K. Vibrational spectroscopy as a green technology for predicting nutraceutical properties and antiradical potential of early-to-late apricot genotypes. *Postharvest Biol. Technol.* **2019**, *155*, 156–166. [CrossRef]

27. Amoriello, T. Multi-criteria approach for land suitability assessment of hop cultivation in Italy. *Int. J. Agric. Environ. Res.* **2019**, *5*, 277–286.

28. Gargani, E.; Ferretti, L.; Faggioli, F.; Haegi, A.; Luigi, M.; Landi, S.; Simoni, S.; Benvenuti, C.; Guidi, S.; Simoncini, S.; et al. A survey on pests and diseases of Italian Hop crops. *Italus Hortus* **2017**, *24*, 1–17.

29. Fongaro, L.; Alamprese, C.; Casiraghi, E. Ripening of salami: Assessment of colour and aspect evolution using image analysis and multivariate image analysis. *Meat Sci.* **2015**, *101*, 73–77. [CrossRef]

30. Cao, X.; Harris, W. Properties of dairy-manure-derived biochar pertinent to its potential use in remediation. *Bioresour. Technol.* **2010**, *101*, 5222–5228. [CrossRef]

31. Hossain, M.K.; Strezov, V.; Chan, K.Y.; Ziolkowski, A.; Nelson, P.F. Influence of pyrolysis temperature on production and nutrient properties of wastewater sludge biochar. *J. Environ. Manag.* **2011**, *92*, 223–228. [CrossRef]

32. Zhang, J.; Wang, Q. Sustainable mechanisms of biochar derived from brewers' spent grain and sewage sludge for ammonia—Nitrogen capture. *J. Clean. Prod.* **2016**, *112*, 3927–3934. [CrossRef]

33. Franciski, M.A.; Peres, E.C.; Godinho, M.; Perondi, D.; Foletto, E.L.; Collazzo, G.C.; Dotto, G.L. Development of CO_2 activated biochar from solid wastes of a beer industry and its application for methylene blue adsorption. *Waste Manag.* **2018**, *78*, 630–638. [CrossRef]

34. Stella Mary, G.; Sugumaran, P.; Niveditha, S.; Ramalakshmi, B.; Ravichandran, P.; Seshadri, S. Production, characterization and evaluation of biochar from pod (*Pisum sativum*), leaf (*Brassica oleracea*) and peel (*Citrus sinensis*) wastes. *Int. J. Recycl. Org. Waste Agric.* **2016**, *5*, 43–53. [CrossRef]

35. Buss, W.; Kammann, C.; Koyro, H. Biochar reduces copper toxicity in chenopodium quinoa willd. in a sandy soil. *J. Environ. Qual.* **2012**, *41*, 1157–1165. [CrossRef] [PubMed]

36. Lei, O.; Zhang, R. Effects of biochars derived from different feedstocks and pyrolysis temperatures on soil physical and hydraulic properties. *J. Soils Sediments* **2013**, *13*, 1561–1572. [CrossRef]

37. Brennan, A.; Jiménez, E.M.; Puschenreiter, M.; Alburquerque, J.A.; Switzer, C. Effects of biochar amendment on root traits and contaminant availability of maize plants in a copper and arsenic impacted soil. *Plant Soil* **2014**, *379*, 351–360. [CrossRef]

38. Donn, S.; Wheatley, R.E.; McKenzie, B.M.; Loades, K.W.; Hallett, P.D. Improved soil fertility from compost amendment increases root growth and reinforcement of surface soil on slopes. *Ecol. Eng.* **2014**, *71*, 458–465. [CrossRef]

39. Lehmann, J.; Rillig, M.C.; Thies, J.; Masiello, C.A.; Hockaday, W.C.; Crowley, D. Biochar effects on soil biota—A review. *Soil Biol. Biochem.* **2011**, *43*, 1812–1836. [CrossRef]

40. Jindo, K.; Martim, S.A.; Navarro, E.C.; Pérez-Alfocea, F.; Hernandez, T.; Garcia, C.; Oliveira Aguiar, N.; Canellas, L.P. Root growth promotion by humic acids from composted and non-composted urban organic wastes. *Plant Soil* **2012**, *353*, 209–220. [CrossRef]

41. Wang, Y.; Ma, Z.; Wang, X.; Sun, Q.; Dong, H.; Wang, G.; Chen, X.; Yin, C.; Han, Z.; Mao, Z. Effects of biochar on the growth of apple seedlings, soil enzyme activities and fungal communities in replant disease soil. *Sci. Hortic.* **2019**, *256*, 108641. [CrossRef]

42. Liu, C.; Liu, F.; Ravnskov, S.; Rubæk, G.H.; Sun, Z.; Andersen, M.N. Impact of Wood Biochar and Its Interactions with Mycorrhizal Fungi, Phosphorus Fertilization and Irrigation Strategies on Potato Growth. *J. Agron. Crop Sci.* **2016**, *203*, 131–145. [CrossRef]

43. Sorrenti, G.; Muzzi, E.; Toselli, M. Root growth dynamic and plant performance of nectarine trees amended with biochar and compost. *Sci. Hortic.* **2019**, *257*, 108710. [CrossRef]

44. Prendergast-Miller, M.; Duvall, M.; Sohi, S. Biochar–root interactions are mediated by biochar nutrient content and impacts on soil nutrient availability. *Eur. J. Soil Sci.* **2014**, *65*, 173–185. [CrossRef]

45. Rellán-Álvarez, R.; Lobet, G.; Dinneny, J.R. Environmental control of root system biology. *Annu. Rev. Plant Biol.* **2016**, *67*, 619–642. [CrossRef] [PubMed]

46. Yu, P.; Li, Q.; Huang, L.; Niu, G.; Gu, M. Mixed Hardwood and Sugarcane Bagasse Biochar as Potting Mix Components for Container Tomato and Basil Seedling Production. *Appl. Sci.* **2019**, *9*, 4713. [CrossRef]

47. Vaughn, S.F.; Eller, F.J.; Evangelista, R.L.; Moser, B.R.; Lee, E.; Wagner, R.E.; Peterson, S.C. Evaluation of biochar-anaerobic potato digestate mixtures as renewable components of horticultural potting media. *Ind. Crop. Prod.* **2015**, *65*, 467–471. [CrossRef]

48. Dunlop, S.J.; Arbestain, M.C.; Bishop, P.A.; Wargent, J.J. Closing the loop: Use of biochar produced from tomato crop green waste as a substrate for soilless, hydroponic tomato production. *HortScience* **2015**, *50*, 1572–1581. [CrossRef]

49. Guo, Y.; Niu, G.; Starman, T.; Volder, A.; Gu, M. Poinsettia growth and development response to container root substrate with biochar. *Horticulturae* **2018**, *4*, 1. [CrossRef]

50. Kumari, K.G.I.D.I.; Moldrup, P.; Paradelo, M.; Elsgaard, L.; Hauggaard-Nielsen, H.; de Jonge, L.W. Effects of Biochar on Air and Water Permeability and Colloid and Phosphorus Leaching in Soils from a Natural Calcium Carbonate Gradient. *J. Environ. Qual.* **2014**, *43*, 647–657. [CrossRef]

Article

Organic Vegetable Crops Managed with Agro-Ecological Practices: Environmental Sustainability Assessment by *DEXi-met* Decision Support System

Francesco Montemurro [1,*], **Alessandro Persiani** [2] **and Mariangela Diacono** [2]

[1] Consiglio per la Ricerca in Agricoltura e l'analisi dell'economia Agraria-Research Centre for Vegetable and Ornamental Crops, Via Salaria 1, 63030 Monsampolo del Tronto (AP), Italy

[2] Consiglio per la Ricerca e l'analisi dell'economia Agraria-Research Centre for Agriculture and Environment, Via Celso Ulpiani 5, 70125 Bari, Italy; alessandro.persiani@crea.gov.it (A.P.); mariangela.diacono@crea.gov.it (M.D.)

* Correspondence: francesco.montemurro@crea.gov.it

Received: 26 August 2019; Accepted: 1 October 2019; Published: 3 October 2019

Abstract: In the last decade, there has been an increasing interest in sustainable agricultural techniques and the environmental evaluation of the effects of agricultural practices. In the present study, we evaluated both the production capacity of organic horticultural systems, and the ex-post sustainability through a new multi-attribute decision model named *"DEXi-met"*. This qualitative model is able to estimate the environmental sustainability of cropping systems managed with different agro-ecological approaches. In particular, we compared the following three horticultural systems: (i) ECO, an organic system with full implementation of agro-ecological strategies (agro-ecological services crops (ASC), strip cultivation, and organic amendment); (ii) GM, an organic system with the introduction of the ASC; (iii) NO ASC, an organic system without ASC. The treatments with ASC presence (ECO and GM) showed similar total energy outputs (substantially higher than the NO ASC), indicating the positive effect of this agro-ecological practice. The findings pointed out that the ECO system, which followed the principles of natural ecosystems, can contribute to building up more complex agro-ecosystems, increasing both resilience and biodiversity. This management strategy reached a good compromise between the production of vegetable cropping systems and environmental sustainability achievement. Then, it is possible to optimize the use of natural resources, support climate adaptation, and reduce greenhouse gas emissions.

Keywords: qualitative multi-attribute model; total energy output; agro-ecological service crops; ex-post sustainability; organic systems

1. Introduction

Sustainable development and environmental sustainability are broadly recognized as global and collective goals because of key issues such as limited resources, environmental pollution, and global warming [1]. These increasing challenges, also considering the local and the global legislative changes, have inevitably involved the agricultural sector. Therefore, agronomists, farmers, and researchers should research, design, and experiment with new agricultural systems that are environmentally friendly, economically viable, socially supportive, and efficiently adapted to a climate change context. In different farming systems, a wide range of cultivation techniques and agro-ecological management strategies that enhance biodiversity in crop fields and support the sustainability of the agro-ecosystems are already practiced and should be further promoted [2,3]. Among them, crop rotations, introduction of agro-ecological service crops (ASC; [4]), utilization of soil amendments [5], and crop/livestock mixing can increase agro-ecosystem diversity and complexity both over space and time [6].

A prerequisite to the systems sustainability implementation is the development, improvement, and/or choice of the best possible assessment methods. In fact, there is a need to determine the reliability of innovative management practices with respect to the conventional ones, and to clarify the benefits and the drawbacks of the full or partial application of an agro-ecological approach. To this end, several approaches to measure, analyze, and assess sustainability have been developed [7]. The different methodologies can be classified on the basis of: (i) the typology of indicators used, from qualitative appraisals to quantitative analytical evaluations [8]; (ii) the scale of analysis, from the single plot to the whole farm or regional scales [9]; (iii) the systems typology, from orchards [10] to arable or horticultural crops [11]; and the timing of analysis, as ex-ante or ex-post evaluation [12].

Within this large number of methodologies, a growing interest focuses on multi-method approaches, which aim at accounting for the complexity of sustainability issues [13]. In this framework, the multi-criteria analysis (MCA) decision-making methods can handle a typical decision-making problem related to sustainability assessment. The MCA are increasingly gaining importance in agriculture, since they can consider multiple and conflicting criteria and, at the same time, they are able to tackle complex decisional problems breaking them down in easily understandable elements [14–16]. In recent years, the scientific community has developed several qualitative MCA tools for the sustainability assessment of different agricultural systems [17,18], based on a computer program for multi-attribute decision-making, defined as "DEXi" by Bohanec et al. [19]. Among them, the *DEXi-met* tool was recently developed, specifically for the ex-post evaluation in organic horticulture, and it was applied to compare different crop rotations in Mediterranean conditions [20]. In this model, a cropping system was considered instead of a single cash crop, in order to have a broader idea of the environmental sustainability of the system.

There is still a lack of knowledge on the sustainability assessment when different levels of the agro-ecological approach are applied, especially in organic horticultural production in the Mediterranean environment. In light of these considerations, the aim of the present research was to evaluate the performance of different cultivation systems managed with agro-ecological practices. To accomplish this aim we evaluated both the production capacity of the systems and the ex-post environmental sustainability by using the *DEXi-met* qualitative multi-attribute model.

2. Materials and Methods

2.1. Study Site

The study was performed in the research farm 'Azienda Sperimentale Metaponto' of the Consiglio per la Ricerca in Agricoltura e l'analisi dell'economia Agraria. The farm is located at Metaponto (MT), in southern Italy (lat. 40°24' N; long. 16°48' E, 8 m above sea level).

The soil, classified as a Typic Epiaquert, has the following properties: low N (1.0 g kg^{-1}) and organic matter (19.0 g kg^{-1}) contents, 759 mg kg^{-1} of exchangeable potassium (K), 31.1 mg kg^{-1} of available phosphorus (P), pH value of 8.4, clay and silt contents of 60 and 36%, respectively, electrical conductivity of 0.48 mS cm^{-1} (at 0 to 30 cm depth), increasing with depth, average bulk density of 1350 kg m^{-3}, cation exchange capacity of 27.1 meq 100 g^{-1} of dry soil and the soil water content (a percentage of soil-dry weight) of 34.5% and 20.1% at field capacity (−0.03 MPa) and permanent wilting point (−1.5 MPa), respectively.

The climate is classified as "accentuated thermo-Mediterranean", considering the UNESCO-FAO classification [21], with mean monthly temperatures of 8.8 °C in the winter, and 24.4 °C in the summer. Winter temperatures can fall below 0 °C, whereas summer temperatures can rise above 40 °C. The total rainfall (on average 490 mm year^{-1}) is concentrated mainly during the winter months and the mean annual potential evaporation rate is 1549 mm.

2.2. Experimental Field Trials, Treatments, and Measurements

The research was carried out during the 2016–2017 cropping season in two different experimental fields. The first one consists in a long-term field trial in organic horticulture, which had been planned to adapt horticultural systems to unfavorable climatic conditions (in particular, extreme rainfall events during autumn and winter periods). In this experimental field, integrated strategies are combined, i.e., soil surface shaping, crop rotations, introduction of agro-ecological services crops (ASC), ASC termination techniques, and fertilization with organic products [4]. The base layer is the soil surface shaping in a "ridge system". Cash crops are planted both on the top of each raised bed 2.5 m wide (ridges) and in the 2.5 m flat areas (strips) between them. The crop rotation is designed to cultivate the cash crop on the ridges and the ASC in the strips during the winter-rainy period of the year. Cover crops are used to prevent soil erosion and provide N to the system via biological fixation, since on the top of the ridges, a leguminous cover crop is intercropped (as living mulch) in the winter as a vegetable crop and maintained as a ground cover. During the winter-rainy period, in the flat soil strips, mixtures of ASC species of different botanical families are cultivated between two consecutive spring–summer cash crops. The used ASC termination methods (before the subsequent cash crops transplant) are green manure (in which the cover crop is chopped and plowed at the end of flowering) vs. cover crop biomass flattening by an in-line roller-crimper, in which the mulch covers the soil surface until the vegetable crop harvest [4]. Finally, the last layer consists of organic fertilization, which is implemented into further horizontal strips, by using commercial and experimental amendments.

The second experimental field was conceived to verify the hypothesis that the use of the in-line roller crimping technology for ASC termination will improve the agronomical performances of the organically managed vegetable cropping systems. A two-year field experiment was carried out to evaluate the effect of ASC termination on tomato, by comparing green manure vs. roller-crimper, NO ASC (control) and plastic mulch (positive control) treatments. Another variability factor consists of three different fertilization treatments (i. on-farm organic fertilizer, ii. commercial organic fertilizer, and iii. unfertilized control).

The experimental design of each field trial was a strip plot with three replications, allowing the ability to calculate the standard deviation of the variables.

In the present research, the following three management systems adopting different agro-ecological approach levels are considered, which have been extrapolated from the two above-defined experimental field trials (Table 1):

1. ECO, an organic system with the full implementation of the described agro-ecological strategies, from the first experimental field. The cultivation area (1 ha) was divided into two parts (0.5 ha ridge furrow and 0.5 ha flat strip) and crops were cultivated both on the ridges and in the strips. On the ridges, with the clover as a living mulch, cauliflower during the winter period (transplanted on 20 October 2016 and harvested on 21 March 2017) and tomato crop during the spring–summer (transplanted on 24 April 2017 and harvested during August 2017) were cultivated. In the strips, the ASC (80% vetch-20% oats) were sown in November 2016 and incorporated as break crops during the following spring. Zucchini, during the spring–summer (transplanted on 27 April 2017 and harvested during July 2017), and lettuce, during the late summer–autumn (transplanted on 31 August 2017 and harvested on 26 October 2017), were then cultivated. A composted anaerobic digestate from cattle manure was used as fertilizer (i.e., on-farm organic fertilizer). The phytosanitary management followed the organic farming rules. The ECO cultivation system is under study in an experimental field in which the adaptation of horticultural systems to extreme climatic events are being tested, since these phenomena are increasing in the Mediterranean area. Consequently, we choose this experimental system to verify the hypothesis that the above-described practices may be used by the farmers as potential adaptation strategies for organic agro-ecosystems.

2. GM, an organic system with the introduction of the ASC, from the second experimental field. The ASC (80% vetch-20% oats) were sown in November 2016 and chopped and plowed into the soil in April 2017. The tomato plants were transplanted in May and harvested during July 2017. The fertilizer, a composted anaerobic digestate from cattle manure (i.e., on-farm organic fertilizer), was spread two times, before ASC sowing (70% of the total amount) and the remaining part (30%) before tomato transplanting. The phytosanitary management followed the organic farming rules. The above-described cultivation technique is becoming more utilized in organic farms, even if it needs further investigation, particularly in horticultural systems.

3. NO ASC, an organic system without ASC, from the second experimental field. The tomato crop was cultivated in the spring–summer period (transplanted on 5 May 2017 and harvested during July–August 2017). The phytosanitary approach followed the organic farming rules and a commercial organic fertilizer (NPK 4-8-12) was spread before transplanting. This cultivation system is still the most commonly used by organic horticultural farmers, even if it does not follow the agro-ecology approach.

Table 1. Description of the three different systems analyzed. ECO = organic system with the full implementation of the agro-ecological strategies; GM = organic system with the introduction of the agro-ecological service crops (ASC); NO ASC = organic system without ASC.

	ECO	GM	NO ASC
Year	2016/2017	2016/2017	2016/2017
Total area	1 ha	1 ha	1 ha
Soil texture	Clay soil	Clay soil	Clay soil
Strip cultivation for agroecological function	Yes (on ridge-flat strips system)	No	No
Cash crop	Ridges: cauliflower/tomato 0.5 ha Strips: zucchini/lettuce 0.5 ha	Tomato 1 ha	Tomato 1 ha
ASC as break crops	Vetch/oats 0.5 ha in strips	Vetch/oats 1 ha	No
ASC as living mulch	Clover 0.5 ha on ridges	No	No
Phytosanitary management	Organic (pyrethrum, Cu, S)	Organic (pyrethrum, Cu, S)	Organic (pyrethrum, Cu, S)
Fertilization management	On-farm organic fertilizers	On-farm organic fertilizers	Commercial organic fertilizers
Amount of N distributed with the fertilization	215 kg ha^{-1}	150 kg ha^{-1}	150 kg ha^{-1}
Soil tillage	Minimum tillage	Minimum tillage	Minimum tillage
Irrigation system and water consumption	Drip irrigation-7320 m^3	Drip irrigation-3300 m^3	Drip irrigation-3300 m^3

In ECO, at the cauliflower, zucchini, and lettuce commercial maturity, five randomly selected plants in each plot were collected to determine both "production quantity" attribute and the most important quality parameters for the calculation of "production quality" attribute. Conversely, at harvest, in GM and NO ASC the tomato fruits were collected from two randomly selected plants (center of the 2 rows in each plot) and both marketable and total yields and quality parameters were recorded to calculate production attributes.

2.3. Sustainability Evaluation

2.3.1. *DEXi-met* Model Application

To assess the sustainability of the agro-ecological practices implemented in the experimental field trials, crops yield and energy outputs were measured. The marketable yields were multiplied by their own coefficient of equivalent energy taken by the literature, to estimate the energy outputs [22]. The data of each agricultural operation were collected in a standardized procedure. All field practices were recorded (human labor as h ha^{-1}, fuels consumption as kg ha^{-1}), during the cover crop management and the cash crop cycles. Moreover, to better understand the systems environmental impact and sustainability, the *DEXi-met* model was used, aiming at assessing the level of sustainability of each considered system [20]. *DEXi-met* was developed for the ex-post assessment in organic horticulture by implementing the original DEXi software, which is utilized in multi-criteria decision analysis [23]. The ex-post assessment carried out with this model includes the basic attributes derived from the field experiment (e.g., productions, organic matter, etc.). In more details, *DEXi-met* is based on a hierarchical decision tree structure that breaks down the sustainability into smaller modules, which can be explained and calculated. Both qualitative and quantitative basic attributes are categorized into a linguistic scale, that is from a three-value scale ("low", "medium", "high"), used for the basic attribute, to a seven-value scale ("very-low", "low", "medium-low", "medium", "medium-high", "high", "very-high") for the "overall sustainability". The evaluation procedure begins with the calculation of the basic attributes, that could be also calculated using a satellite tree [24]. Their homogenization into the rank-ordered qualitative scale and the pyramidal aggregation of attributes contributed to the calculation of the aggregated final sustainability. The aggregation procedure is based on decision rules and relative weightings, that were given to each attribute, according to their alleged significance and contribution to sustainability. The weightings were defined involving both decision analysts and experts (i.e., researchers, agronomists, and farmers) and considering the literature, as indicated in Montemurro et al. [20]. The *DEXi-met* model tree structure is reported in Figure 1. All the attributes (basic and aggregate) from the bottom to the top, their aggregation weights and the corresponding scales are presented, to understand the calculation of the final "overall environmental sustainability".

2.3.2. *DEXi-met* Sensitivity Analysis

In order to identify the most significant variables that affected the sustainability of the systems, a sensitivity analysis (SA) of the *DEXi-met* model was also performed. According to the suggestions of Iocola et al. [17], the SA was performed utilizing the IZIEval tool (http://wiki.inra.fr/wiki/deximasc/Interface+IZI-EVAL/Accueil). The IZIEval is an interface shaped to facilitate the multi-criteria sustainability assessment of cropping systems based on models developed with the DEXi software, supplementing the existing features of DEXi. The Algdesign and XML packages, of the open-source R software [25], were used for the SA.

Through the IZIEval interface, both the sensitivity indexes (SI) and Monte Carlo (MC) analyses were performed, to gain the SA. In particular, according to Carpani et al. [15] we used the same basic attributes utilized for the "overall environmental sustainability" in the sensitivity indexes computation. The software automatically attributed an equal weight or probability of occurrence to all possible values of each variable. The SI highest values corresponded to the most important effect for a specific variable within the "overall environmental sustainability". The SI used the hierarchical model tree structure to obtain the results. Aggregation weights and number of the basic variables at the same level, aggregation weights of the aggregated variables, and depth levels influenced the findings.

To model the probability of different outcomes when random variables are involved, the Monte Carlo simulations are a possible tool. They allow obtaining the relative frequency distribution of the output values of an aggregated variable. In our study, according to Iocola et al. [17] this analysis was carried out by using IZIeval, randomly sampling and simulating a large number of values (5000) of each variable, to obtain the frequency distribution of the overall sustainability values and their main components.

BASIC ATTRIBUTES	Weight %	Aggregate attributes	Weight %	Aggregate attributes	Weight %	Aggregate attributes	Weight %	Overall sustainability
Insect pests and diseases		50		Control pests	33	Production capacity	25	Sustainability
Weeds		50						
Soil structure		25		Physico chemical fertility	33			
P and K Fertility		25						
Acido basic balance		25						
N balance		25						
Production quantity		60		Production	33			
Production quality		40						
Water consumption		33		Water	50	Soil and water preservation	25	
Groundwater utilization		33						
Irrigation technology		33						
Tillage diversification	50	soil treatments	32	Soil	50			
Tillage typology and depth	50							
Soil erosion control		32						
Organic matter balance		37						
External energy input		40		Energy	33	Resources preservation	25	
No-renewable input dependence		30						
Reuse input		30						
Fertilizer C/N		70		Fertilization	33			
On-farm fertilizers		30						
Cause of phytosanitary treatment		30		Phytosanitary	33			
Impact of phytosanitary treatment		40						
Approach of phytosanitary treatment		30						
Crop rotation	60	specific variability	50	Flora	50	Biodiversity conservation	25	
Strip cultivation agroecological function	40							
Floristic abounance	50	flora conservation	50					
Floristic diversity	50							
Macrofauna preservation		60		Fauna	25			
Flying insects preservation		40						
Microorganism preservation		100		Microorganism	25			
(low, medium low, medium high, high)/(low,medium,high)		(low, medium low, medium high, high)		(low, medium low, medium high, high)		(low, medium low, medium, medium high, high)		(very low, low, medium low,medium, medium high, high, very high)

SCALE

Figure 1. The *DEXi-met* model decision tree. The model includes 30 basic attributes, aggregate attributes at different levels, four nodal attributes, and the overall sustainability. The numbers between attribute levels represent the default aggregation weights (expressed in %). For each attribute level (basic, aggregate, and overall) the scale is reported at the bottom of the figure.

3. Results

3.1. Yields Performance and Energetic Outputs

The highest absolute value of tomato marketable yield was found in GM, whereas the ECO treatment showed the lowest one with a reduction of −58 and −24% in comparison with GM and NO ASC systems, respectively (Table 2).

Table 2. Effects of management strategies on marketable yields (Mg ha^{-1}, values ± standard deviation). ECO = organic system with the full implementation of the agro-ecological strategies; GM = organic system with the introduction of the agro-ecological service crops (ASC); NO ASC = organic system without ASC.

Cash Crops	ECO			GM			NO ASC		
	Mg ha^{-1}		St. dev.	Mg ha^{-1}		St. dev.	Mg ha^{-1}		St. dev.
Cauliflower	0.96	±	0.05	-	-	-	-	-	-
Zucchini	13.21	±	3.73	-	-	-	-	-	-
Lettuce	24.69	±	1.99	-	-	-	-	-	-
Tomat	13.83	±	3.23	30.88	±	13.83	18.13	±	12.24

The ECO treatment also showed a very low marketable yield in cauliflower cultivation. On the whole, considering that in ECO the cultivation area of each crop was 0.5 ha, while in GM and NO ASC it was doubled, the GM treatment determined higher total energy output by 70.3% and 14.4% as compared to NO ASC and ECO treatments, respectively (Table 3). Furthermore, the difference between GM and ECO was due to the low energy output occurred in cauliflower. In any case, all values of energy output were characterized by a high variation.

Table 3. Crop ((MJ ha^{-1} values ± standard deviation) and total energy output divided by the management strategies. ECO = organic system with the full implementation of the agro-ecological strategies; GM = organic system with the introduction of the agro-ecological service crops (ASC); NO ASC = organic system without ASC.

Crop	Energy Equivalent	ECO			GM			NO ASC		
	MJ kg^{-1} (USDA, 2019)	MJ ha^{-1}		St. dev.	MJ ha^{-1}		St. dev.	MJ ha^{-1}		St. dev.
Cauliflower	1	480	±	48	-		-	-		-
Zucchini	0.9	5944	±	3361	-		-	-		-
Lettuce	0.7	8643	±	1392	-		-	-		-
Tomato	0.75	5185	±	2422	23,162	±	10,376	13,599	±	9178
Total Energy output (MJ ha^{-1})		**20,252**			**23,162**			**13,599**		

3.2. Environmental Sustainability Evaluation

The overall environmental sustainability of the tested cropping systems varied considering the different crop management (Figure 2). In particular, the "high" score was reached by ECO treatment, while a "medium-high" and "medium-low" score was obtained for GM and NO ASC, respectively.

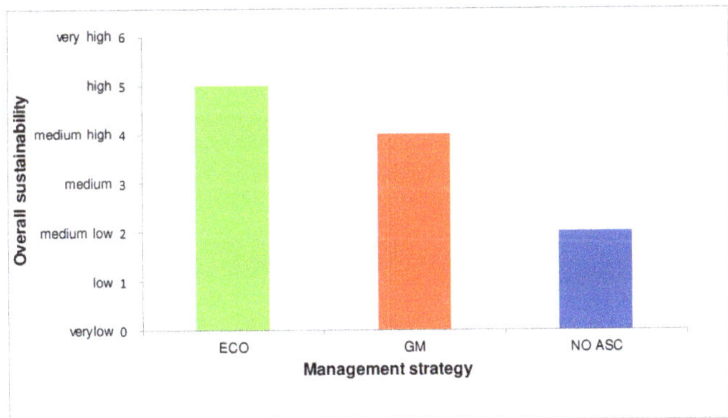

Figure 2. Comparison among the different crop management strategies: evaluation results of the multi-criteria decision model *DEXi-met* on the overall sustainability. ECO = organic system with the full implementation of the agro-ecological strategies; GM = organic system with the introduction of the agro-ecological service crops (ASC); NO ASC = organic system without ASC.

The nodal aggregate attribute "production capacity" resulted in "medium" in ECO and GM treatments and "medium-low" in NO ASC (Figure 3).

The ECO treatment scored "high" value for the aggregate attributes "soil and water preservation" and "resource preservation", while GM and NO ASC reached "medium-high" and "medium" scores, respectively, for these same aggregates. The "biodiversity conservation" ranged from "high" in ECO to "low" in NO ASC.

The sustainability evaluation of all the components (from the overall sustainability to the basic attributes) for the three scenarios is reported in Figure 4, as a comparison among the different tested systems. It is also reported the level of sustainability (from "sustainable" to "not sustainable") of each item, according to the specific linguistic scale (from three to seven) described in Figure 1.

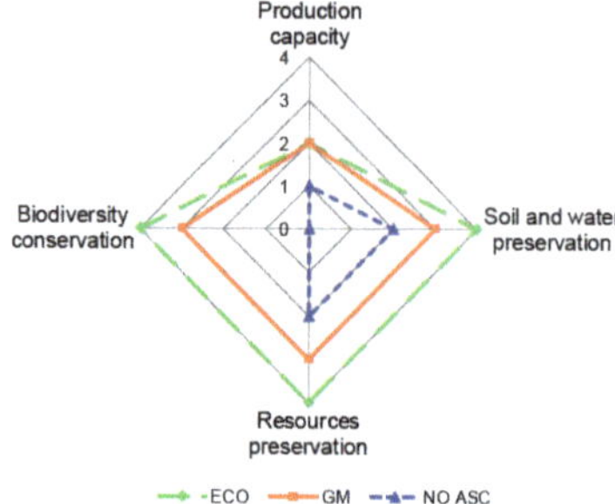

Figure 3. Comparison among the different crop management strategies: evaluation results of the multi-criteria decision model *DEXi-met* on the four main aggregate attributes ("production capacity", "soil and water preservation", "resources preservation", and "biodiversity conservation"). ECO = organic system with the full implementation of the agro-ecological strategies; GM = organic system with the introduction of the agro-ecological service crops (ASC); NO ASC = organic system without ASC.

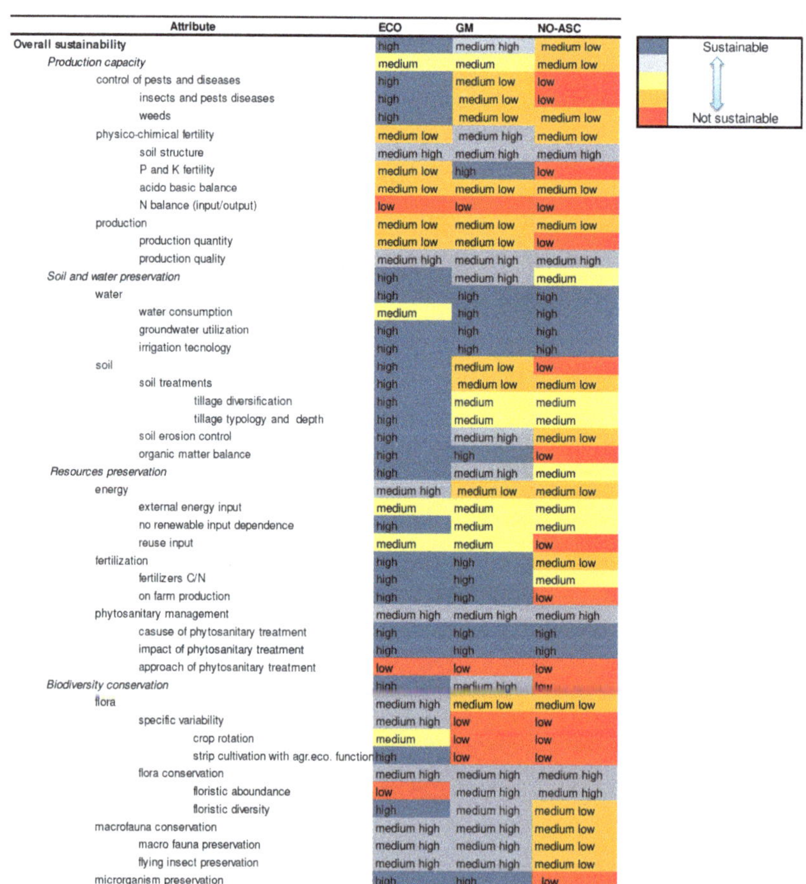

Attribute	ECO	GM	NO-ASC
Overall sustainability	high	medium high	medium low
Production capacity	medium	medium	medium low
control of pests and diseases	high	medium low	low
insects and pests diseases	high	medium low	low
weeds	high	medium low	medium low
physico-chimical fertility	medium low	medium high	medium low
soil structure	medium high	medium high	medium high
P and K fertility	medium low	high	low
acido basic balance	medium low	medium low	medium low
N balance (input/output)	low	low	low
production	medium low	medium low	medium low
production quantity	medium low	medium low	low
production quality	medium high	medium high	medium high
Soil and water preservation	high	medium high	medium
water	high	high	high
water consumption	medium	high	high
groundwater utilization	high	high	high
irrigation tecnology	high	high	high
soil	high	medium low	low
soil treatments	high	medium low	medium low
tillage diversification	high	medium	medium
tillage typology and depth	high	medium	medium
soil erosion control	high	medium high	medium low
organic matter balance	high	high	low
Resources preservation	high	medium high	medium
energy	medium high	medium low	medium low
external energy input	medium	medium	medium
no renewable input dependence	high	medium	medium
reuse input	medium	medium	low
fertilization	high	high	medium low
fertilizers C/N	high	high	medium
on farm production	high	high	low
phytosanitary management	medium high	medium high	medium high
casuse of phytosanitary treatment	high	high	high
impact of phytosanitary treatment	high	high	high
approach of phytosanitary treatment	low	low	low
Biodiversity conservation	high	medium high	low
flora	medium high	medium low	medium low
specific variability	medium high	low	low
crop rotation	medium	low	low
strip cultivation with agr.eco. function	high	low	low
flora conservation	medium high	medium high	medium high
floristic aboundance	low	medium high	medium high
floristic diversity	high	medium high	medium low
macrofauna conservation	medium high	medium high	medium low
macro fauna preservation	medium high	medium high	medium low
flying insect preservation	medium high	medium high	medium low
microrganism preservation	high	high	low

Figure 4. Evaluation results from *DEXi-met* model from the overall environmental sustainability to the basic attributes. ECO = organic system with the full implementation of the agro-ecological strategies; GM = organic system with the introduction of the agro-ecological service crops (ASC); NO ASC = organic system without ASC.

The "production capacity" aggregate attribute is generated from the first order aggregate attributes "control of pests and diseases", "physical-chemical fertility", and "production". In the ECO management, most of these attributes scored "high", "medium-high" or "medium-low", with one only basic attribute (N balance) with "low" value. Conversely, in NO ASC, the most frequent score was "medium-low", while GM showed intermediate values between the other two treatments.

The "soil and water preservation" aggregate attribute showed small differences among water management options, while the first order attribute "soil" was "high", "medium-low" and "low" for ECO, GM, and NO ASC, respectively. These differences are generated by the basic attributes "tillage diversification" and "tillage typology and depth" ("high" in ECO and "medium" in GM and NO ASC), "soil erosion control ("high" in ECO and "medium-high" and "medium-low" for GM and NO ASC, respectively) and "organic matter balance" ("high" in ECO and GM and "low" in NO ASC).

The "resources preservation" aggregate attribute differed for the attributes related to the "energy" and "fertilization", which scored frequently "high" and "medium" in ECO treatment, "medium" in GM and "medium-low" and "low" in NO ASC. No differences were found in the first order "phytosanitary management" attribute and in the basic attributes.

Finally, a large number of differences were recorded in the "biodiversity conservation" component. In particular, the ECO treatment scored "high" and "medium-high" in most of the basic attributes, GM showed frequently "medium-high" values, while NO ASC scored "medium-low" and "low" values.

The results of the SI calculation for the basic attributes are reported in Figure 5.

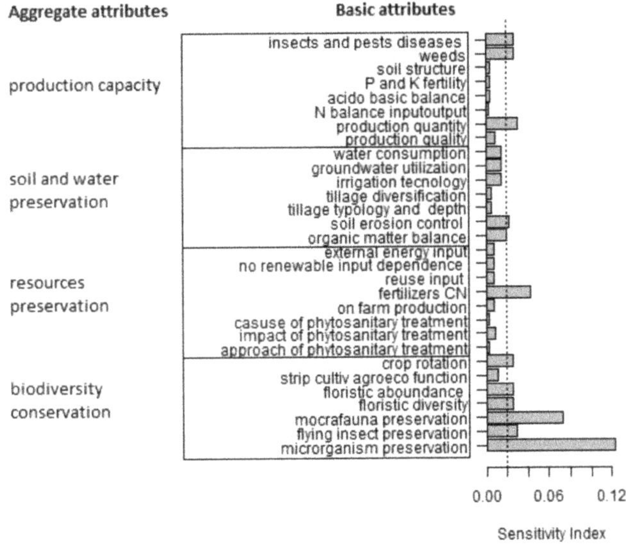

Figure 5. Sensitivity index values obtained with IZIEval tool for each basic variable of *DEXi-met* referred to the overall sustainability. The vertical line distinguishes the more sensitive variables (right side of the line) from the others.

The "microorganism preservation" and the "macrofauna preservation" reached the highest (0.12 and 0.08, respectively) SI values, being the most influential variables of the first order "biodiversity conservation" attribute. Within the "resources preservation" aggregate attribute, the "fertilizer C/N" basic attribute reached the highest SI value, while in the "production capacity" attribute, the "insects and pest diseases" and the "weeds" and "production quantity" showed higher values compared with the other basic attributes. Within the "soil and water preservation" component, the only basic attribute "soil erosion control" overtakes the 0.2 sensitivity index.

Table 4 reports the frequency distributions of the 5000 simulated outputs of the Monte Carlo (MC) analysis for the overall sustainability and for the main model components (nodal attributes). The "overall environmental sustainability" showed the qualitative value "medium" more frequently (MC = 0.543) than the other values. Among the nodal attributes, the "production capacity" recorded the value "medium" (MC = 0.496) that occurred more frequently than the other modalities. The "medium-low" and "medium" values were the most frequent for the "soil and water preservation" (MC = 0.571 and MC = 0.275, respectively), while the "resources preservation" principally scored the "medium" and "medium-high" values (MC = 0.475 and MC = 0.399, respectively). In the nodal attribute "biodiversity conservation", the "medium-low" and "medium-high" were the most frequent values.

Table 4. Relative frequency distributions of the results of 5000 Monte Carlo simulations among the seven different qualitative values ("very low", "low", "medium-low", "medium", "medium-high", "high", "very high") for the overall sustainability and among the five different qualitative values (low", "medium-low", "medium", "medium-high", "high") for the main aggregate attributes ("production capacity", "soil and water preservation", "resources preservation", and "biodiversity conservation") obtained with *DEXi-met*.

	Very Low	Low	Medium-Low	Medium	Medium-High	High	Very High
Overall sustainability	0.000	0.003	0.257	0.543	0.190	0.005	0.000
Production capacity	-	0.014	0.237	0.496	0.237	0.016	-
Soil and water preservation	-	0.020	0.571	0.265	0.136	0.008	-
Resources preservation	-	0.004	0.034	0.475	0.399	0.089	-
Biodiversity conservation	-	0.246	0.309	0.000	0.338	0.107	-

4. Discussion

4.1. Yield Performances and Energetic Output

The values of the tomato marketable yields in GM were higher by 70 and 123% compared with NO ASC and ECO treatments, respectively. This result was probably due to higher availability of readily available N, which derives from decomposition of the aboveground biomass of the ASC plowed into the soil [4,11]. Conversely, in the ECO plots, in which the clover was used as living mulch [26,27], and in the NO ASC, the tomato yield did not reach the standard level of organic production [28]. Moreover, in the ECO system, the cauliflower marketable yield was very low, likely because adverse climatic conditions occurred. In fact, during the growing season there was an extreme adverse event, unusual in the experimental area, showing low mean month temperature (4.5 °C) and several days of values below −4 °C on January 2017, associated with high rainfall intensity (117 mm). The zucchini and lettuce marketable yields, which benefited from the residual fertility of the ASC, were comparable with other experimental results on organic crops production [29–31].

The total energy output per hectare was higher in GM by 14% and 70% than ECO and NO ASC, respectively (Table 3). In particular, the treatments with the presence of ASC showed similar total energy output productions (about 20,000 MJ ha^{-1}), which were substantially different in comparison with the management without cover crops, thus indicating the positive effect of such agro-ecological practice [32]. This result was probably due to the large difference generated in the total tomato output for GM and NO ASC treatments. Conversely, the ECO treatment showed a slightly reduced energy output in comparison with GM, because of the very low productions in the cauliflower cropping cycle [33], even if the four different crops contributed to the overall production by half a hectare. According to the USDA indications [22], the coefficient of equivalent energy of the cauliflower was the highest (value equal to 1), therefore, the low production of this crop influenced the total energy output of the ECO treatment. In any case, we must take into account that the differences among treatments showed a high standard deviation, which was generated by the huge variability of the data.

4.2. Environmental Sustainability Evaluation by DEXi-met

The "overall environmental sustainability" of the cropping systems varied among the three evaluated management systems, passing from "medium-low" showed in NO ASC to the "high" in ECO. This response can be explained by the scores both of the aggregate nodal attributes (Figure 3) and the basic attributes (Figure 4), and it was a consequence of the intensification of the agro-ecological strategies adopted. In particular, the *DEXi-met* output showed that the ECO strategy was the most sustainable one, mainly due to the differences detected in the nodal attribute "biodiversity conservation". In fact, the study of Depalo et al. [34] pointed out a general positive influence of the living mulch techniques on arthropods in plant/soil systems, as shown by a high level of soil biodiversity and a lack of negative impacts on the density of canopy insects. Also, the presence of ASC in the rotation as break crops enhances the "biodiversity conservation" and, at the same time, it may have impact on occurrence of weeds, diseases, and pests [4,35]. Our results confirmed these findings, being the ECO system characterized by ASC presence both as break crops and living mulch, compared with the other two treatments (Table 1).

For the "production capacity", differently from the yield performance and the energetic output, the *DEXi-met* model considered not only the crop productions, but also the physico-chemical fertility and the systems control on pests and disease. The differences between ECO and GM were not perceivable by the model and scored "medium" value in both systems (Figure 3). Conversely, the NO ASC scored "medium-low", due to the "low" value of pests and disease control (Figure 4).

The systems with the introduction of the agro-ecological service crops (ECO and GM) scored high soil erosion control, in agreement with the study of De Benedetto et al. [36]. This result was generated by a better soil cover, in particular during the winter/heavy rainy period. Therefore, the nodal attribute "soil and water preservation" was the highest in ECO, followed by the GM system. The differences in "resources preservation" was mainly due to the different fertilizers used. In fact, the composted anaerobic digestate, which was utilized in the ECO and GM plots, is a renewable, more sustainable fertilizer than the commercial organic one, and its application did not compromise the systems production capacity, thus confirming the findings of previous studies [11,37].

Finally, the *DEXi-met* output showed substantial differences in the aggregate nodal attribute "biodiversity conservation" among the systems, as explained above. In particular, the introduction of the ASC increased the score both in ECO and GM ("high" and "medium-high", respectively), compared to the NO ASC management strategy. Similarly, other studies indicated that the presence of ASC enhances the insect and arthropods communities [38], as well as the soil microbial activities [39].

Even if the *DEXi-met* presents some aspects that should be improved, it showed some strengths and, therefore, its application gave us the possibility to analyze in detail the general structure of the overall sustainability, as well as the components and the single variables of the systems considered. We should also take into account that *DEXi-met* model is one of the new ex-post tools, which considers some attributes derived from the field experiment. However, to better understand how the model tree structure affects the results and to find the most significant variables that contributed most to the output variability, a sensitivity analysis was necessary. The sensitivity index results were affected by the level of complexity of each component and by the number of variables. Carpani et al. [15] indicated that a simpler component structure has a greater influence on the overall sustainability, whereas a higher number of variables, that individually could have no significant impact, become more sensitive if they are considered together. In our study, the *DEXi-met* produced both the "microorganism preservation" and the "macrofauna preservation" attributes as the most influential variables of our sensitivity analysis. This last result was due to the difference in the systems analyzed. In particular, it is a consequence of using agro-ecological practices, especially in the ECO system. Except for "strip cultivation with agro-ecological functions", all the other basic attributes in the nodal aggregate "biodiversity conservation" showed the SI higher than 0.2, indicating the positive influence of the systems on the environmental sustainability [40]. The basic attributes "fertilizer C/N" (within the "resources preservation" nodal attribute), the "insects and pests diseases", "weeds", and "production quantity" (within the "production capacity"), and "soil erosion control" (within the "soil and water preservation") reached the highest values. According to Carpani et al. [15], when the SI is high, the effect of each variable on the overall sustainability is more relevant.

The detailed analysis of the *Dexi-met* model structure through the distribution of frequencies in the overall sustainability and the nodal attributes (showed by the Monte Carlo analysis) revealed that the model adequately represents the diversity of the systems evaluated. In fact, the frequencies obtained in the "overall sustainability" showed the highest value for the "medium" modality, following a normal Gaussian pattern. This behavior is due to the use of till to seven qualitative classes at the "overall sustainability" level, allowing to distinguish the different scenarios. However, the number of the qualitative classes (from "very low" to "very high") was not so large to generate unnecessary complications in the use of the model, and to reduce its ability to distinguish differences between systems. Besides, to avoid further complications, in agreement with Craheix et al. [24], the aggregate attributes were composed by five qualitative classes (from "low" to "high") and the basic attributes were composed by only three classes ("high", "medium", and "low").

5. Conclusions

Sustainability in agriculture is a complex concept and there are no common viewpoints among scientists about its dimension. Nonetheless, various parameters for measuring agricultural sustainability have been proposed, since the measure of the mere production capacity of an agro-system is not enough to evaluate it. This study clearly highlighted the relevance of considering different criteria, when we assess the advances in sustainability achievement that could be obtained introducing agro-ecological management practices and innovations. The findings also demonstrated that applying principles and practices which tend to follow the natural ecosystems can contribute to building up more complex agro-ecosystems, increasing resilience, and optimizing and maintaining biodiversity. In particular, the agro-ecological approach (ECO) both reduces the use of and dependency on external synthetic inputs by enabling to control pests, weeds, and improving fertility with ecological management. This management strategy could optimize and close resource loops (nutrients, biomass, etc.), by recycling nutrients and biomass in the farm. It may also support climate adaptation and resilience and contribute to greenhouse gas emissions mitigation, through lower use of fossil fuels and higher carbon sequestration in soils.

As revealed by our results, the introduction of the agro-ecological management practices such as ASC, use of on-farm produced fertilizers (composts), intercropping, etc., is an interesting way to improve the sustainability of the system. In any case, the results found here could not be fully generalized, since the *Dexi-met* model did not take into account some other aspects (e.g., the economic sustainability, the length of the study period, etc.). Moreover, when these strategies are applied, agronomic and productions difficulties should be kept in mind, at least in the short transition period between conventional and agro-ecological systems.

The proposed modeling approach provides a simple method of decisional support to farmers to efficiently select different crop management strategies, by assessing the environmental sustainability of the cultivation systems. An interesting topic of further research could be testing the considered agro-ecological management practices in different environmental conditions.

Author Contributions: Conceptualization, F.M., A.P., and M.D.; formal analysis, A.P.; data curation, A.P.; writing—original draft preparation, F.M., A.P., and M.D.; writing—review and editing, F.M., A.P., and M.D.; funding acquisition, F.M.

Funding: This paper is a result of the research projects RETIBIO (*Attività di supporto nel settore dell'agricoltura biologica per il mantenimento dei dispositivi sperimentali di lungo termine e il rafforzamento delle reti di relazioni esistenti a livello nazionale e internazionale*), funded by the Organic Farming Office of the Italian Ministry of Agriculture, and SOILVEG (*Improving soil conservation and resource use in organic cropping systems for vegetable production through introduction and management of Agro-ecological Service Crops (ASC)*) funded by ERA-Net CORE Organic Plus Funding Bodies partners of the European Union's FP7 research and innovation programme under the grant agreement No. 618107.

Conflicts of Interest: The authors declare no conflict of interest. The funders had no role in the design of the study; in the collection, analyses, or interpretation of data; in the writing of the manuscript, or in the decision to publish the results.

References

1. Bitter, J.; Janssen, D.; Vossen, R. Requirements for multi-method approaches to sustainability assessment—A theoretical and empirical study. In *The European Conference on Sustainability, Energy & the Environment. Official Conference Proceedings*; The International Academic Forum: Brighton, UK, 2018; pp. 239–258.
2. Diacono, M.; Persiani, A.; Canali, S.; Montemurro, F. Agronomic performance and sustainability indicators in organic tomato combining different agro-ecological practices. *Nutr. Cycl. Agroecosyst.* **2018**, *112*, 101–117. [CrossRef]
3. Diacono, M.; Baldivieso-Freitas, P.; Sans Serra, F.X. Nitrogen utilization in a cereal-legume rotation managed with sustainable agricultural practices. *Agronomy* **2019**, *9*, 113. [CrossRef]
4. Diacono, M.; Fiore, A.; Farina, R.; Canali, S.; Di Bene, C.; Testani, E.; Montemurro, F. Combined agro-ecological strategies for adaptation of organic horticultural systems to climate change in Mediterranean environment. *Ital. J. Agron.* **2016**, *11*, 85–991. [CrossRef]
5. Diacono, M.; Persiani, A.; Testani, E.; Montemurro, F.; Ciaccia, C. Recycling agricultural wastes and by-products in organic farming: Biofertilizers production, yield performance and carbon footprint analysis. *Sustainability* **2019**, *11*, 3824. [CrossRef]
6. Altieri, M.A.; Koohafkan, P. Strengthening resilience of farming systems: A key prerequisite for sustainable agricultural production. In *Wake Up before It is Too Late: Make Agriculture Truly Sustainable Now for Food Security in a Changing Climate*; UNCTAD, TER13 Report; UNCTAD: Geneva, Switzerland, 2013; pp. 56–60.
7. Cinelli, M.; Coles, S.R.; Kirwan, K. Analysis of the potentials of multi criteria decision analysis methods to conduct sustainability assessment. *Ecol. Indic.* **2014**, *46*, 138–148. [CrossRef]
8. Waas, T.; Hugé, J.; Block, T.; Wright, T.; Benitez-Capistros, F.; Verbruggen, A. Sustainability Assessment and Indicators: Tools in a Decision-Making Strategy for Sustainable Development. *Sustainability* **2014**, *6*, 5512–5534. [CrossRef]
9. Rodrigues, G.A.; Rodrigues, I.A.; Buschinelli, C.C.; de Barros, I. Integrated farm sustainability assessment for the environmental management of rural activities. *Environ. Impact Assess. Rev.* **2010**, *30*, 229–239. [CrossRef]
10. Pergola, M.; Persiani, A.; Pastore, V.; Palese, A.M.; Arous, A.; Celano, G. A comprehensive Life Cycle Assessment (LCA) of three apricot orchard systems located in Metapontino area (Southern Italy). *J. Clean. Prod.* **2017**, *142*, 4059–4407. [CrossRef]
11. Diacono, M.; Persiani, A.; Fiore, A.; Montemurro, F.; Canali, S. Agro-ecology for potential adaptation of horticultural systems to climate change: Agronomic and energetic performance evaluation. *Agronomy* **2017**, *7*, 35. [CrossRef]
12. Bockstaller, C.; Feschet, P.; Angevin, F. Issues in evaluating sustainability of farming systems with indicators. *Oléagineux Corps Gras Lipides John Libbey Eurotext* **2015**, *22*, 12. [CrossRef]
13. Bond, A.J.; Morrison-Saunders, A.; Pope, J. Sustainability assessment: The state of the art. *Impact Assess. Proj. Apprais.* **2012**, *30*, 53–62. [CrossRef]
14. Bohanec, M. *DEXi: Program for Multi-Criteria Decision Making, User's Manual, Version 5.00*; IJS Report DP-11897; Jožef Stefan Institute: Ljubljana, Slovenia, 2015; Available online: http://kt.ijs.si/MarkoBohanec/DEXi/html/DEXiDoc.htm (accessed on 5 January 2018).
15. Carpani, M.; Bergez, J.E.; Monod, H. Sensitivity analysis of a hierarchical qualitative model for sustainability assessment of cropping systems. *Environ. Model. Softw.* **2012**, *27*, 15–22. [CrossRef]
16. Sadok, W.; Angevin, F.; Bergez, J.; Bockstaller, C.; Colomb, B.; Guichard, L.; Reau, R.; Messéan, A.; Doré, T. MASC, a qualitative multi-attribute decision model for ex ante assessment of the sustainability of cropping systems. *Agron. Sustain. Dev.* **2009**, *29*, 447–461. [CrossRef]
17. Iocola, I.; Campanelli, G.; Diacono, M.; Leteo, F.; Montemurro, F.; Persiani, A.; Canali, S. Sustainability Assessment of organic vegetable production using a qualitative multi-attribute model. *Sustainability* **2018**, *10*, 3820. [CrossRef]
18. Nikoloski, T.; Udovč, A.; Pavlovič, M.; Rajkovič, U. Farm reorientation assessment model based on multi-criteria decision making. *Comput. Electron. Agric.* **2017**, *140*, 237–243. [CrossRef]

19. Bohanec, M.; Messean, A.; Scatasta, S.; Angevin, F.; Griffiths, B.; Krogh, P.H.; Žnidaršič, M.; Džeroski, S. A qualitative multi-attribute model for economic and ecological assessment of genetically modified crops. *Ecol. Model.* **2008**, *215*, 247–261. [CrossRef]

20. Montemurro, F.; Persiani, A.; Diacono, M. Environmental sustainability assessment of horticultural systems: A multi-criteria evaluation approach applied in a case study in Mediterranean conditions. *Agronomy* **2018**, *8*, 98. [CrossRef]

21. Unesco-FAO Etude écologique de la Zone Méditerranéenne. Carte Bioclimatique de la Zone Méditerranéenne. 1963. Available online: http://unesdoc.unesco.org/images/0013/001372/137255fo.pdf (accessed on 5 April 2017).

22. USDA. United States Department of Agriculture Agricultural Research Service National Nutrient Database for Standard Reference Legacy Release. Available online: https://ndb.nal.usda.gov/ndb/ (accessed on September 2018).

23. Bohanec, M.; Dzeroski, S.; Znidarsi, M.; Messean, A.; Scatasta, S.; Wesseler, J. Multi-attribute modeling of economic and ecological impacts of cropping systems. *Informatica* **2004**, *28*, 387–392.

24. Craheix, D.; Bergez, J.E.; Angevin, F.; Bockstaller, C.; Bohanec, M.; Colomb, B.; Doré, T.; Fortino, G.; Guichard, L.; Pelzer, E.; et al. Guidelines to design models assessing agricultural sustainability, based upon feedbacks from the DEXi decision support system. *Agron. Sustain. Dev.* **2015**, *35*, 1431–1447. [CrossRef]

25. R Core Team. *R: A Language and Environment for Statistical Computing*; R Foundation for Statistical Computing: Vienna, Austria, 2013; Available online: http://www.R-project.org/ (accessed on February 2018).

26. Chase, C.A.; Mbuya, O.S. Greater interference from living mulches than weeds in organic broccoli production. *Weed Technol.* **2008**, *22*, 280–285. [CrossRef]

27. Canali, S.; Campanelli, G.; Ciaccia, C.; Diacono, M.; Leteo, F.; Fiore, A.; Montemurro, F. Living mulch strategy for organic cauliflower (Brassica oleracea L.) production in central and southern Italy. *Ital. J. Agron.* **2015**, *10*, 90–96. [CrossRef]

28. Adamczewska-Sowińska, K.; Kołota, E.; Winiarska, S. Living mulches in field cultivation of vegetables. *Veg. Crops Res. Bull.* **2009**, *70*, 19–29. [CrossRef]

29. Ciaccia, C.; Montemurro, F.; Campanelli, G.; Diacono, M.; Fiore, A.; Canali, S. Legume cover crop management and organic amendments application: Effects on organic zucchini performance and weed competition. *Sci. Hortic.* **2015**, *185*, 48–58. [CrossRef]

30. Montemurro, F.; Fiore, A.; Campanelli, G.; Tittarelli, F.; Ledda, L.; Canali, S. Organic fertilization, green manure, and vetch mulch to improve organic zucchini yield and quality. *HortScience* **2013**, *48*, 1027–1033. [CrossRef]

31. Montemurro, F.; Ciaccia, C.; Leogrande, R.; Ceglie, F.; Diacono, M. Suitability of different organic amendments from agro-industrial wastes in organic lettuce crops. *Nutr. Cycl. Agroecosyst.* **2015**, *102*, 243–252. [CrossRef]

32. Gomiero, T.; Pimentel, D.; Paoletti, M.G. Energy and environmental issues in organic and conventional agriculture. *Crit. Rev. Plant Sci.* **2008**, *27*, 239–254. [CrossRef]

33. Persiani, A.; Diacono, M.; Monteforte, A.; Montemurro, F. Agronomic performance, energy analysis and carbon balance comparing different fertilization strategies in horticulture under Mediterranean conditions. *Environ. Sci. Pollut. Res.* **2019**, *26*, 19250–19260. [CrossRef]

34. Depalo, L.; Burgio, G.; von Fragstein, P.; Kristensen, H.L.; Bavec, M.; Robačer, M.; Campanelli, G.; Canali, S. Impact of living mulch on arthropod fauna: Analysis of pest and beneficial dynamics on organic cauliflower (Brassica oleracea L. var. botrytis) in different European scenarios. *Renew. Agric. Food Syst.* **2017**, *32*, 240–247. [CrossRef]

35. Campiglia, E.; Mancinelli, R.; Radicetti, E.; Caporali, F. Effect of cover crops and mulches on weed control and nitrogen fertilization in tomato (Lycopersicon esculentum Mill.). *Crop Prot.* **2010**, *29*, 354–363. [CrossRef]

36. De Benedetto, D.; Montemurro, F.; Diacono, M. Impacts of Agro-Ecological Practices on Soil Losses and Cash Crop Yield. *Agriculture* **2017**, *7*, 103. [CrossRef]

37. Montemurro, F.; Vitti, C.; Diacono, M.; Canali, S.; Tittarelli, F.; Ferri, D. A three-year field anaerobic digestates application: Effects on fodder crops performance and soil properties. *Fresenius Environ. Bull.* **2010**, *19*, 2087–2093.

38. Navarro-Miró, D.; Caballero-López, B.; Blanco-Moreno, J.M.; Pérez, A.; Depalo, L.; Masetti, A.; Burgio, G.; Canali, S.; Sans, F.X. Agro-ecological Service Crops with roller crimper termination enhance ground-dwelling predator communities and pest regulation. In Proceedings of the 5th ISOFAR Scientific Conference "Innovative Research for Organic 3.0" 19th Organic World Congress, New Dehli, India, 9–11 November 2017.

39. Hartwig, N.L.; Ammon, H.U. Cover crops and living mulches. *Weed Sci.* **2002**, *50*, 688–699. [CrossRef]
40. Scherr, S.J.; McNeely, J.A. Biodiversity conservation and agricultural sustainability: Towards a new paradigm of "ecoagriculture" landscapes. *Philos. Trans. R. Soc. Lond. B Biol. Sci.* **2008**, *363*, 477–494. [CrossRef] [PubMed]

applied sciences

Article

Stomatal Response of Maize (*Zea mays* L.) to Crude Oil Contamination in Soils

Chaolan Zhang [1,2]**, He Huang** [1]**, Yongxin Zhou** [3]**, Haiying Lin** [2]**, Tian Xie** [3] **and Changjun Liao** [3,*]

[1]   College of Life Science and Technology, Guangxi University, Nanning 530004, China; zhangcl@gxu.edu.cn (C.Z.); river_63@163.com (H.H.)
[2]   School of Resources, Environmental and Materials, Guangxi University, Nanning 530004, China; linhaiying@gxu.edu.cn
[3]   Guangxi Bossco Environmental Protection Technology Co., Ltd, Nanning 530004, China; zhouyx@bossco.cc (Y.Z.); xiet@bossco.cc (T.X.)
*   Correspondence: hjxfyf@bossco.cc

Received: 9 August 2019; Accepted: 27 September 2019; Published: 29 September 2019

Abstract: In this study, maize plant was cultured in soil contaminated with different levels of crude oil. The purpose was to investigate the change of soil properties, leaf physiological and chemical parameters, and phenanthrene content in the leaf. Results showed that soil water content significantly increased when the levels of total petroleum hydrocarbons were 3700–17,800 mg/kg in soil, and soil electrical conductivity significantly increased compared with the control. In maize leaf, stomatal length and density, as well as K and Na contents decreased in contaminated treatments compared with the control. Stomatal length has a significant positive correlation with K content in leaf (r = 0.92, $p < 0.01$), while stomatal density was negatively correlated to the crude oil level in soil (r = −0.91, $p < 0.05$). Accumulation of phenanthrene in maize leaf was mainly through the foliar uptake pathway. Phenanthrene concentrations of maize leaf in oil-treated soil were less than that of the control, which exhibited a significant positive relationship with stomatal length (r = 0.98, $p < 0.01$). This study demonstrated that the stomata structure of maize could be influenced by crude oil and thus possibly controlling the accumulation of polycyclic aromatic hydrocarbons in aerial tissues. Based on these results, controlling stomata movement will be beneficial to phytoremediation of contaminated soil.

Keywords: maize; stomata; soil; phenanthrene; remediation

---

1. Introduction

Petroleum oil is the main energy source and plays an important role in modern society. Soil contaminated with petroleum hydrocarbons is an increasing environmental concern as oil consumption increases dramatically around the world [1]. Among the numerous components of petroleum hydrocarbons, polycyclic aromatic hydrocarbons (PAHs) are persistent and carcinogenic in the environment, and thus threaten human health through contaminated food chain [2]. In general, compared with inhalation or skin contact, ingestion of contaminated food is the primary pathway for human to exposure to PAHs [3,4].

Oil residuals can cause some major changes in the soil chemical properties, such as decreased total nitrogen content, and increased pH value to some extent [5], which influences the growth of plants in soil. The toxic effects of crude oil contamination prevent germination from occurring and provide unsatisfactory soil conditions [6,7]. Such poor soil conditions may result from insufficient aeration caused by decreased air-filled pore space, low water content, as well as a reduction of available nutrients [8]. Therefore, it is necessary to fully understand the changed properties of petroleum hydrocarbon contaminated soils that are closely related to plant growth.

Plants adapt to environmental stress by adjusting their external morphology, internal structure, and physiological and ecological characteristics [9]. Crude oil is likely to directly affect plants after contact and provide a resulting intake of contaminants [10,11]. Plants take in PAHs mainly through soil-to-plant and air-to-plant, i.e., root uptake and atmospheric deposition from gaseous or particulate forms [12,13]. In the former pathway, PAHs can be taken in the aerial plant tissues from the root through the transpiration stream within the xylem, while in the latter one, they can be diffused into plant leaves via the cuticle or the stomata [14]. Since stomatal closure or opening is vital to the transpiration and gas exchange, it should be considered in the study of PAHs uptake by plants.

In botany, a stoma is a pore found in the epidermis of leaves, stems, and other organs controlling gas exchange between the atmosphere and plants [15]. Stomatal density and aperture (length of stomata) vary under many environmental factors such as atmospheric CO_2 concentration, light intensity, air temperature [16,17], potassium and sodium concentration [18], air pollution, and environmental stress [19]. For example, stomatal size obviously decreases with water deficit, and stomatal density is positively correlated with stomatal conductance, net CO_2 assimilation rate, and water use efficiency [20]. Thus, the stomata can adapt to local and global changes on all time scales from minutes to millennia [15].

The edible plants grown in contaminated soils are of great concern to human health for its potential risk. For instance, maize plant has been reported to be a candidate for phytoremediation of hydrocarbon-contaminated soil [21,22]. In our previous work, maize plant was also applied for phytoremediation of crude oil contaminated soil, in which several PAHs had been detected in maize plants, and phenanthrene (PHE) had the highest level [23,24]. Stomata play an important role in the air-to-plant uptake pathway of PAHs. However, the information about the changes in leaf stomata of maize plants grown in crude oil contaminated soils is scarce. Therefore, it is necessary to investigate the changes of leaf stomata of plants grown in contaminated soil and the influence factors. The main objectives of this study were (1) to understand the changes of stomata structure and the concentrations of potassium (K) and sodium (Na) in maize leaf in response to crude oil contaminated soil; (2) to evaluate the relationships between the PHE (a representative of PAHs) leaf concentration and leaf parameters/soil properties.

2. Materials and Methods

2.1. Chemicals, Seed of Maize and Soil

Crude oil without refining was obtained from Guangzhou Department, Sinopec Corporation, China. All other agents used in this study were analytical grade. Seed of CT 38 was purchased from Research Institution of Crop, Guangdong Academy of Agricultural Sciences, China.

The soil in the experiment was collected from the upper layer (0–20 cm) of an abandoned farm in Guangzhou Higher Education Mega Center, Guangzhou, China. After stones and roots were removed, the soil was air-dried, smashed, and passed through a 4 mm sieve. The organic matter content and pH of the soil were 1.3% and 6.54, respectively. Nutrient levels were 24.5 g/kg ammoniac nitrogen, 4.32 g/kg total phosphorus (P), and 0.40 g/kg total K.

2.2. Experimental Design and Management

The soil (1.5 kg) was placed in a plastic crate, spiked with different amounts of crude oil, and stirred for homogeneity with a wood spoon. The soil was then put into plastic pots and placed outdoors for 4 months to adequately evaporate the volatile fractions of crude oil. And then the total petroleum hydrocarbon (TPH) levels were measured to be 0, 2600, 3700, 6500, 17,800, and 48,800 mg/kg, respectively, using the method of our previous work [23]. Each treatment was replicated three times. Three maize seeds were placed into soils at 2 cm depth in each pot. After the maize seedlings grow out of the soils with three expanded leaves, one seedling was left in each pot. The pots were placed outdoors at the top of our laboratory building. The experiment was started in September. The average temperature was 23.2 °C during the experiment. Water was added into potholders for soil moisture. After two months, soil properties and maize leaf parameters were determined.

2.3. Analytical Methods

2.3.1. Soil Water Content and Soil Electrical Conductivity

To understand the water content of soil contaminated with different levels of crude oil, soil samples were collected in pots with a core sampler when water sufficiently infiltrated into the soil from a potholder. Soil water content was determined gravimetrically by weighing, after drying in an oven at 105 °C for 12 h according to the method described by Liu et al. [25]. Soil electrical conductivity was measured by a portable electrical conductivity meter (Hanna HI-993310D).

2.3.2. Determination of Stomatal Traits

Stomatal traits in maize leaf were determined according to the method described by Zheng et al. [26]. The first leaf fully expanded on the main stem was sampled for each plant. Colorless nail polish was carefully smeared on leaf samples for about half an hour. Then the thin film was immediately covered with a cover slip and pressured lightly with a fine-point tweezer. Leaf stomatal length and density were measured from the base, middle, and tip sections on leaves of maize. Three slides were prepared for each taxon. Stomatal length was determined by micro-morphological observations carried out on 1 cm^2 portion per leaf (excised from similar areas) with a microscope (Carl Zeiss Micro-imaging, GER) equipped with a spot insight color camera (Diagnostic Instruments, Sterling Heights, Sterling Heights, MI, USA). Stomatal density (NO/cm^2) was calculated on 10 representative fields of leaves according to the method described by Orsini et al. [27].

The gravimetric measurement of water loss after leaf excision is a rapid method to evaluate the transpiration rate. The initial fresh weight (FW) and the weight after 5 min (W) were recorded. Water loss in 5 min was the difference between FW and W, which were used to calculate the transpiration rate [28]. The leaf water content was also determined gravimetrically by the method of soil water content mentioned above. The leaves were cut into pieces and dried in an oven at 105 °C until they reached a constant weight.

2.3.3. Determination of K and Na Concentration in Maize Leaves

To determine the concentration of K and Na in maize leaves, samples were collected and dried, followed by digestion with HNO_3 and oxidation by H_2O_2 with a heating plate. The residual was dissolved in 5% (*V/V*) HNO_3 solution, the concentrations were measured by atomic absorption spectrophotometry (AAS, Z-2000, Hitachi, Tokyo, Japan) as previously described by Cicek and Cakirlar [29].

2.3.4. Determination of PHE in Maize Leaves

Concentration determination of PHE in maize leaf was conducted according to the previous method described by Tao et al. [30] with some modification. Maize samples (1.00 g) homogenized with about 1 g of anhydrous sodium sulfate were put in glass tubes. The samples were extracted with 10 mL hexane/dichloromethane (1:1) under ultrasonic conditions for 30 min. Then the extract was collected in a beaker. This process was replicated three times. The collected extract was purified by passage through a silica gel column and vacuum concentrated with a rotary evaporator at 40 °C. The samples were re-suspended in *n*-hexane to a final volume of 1 mL for further analysis by gas chromatography mass spectrometry (GC-MS).

Analysis of plant samples was conducted using a GC–MS with Thermo Trace GC Ultra instrument coupled to a Thermo DSQ II mass spectrometer (Thermo Electron Corporation, Waltham, MA, USA). Compounds were separated in a 30 m 0.25 mm id capillary column coated with 0.25 μm film (HP-5MS, Agilent, USA). GC temperature was programmed from an initial 80 °C before commencing at 10 °C/min up to 290 °C, with a final holding time of 10 min. Helium was used as carrier gas. A 1.0 µL aliquot of the extract was injected while the injector port was held at 280 °C and operated in a splitless mode at a flow rate of 1.0 mL/min. The head column pressure was 30 kPa. The mass spectrometer was operated in scan mode with an electron impact ionization of 70 eV and an ion source temperature of 230 °C. Solvent delay was set at 4 min. Selective ion monitoring model was used. The target ions and retention time was 178 and 14.84 min for PHE, respectively [31].

2.4. Statistical Analysis

Statistical Product and Service Solutions statistic software 17.0 (SPSS company, Chicago, IL, USA) was used for the statistical evaluation of the results designed as completely randomized with three replicates of each parameter. Mean values followed by the same letter were not significantly different, as determined by an analysis of variance (ANOVA). The differences were compared by Duncan's range at a significance level of $p < 0.05$. The relationships between parameters were evaluated by Pearson correlation analysis.

3. Results

3.1. Changes in Soil Properties

The changes in soil water content and soil electrical conductivity in different treatments were shown in Figure 1. Soil water content was significantly increased when the TPH levels rose from 3700 to 17,800 mg/kg, but dramatically decreased at the extremely high level of 48,800 mg/kg, compared to the control soil (Figure 1A). At the low-level contaminated soil (2600 mg/kg), water content was similar to that of the control. The values of soil electrical conductivity in the contaminated treatments were significantly higher than that of the control (Figure 1B), but it exhibited no regular tendency.

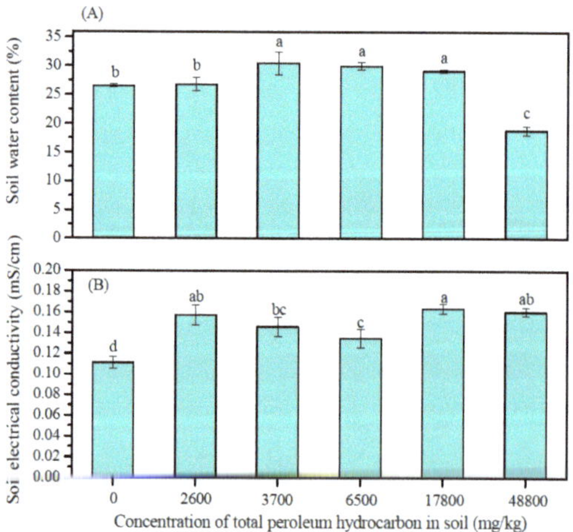

Figure 1. Effect of soil contamination with crude oil on soil water content (**A**) and soil electrical conductivity (**B**). ($p < 0.05$). Different letters on top of the bar indicate they are significantly different at $p < 0.05$.

3.2. Leaf Growth and Stomatal Density and Length

Stomatal length of maize leaf in contaminated treatments significantly decreased in comparison with that in the control (Figure 2A), but there were no significant differences among 3700–48,800 mg/kg treatments. Stomatal density decreased with increasing TPH levels in soil (Figure 2B). In the highest-level contaminated soil, stomatal density was decreased by 46% compared with the control. In comparison, the downtrend showed that stomatal length was more sensitive than stomatal density to contaminated soil.

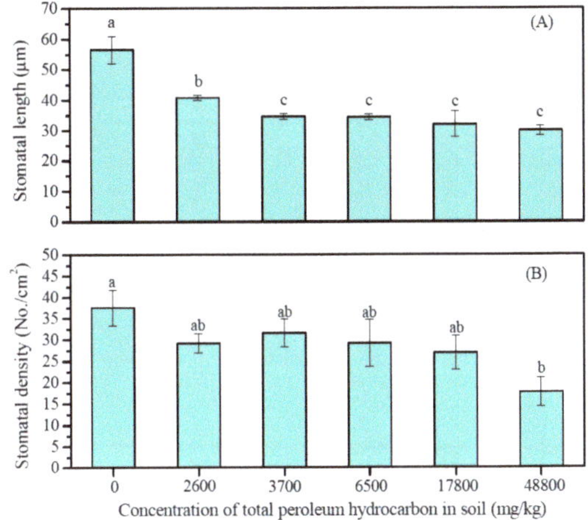

Figure 2. Effect of soil contamination with crude oil on stomatal length (**A**) and stomatal density (**B**) of maize leaf. ($p < 0.05$). Different letters on top of the bar indicate they are significantly different at $p < 0.05$.

3.3. Water Content and Transpiration Rate

Water content and transpiration rate are important physiological functions of plants, which may be influenced by soil conditions. As shown in Figure 3A, leaf water contents in all samples grown in contaminated soil were similar, suggesting the crude oil contaminated soil with different concentration did not have a remarkable effect on the water transport from soil to plant tissues, but did change the water content in the maize leaf. As shown in Figure 3B, transpiration rates of maize leaf in contaminated soil did not exhibit a significant difference, but were slightly higher than the control. This indicated the transpiration rate of maize plant could be affected by crude oil contaminated soil to some extent.

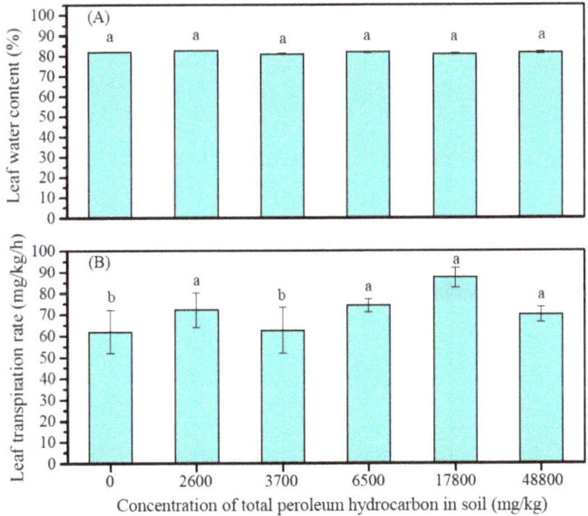

Figure 3. Effect of soil contamination with crude oil on water content (**A**) transpiration rate (**B**) of maize leaf ($p < 0.05$). Different letters on top of the bar indicate they are significantly different at $p < 0.05$.

3.4. Concentrations of K and Na in Maize Leaf

As the major mineral elements in plant tissues, K and Na play important roles in maintaining the physiological functions, especially in regulating the opening and closure of stomata. It is necessary to investigate the effect of crude oil contaminated soil on the K and Na assimilation of maize. As shown in Figure 4, both K and Na concentrations in maize leaf significantly decreased with increasing TPH levels of soils, indicating the crude oil contaminated soil had a significant effect on the assimilation of K and Na in maize plant. Additionally, K and Na concentrations in maize leaf at high levels of contaminated soil (above 6500 mg/kg) were much lower than those of the control.

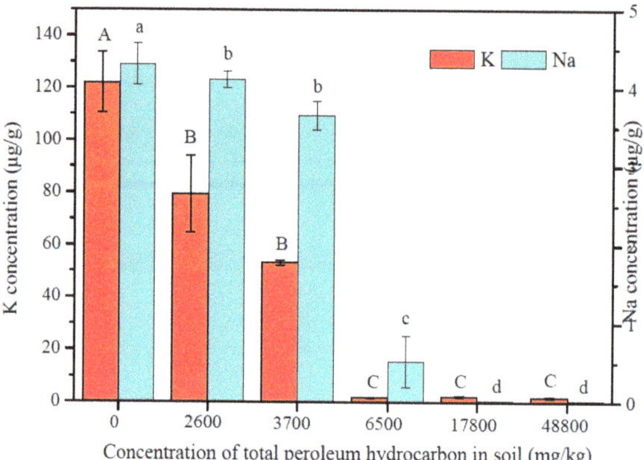

Figure 4. Effect of soil contamination with crude oil on K and Na concentrations in maize leaf. ($p < 0.05$). Different letters on top of the bar indicate they are significantly different at $p < 0.05$.

3.5. Phenanthrene Concentration in Maize Leaf

As shown in Figure 5, PHE concentrations of maize leaf in contaminated treatments were lower than that in the control group, and there were significant differences when TPH levels reached 3700 mg/kg. Furthermore, in the soil treatments with TPH levels ranging from 3700 to 48,800 mg/kg, PHE concentrations in maize did not exhibit significant changes.

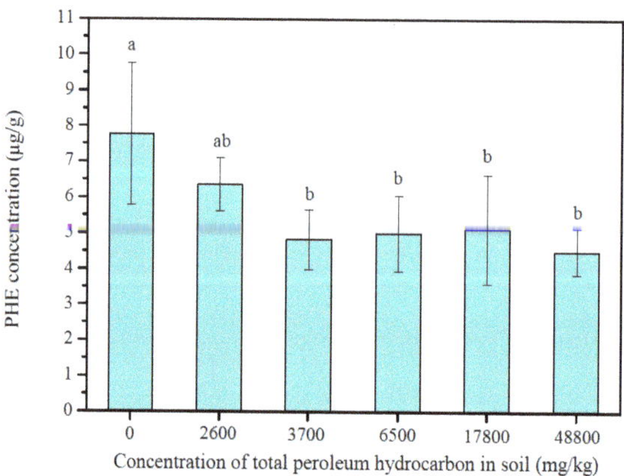

Figure 5. Phenanthrene concentration of maize leaf in different treatments. ($p < 0.05$). Different letters on top of the bar indicate they are significantly different at $p < 0.05$.

3.6. Relationship

The relationships between different parameters were presented in Table 1. TPH level in soil was negatively correlated to soil water content, leaf water content, stomatal length and density, K, Na and PHE contents in maize leaf, but positively correlated to soil electrical conductivity and transpiration rate of maize leaf. Especially, significant negative correlations were observed between soil TPH level and soil water content ($r = -0.82$, $p < 0.05$), as well as stomatal density ($r = -0.91$, $p < 0.05$). In this study, correlation analysis also covered the stomata structure, ion contents, and PHE concentration in maize leaf. Stomatal length was significantly positively correlated to leaf K content ($r = 0.92$, $p < 0.01$). Besides, PHE concentration had a significantly positive correlation with stomatal length ($r = 0.98$, $p < 0.01$).

Table 1. Simple correlation coefficient (r) between parameters.

	TPH	SW	EC	LW	TR	SL	SD	K	Na	PHE
TPH	1.00									
SW	−0.82 *	1.00								
EC	0.55	−0.22	1.00							
LW	−0.10	−0.25	0.23	1.00						
TR	0.27	0.10	0.65	−0.28	1.00					
SL	−0.59	0.10	−0.83 *	0.05	−0.55	1.00				
SD	−0.91 *	0.65	−0.78	−0.10	−0.46	0.81	1.00			
K	−0.62	0.11	−0.66	0.30	−0.67	0.92 **	0.79	1.00		
Na	−0.71	0.27	−0.54	0.44	−0.72	0.75	0.78	0.94 **	1.00	
PHE	−0.59	0.09	−0.71	0.11	−0.40	0.98 **	0.76	0.89 *	0.72	1.00

TPH: total petroleum hydrocarbon in soil; SW: soil water content; EC: soil electrical conductivity; LW: leaf water content; TR: transpiration rate; SL: stomatal length; SD: stomatal density; K: leaf K content; Na: leaf Na content; PHE: phenanthrene concentration in leaf. * indicate the r values are significant at $p < 0.05$. ** indicate the r values are significant at $p < 0.01$.

4. Discussion

4.1. Soil Properties in Crude Oil Contaminated Soil

Soil properties play an important role in soil microorganism activity and plant growth. According to previous work, plants exposed to soils contaminated with petroleum hydrocarbons were subjected to growth limitations, due to low water uptake and reduced nutrient availability [8]. Mineral nutrient availability can be reflected by soil electrical conductivity. Therefore, soil water content and soil electrical conductivity need to be well understood in phytoremediation of soil contaminated with petroleum hydrocarbons. In this study, water contents increased in soil contaminated with certain TPH levels (3700–17,800 mg/kg) but decreased significantly when TPH levels reached 48,800 mg/kg. Water-stable aggregates in soil are related to the content of soil organic matter [32]. The addition of crude oil to soil increased the soil organic matter content, possibly resulting in enhanced water holding capacity at a certain limited range, as soil organic matter was an important determinant of the available water capacity [33]. However, the high concentration of crude oil in soil might prevent water from entering the pores of soil, which decreases water holding capacity.

Soil electrical conductivity increased in crude oil contaminated soils compared to the control soil in this study. This result was in agreement with previous work, which showed that the value of electrical conductivity in contaminated soil was higher than that of the control site [5]. Soil electrical conductivity represents soil salinity, which is mainly composed of cation ions such as Na^+, K^+, and Ca^{2+}. The result of this study illustrated that availability of these ions in soil was not the limiting factors for plant uptake. The addition of crude oil in soil leads to higher soil electrical conductivity, possibly resulting from the production of metabolites from crude oil biodegradation [34].

4.2. Stomata and Other Leaf Parameters

Stomata are the pores in the epidermis of botany leaf controlling gas exchange, mainly CO_2 and water vapor, between the atmosphere and plants [15]. In the present work, stomatal length and density of maize leaf in contaminated soil treatments decreased compared with those in the control, indicating that crude oil contaminated soil harmed stomatal structures. And the downtrend showed stomatal length to be more sensitive than stomatal density to contaminated soil. Previous work showed stomata in plants were not only influenced by air pollution of automobile emissions [35], but also by soil or water contamination of environmental stress [19,36]. It is interesting to observe the changes in maize stomata induced by the contamination of air or soil in the present work.

Plasticity of stomatal development may be determined by many exogenous and environmental cues, of which abscisic acid (ABA) is considered as a vital regulator of environmentally determined stomatal development [37]. According to previous work, ABA can increase progressively in the root with responses to abiotic environmental stress [38,39]. In particular, the level of endogenous ABA significantly increased in pea (*Pisum sativum* L.) plant with increasing fluoranthene concentrations [40]. Therefore, crude oil contaminated soil might stimulate the synthesis of ABA in maize root and then increase the amount of ABA, thus decreased stomatal length and density.

In addition, the changed stomata might affect other leaf parameters. Stomatal length has an extremely positive correlation with leaf K content ($r = 0.92$, $p < 0.01$), and positive correlation with leaf Na content ($r = 0.75$, $p < 0.01$). It seems that K and Na content in leaf may be influenced by the stomata structure. But on the other hand, K^+ is considered to involve in controlling stomatal movements, in which guard cell K^+ uptake from the apoplast is mediated by a proton-extruding adenosine triphosphatase on the plasmalemma [41]. Moreover, tonoplast-localized NHX proteins as Na^+, K^+/H^+ antiporters are essential for active K^+ uptake at the tonoplast for stomatal function [42], so that K^+ and Na^+ in plant are considered as twins [43]. The contents of K and Na in maize leaf might also affect the length and density of stomata. The relation between those ions and stomata structure still needs further study. Additionally, leaf water content and transpiration rate did not significantly change in different treatments, indicating that the response of water balance in maize plant was different from nutrient ions in contaminated soil.

4.3. PHE Uptake and Stomata

Accumulation of PAHs in aerial plant tissues may be from root through transpiration stream and from diffusion via leaf stomata [14]. In the present work, transpiration rates in all plants were similar (Figure 3B). Besides, PHE concentrations of maize leaf in contaminated treatments were lower than that of the control. PHE could volatilize from contaminated soil. These results indicated PHE accumulation in maize aerial tissues might be from foliar uptake pathway which was controlled by stomata. This suggestion was also confirmed by the fact that PHE concentration in maize leaf was significantly positively correlated with stomatal length ($r = 0.98$, $p < 0.01$). Besides, previous studies also confirmed that foliar uptake was the dominant pathway of PHE accumulation by plant [44–46].

Since stomata play an important role in uptaking pollutants by plants grown in contaminated soils, measurements influencing stomatal movement can be applied in phytoremediation of contaminated soil for different purposes. For example, ABA can be used on maize husk for inhibition of PAH accumulation by grain due to being able to induce stomatal closure and inhibit a light-induced stomatal opening [47], when maize plant is considered for phytoremediation of PAHs-contaminated soil. Thus, safe food will be produced. In contrast, since fusicoccin can prevent dark-induced stomatal closure [48], it can be used on hyperaccumulators for extracting more pollutants in remediation of soil contaminated with heavy metals, which are taken up by plant root and transferred to aboveground tissues with a transpiration stream that closely relates to a stomatal opening. Therefore, phytoremediation of soils contaminated with organic pollutants or heavy metals can benefit from the controlling of stomatal movement. However, the data for the effect of phytohormone on maize plant was not provided in this study, which needs to be investigated in future work.

5. Conclusions

Stomatal response and the change of related parameters of the maize plant (*Zea mays* L.) to crude oil contaminated soil were investigated in this study. Soil water content and electrical conductivity increased to a certain extent in contaminated soil, whereas the TPH level exhibited a negative relationship with soil water content ($r = -0.82$, $p < 0.05$). Stomatal length and density, leaf K, and Na contents decreased in contaminated soil compared with that of the control group. Stomatal length is positively correlated to leaf K content ($r = 0.92$, $p < 0.01$), while stomatal density is negatively correlated to soil TPH level ($r = -0.91$, $p < 0.05$). Moreover, it is found that the accumulation of PAHs in maize mainly occurred through the foliar uptake pathway. And PHE concentration exhibits a significantly positive relationship with stomatal length ($r = 0.98$, $p < 0.01$). Based on this study, measurements should be applied to control stoma closure or opening for different purposes in phytoremediation of contaminated soils.

Author Contributions: C.L. and C.Z. conceived and designed the experiments; C.Z. and H.H. performed the experiments; H.H., Y.Z. and T.X. analyzed the data; C.L. and C.Z. written original draft; H.L. and H.H. reviewed and edited the draft.

Funding: The research was financially supported by the Program for the National Key R&D Program of China (2018YFD0800700, 2017YFD0801300).

Conflicts of Interest: The authors declare no conflict of interest.

References

1. Klamerus-Iwan, A.; Błońska, E.; Lasota, J.; Kalandyk, A.; Waligórski, P. Influence of oil contamination on physical and biological properties of forest soil after chainsaw use. *Water Air Soil Poll.* **2015**, *226*, 389. [CrossRef] [PubMed]

2. Xue, W.; Warshawsky, D. Metabolic activation of polycyclic and heterocyclic aromatic hydrocarbons and DNA damage: A review. *Toxicol. Appl. Pharmacol.* **2005**, *206*, 73–93. [CrossRef] [PubMed]

3. Cocco, P.; Moore, P.S.; Ennas, M.G.; Tocco, M.G.; Ibba, A.; Mattuzzi, S.; Meloni, M.; Monne, M.; Piras, G.; Collu, S. Effect of urban traffic, individual habits, and genetic polymorphisms on background urinary 1-hydroxypyrene excretion. *Ann. Epidemiol.* **2007**, *17*, 1. [CrossRef] [PubMed]

4. Desalme, D.; Binet, P.; Chiapusio, G. Challenges in Tracing the Fate and Effects of Atmospheric Polycyclic Aromatic Hydrocarbon Deposition in Vascular Plants. *Environ. Sci. Technol.* **2013**, *47*, 3967–3981. [CrossRef]

5. Wang, X.Y.; Feng, J.; Zhao, J.M. Effects of crude oil residuals on soil chemical properties in oil sites, Momoge Wetland, China. *Environ. Monit. Assess.* **2010**, *161*, 271–280. [CrossRef] [PubMed]

6. Kyunghwa, B.; Heesik, K.; Heemock, O.; Byungdae, Y.; Jaisoo, K.; Insook, L. Effects of crude oil, oil components, and bioremediation on plant growth. *J. Environ. Sci. Health Part A* **2004**, *39*, 2465–2472.

7. Kisic, I.; Mesic, S.; Basic, F.; Brkic, V.; Mesic, M.; Durn, G.; Zgorelec, Z.; Bertovic, L. The effect of drilling fluids and crude oil on some chemical characteristics of soil and crops. *Geoderma* **2009**, *149*, 209–216. [CrossRef]

8. Merkl, N.; Schultzekraft, R.; Infante, C. Assessment of tropical grasses and legumes for phytoremediation of petroleum-contaminated soils. *Water Air Soil Pollut.* **2005**, *165*, 195–209. [CrossRef]

9. Frei, E.R.; Ghazoul, J.; Matter, P.; Heggli, M.; Pluess, A.R. Plant population differentiation and climate change: Responses of grassland species along an elevational gradient. *Glob. Chang. Biol.* **2014**, *20*, 441–455. [CrossRef] [PubMed]

10. Baruah, P.; Saikia, R.R.; Baruah, P.P.; Deka, S. Effect of crude oil contamination on the chlorophyll content and morpho-anatomy of Cyperus brevifolius (Rottb.) Hassk. *Environ. Sci. Pollut. Res.* **2014**, *21*, 12530. [CrossRef] [PubMed]

11. Balasubramaniyam, A.; Harvey, P.J. Scanning electron microscopic investigations of root structural modifications arising from growth in crude oil-contaminated sand. *Environ. Sci. Pollut. Res.* **2014**, *21*, 12651–12661. [CrossRef] [PubMed]

12. Fismes, J.; Perrin-Ganier, C.; Empereur-Bissonnet, P.; Morel, J.L. Soil-to-root transfer and translocation of polycyclic aromatic hydrocarbons by vegetables grown on industrial contaminated soils. *J. Environ. Qual.* **2002**, *31*, 1649–1656. [CrossRef] [PubMed]

13. Tao, Y.Q.; Zhang, S.Z.; Zhu, Y.G.; Christie, P. Uptake and acropetal translocation of polycyclic aromatic hydrocarbons by wheat (Triticum aestivum L.) grown in field-contaminated soil. *Environ. Sci. Technol.* **2009**, *43*, 3556–3560. [CrossRef] [PubMed]

14. Collins, C.; Fryer, M.; Grosso, A. Plant uptake of non ionic organic chemicals. *Environ. Sci. Technol.* **2006**, *40*, 45–52. [CrossRef]

15. Hetherington, A.M.; Woodward, F.I. The role of stomata in sensing and driving environmental change. *Nature* **2003**, *424*, 901–908. [CrossRef]

16. Buckley, T.N.; Mott, K.A. Modelling stomatal conductance in response to environmental factors. *Plant Cell Environ.* **2013**, *36*, 1691. [CrossRef]

17. Rogiers, S.Y.; Hardie, W.J.; Smith, J.P. Stomatal density of grapevine leaves (*Vitis vinifera L.*) responds to soil temperature and atmospheric carbon dioxide. *Aust. J. Grape Wine Res.* **2011**, *17*, 147–152. [CrossRef]

18. Battie-Laclau, P.; Laclau, J.-P.; Beri, C.; Mietton, L.; Almeida Muniz, M.R.; Arenque, B.C.; Piccolo, M.d.C.; Jordan-Meille, L.; Bouillet, J.-P.; Nouvellon, Y. Photosynthetic and anatomical responses of Eucalyptus grandis leaves to potassium and sodium supply in a field experiment. *Plant Cell Environ.* **2014**, *37*, 70–81. [CrossRef]

19. Ahmad, S.H.; Reshi, Z.; Ahmad, J.; Iqbal, M. Morpho-anatomical responses of Trigonella foenum graecum Linn. to induced cadmium and lead stress. *J. Plant Biol.* **2005**, *48*, 64–84. [CrossRef]

20. Xu, Z.; Zhou, G. Responses of leaf stomatal density to water status and its relationship with photosynthesis in a grass. *J. Exp. Bot.* **2008**, *59*, 3317–3325. [CrossRef]

21. Shahsavari, E.; Adetutu, E.M.; Anderson, P.A.; Ball, A.S. Tolerance of selected plant species to petrogenic hydrocarbons and effect of plant rhizosphere on the microbial removal of hydrocarbons in contaminated soil. *Water Air Soil Poll.* **2013**, *224*, 1–14. [CrossRef]

22. Wyszkowski, M.; Ziółkowska, A. Content of polycyclic aromatic hydrocarbons in soils polluted with petrol and diesel oil after remediationwith plants and various substances. *Plant Soil Environ.* **2013**, *59*, 287–294. [CrossRef]

23. Liao, C.; Xu, W.; Lu, G.; Deng, F.; Liang, X.; Guo, C.; Dang, Z. Biosurfactant-enhanced phytoremediation of soils contaminated by crude oil using maize (*Zea mays. L*). *Ecol. Eng.* **2016**, *92*, 10–17. [CrossRef]

24. Liao, C.; Xu, W.; Lu, G.; Liang, X.; Guo, C.; Yang, C.; Dang, Z. Accumulation of hydrocarbons by maize (*Zea mays L.*) in remediation of soils contaminated with crude oil. *Int. J. Phytoremediat.* **2015**, *17*, 693–700. [CrossRef] [PubMed]

25. Liu, Z.; Fu, B.; Zheng, X.; Liu, G. Plant biomass, soil water content and soil N:P ratio regulating soil microbial functional diversity in a temperate steppe: A regional scale study. *Soil Biol. Biochem.* **2010**, *42*, 445–450. [CrossRef]

26. Zheng, Y.; Xu, M.; Hou, R.; Shen, R.; Qiu, S.; Ouyang, Z. Effects of experimental warming on stomatal traits in leaves of maize (*Zea may L.*). *Ecol. Evol.* **2013**, *3*, 3095–3111. [CrossRef] [PubMed]

27. Orsini, F.; Alnayef, M.; Bona, S.; Maggio, A.; Gianquinto, G. Low stomatal density and reduced transpiration facilitate strawberry adaptation to salinity. *Environ. Exp. Bot.* **2012**, *81*, 1–10. [CrossRef]

28. Tanaka, Y.; Sugano, S.S.; Shimada, T.; Hara-Nishimura, I. Enhancement of leaf photosynthetic capacity through increased stomatal density in Arabidopsis. *New Phytol.* **2013**, *198*, 757–764. [CrossRef] [PubMed]

29. Çiçek, N.; Çakirlar, H. The effect of salinity on some physiological parameters in two maize cultivars. *Bulg. J. Plant Physiol.* **2002**, *28*, 66–74.

30. Tao, S.; Jiao, X.C.; Chen, S.H.; Liu, W.X.; Jr, C.R.; Zhu, L.Z.; Luo, Y.M. Accumulation and distribution of polycyclic aromatic hydrocarbons in rice (*Oryza sativa*). *Environ. Pollut.* **2006**, *140*, 406–415. [CrossRef]

31. Chen, Z.L.; Zhang, M.; Liu, H.Y. Micro-Column convenient chromatography for separation of aromatic hydrocarbon compound and Gc/Irms analysis. *Pet. Geol. Exp.* **2013**, *35*, 347–350. (In Chinese)

32. Tisdall, J.M.; Oades, J.M. Organic matter and water-stable aggregates in soils. *J. Soil Sci.* **1982**, *33*, 141–163. [CrossRef]

33. Hudson, B. Soil organic matter and available water capacity. *J. Soil Water Conserv.* **1994**, *49*, 189–194.

34. Aal, G.Z.A.; Slater, L.D.; Atekwana, E.A. Induced-polarization measurements on unconsolidated sediments from a site of active hydrocarbon biodegradation. *Geophysics* **2006**, *71*, H13–H24. [CrossRef]

35. Pal, A.; Kulshreshtha, K.; Ahmad, K.J.; Behl, H.M. Do leaf surface characters play a role in plant resistance to auto-exhaust pollution? *Flora-Morphol. Distrib. Funct. Ecol. Plants* **2002**, *197*, 47–55. [CrossRef]

36. Rai, V.; Khatoon, S.; Bisht, S.S.; Mehrotra, S. Effect of cadmium on growth, ultramorphology of leaf and secondary metabolites of Phyllanthus amarus Schum. and Thonn. *Chemosphere* **2005**, *61*, 1644–1650. [CrossRef] [PubMed]

37. Chater, C.C.; Oliver, J.; Casson, S.; Gray, J.E. Putting the brakes on: Abscisic acid as a central environmental regulator of stomatal development. *New Phytol.* **2014**, *202*, 376–391. [CrossRef]

38. Rai, M.K.; Shekhawat, N.S.; Gupta, A.K.; Phulwaria, M.; Ram, K.; Jaiswal, U. The role of abscisic acid in plant tissue culture: A review of recent progress. *Plant Cell Tissue Organ Cult.* **2011**, *106*, 179–190. [CrossRef]

39. Serna, L. The role of brassinosteroids and abscisic acid in stomatal development. *Plant Sci.* **2014**, *225*, 95–101. [CrossRef]

40. Váňová, L.; Kummerová, M.; Klemš, M.; Zezulka, Š. Fluoranthene influences endogenous abscisic acid level and primary photosynthetic processes in pea (*Pisum sativum L.*) plants in vitro. *Plant Growth Regul.* **2009**, *57*, 39–47. [CrossRef]

41. Outlaw, W.H., Jr. Current concepts on the role of potassium in stomatal movements. *Physiol. Plant.* **2010**, *59*, 302–311. [CrossRef]

42. Verónica, B.; Leidi, E.O.; Zaida, A.; Lourdes, R.; Anna, D.L.; Fernández, J.A.; Beatriz, C.; Pardo, J.M. Ion exchangers NHX1 and NHX2 mediate active potassium uptake into vacuoles to regulate cell turgor and stomatal function in Arabidopsis. *Plant Cell* **2012**, *24*, 1127–1142.

43. Benito, B.; Haro, R.; Amtmann, A.; Cuin, T.A.; Dreyer, I. The twins K^+ and Na^+ in plants. *J. Plant Physiol.* **2014**, *171*, 723–731. [CrossRef] [PubMed]

44. Kipopoulou, A.M.; Manoli, E.; Samara, C. Bioconcentration of polycyclic aromatic hydrocarbons in vegetables grown in an industrial area. *Environ. Pollut.* **1999**, *106*, 369–380. [CrossRef]

45. Lin, H.; Tao, S.; Zuo, Q.; Coveney, R.M. Uptake of polycyclic aromatic hydrocarbons by maize plants. *Environ. Pollut.* **2007**, *148*, 614–619. [CrossRef] [PubMed]

46. Simonich, S.L.; Hites, R.A. Vegetation-atmosphere partitioning of polycyclic aromatic hydrocarbons. *Environ. Sci. Technol.* **1994**, *28*, 939–943. [CrossRef] [PubMed]

47. Yin, Y.; Adachi, Y.; Ye, W.; Hayashi, M.; Nakamura, Y.; Kinoshita, T.; Mori, I.C.; Murata, Y. Difference in abscisic acid perception mechanisms between closure induction and opening inhibition of stomata. *Plant Physiol.* **2013**, *163*, 600–610. [CrossRef] [PubMed]

48. She, X.P.; Huang, A.X.; Li, J.; Han, X.Z. Inhibition of dark-induced stomatal closure by fusicoccin involves a removal of hydrogen peroxide in guard cells of Vicia faba. *Physiol. Plant.* **2010**, *140*, 258–268. [CrossRef] [PubMed]

applied
sciences

Technical Note

A Standardized Method for Estimating the Functional Diversity of Soil Bacterial Community by Biolog® EcoPlates™ Assay—The Case Study of a Sustainable Olive Orchard

Adriano Sofo [1,*] and **Patrizia Ricciuti** [2]

[1] Department of European and Mediterranean Cultures: Architecture, Environment and Cultural Heritage (DiCEM), University of Basilicata, Via San Rocco, 3–75100 Matera, Italy
[2] Department of Soil, Plant and Food Sciences (DiSSPA), University of Bari 'Aldo Moro', Via Amendola, 165–70126 Bari, Italy; patrizia.ricciuti@uniba.it
* Correspondence: adriano.sofo@unibas.it

Received: 23 August 2019; Accepted: 24 September 2019; Published: 26 September 2019

Featured Application: Soil bacteria are of paramount importance for determining soil quality and fertility but evaluating the microbiological status of the soils in a simple and reliable way is not easy. For doing this, specific plates are commercially available but, in order to have uniform results, a common procedure should be followed by everyone. We tried to fill this gap. In addition to agricultural production, our results are useful for mitigating soil pollution and climate change, where bacteria play a key role.

Abstract: Biolog® EcoPlates™ (Biolog Inc., Hayward, CA, USA) were developed to analyse the functional diversity of bacterial communities by means of measuring their ability to oxidize carbon substrates. This technique has been successfully adopted for studying bacterial soil communities from different soil environments, polluted soils and soils subjected to various agronomic treatments. Unfortunately, Biolog® EcoPlates™ assay, especially working on soil, can be difficult to reproduce and hard to standardize due to the lack of detailed procedures and protocols. The main problems of this technique mainly regard soil preparation, bacterial inoculum densities and a correct definition of blank during the calculation of the diversity indices. On the basis of our previous research on agricultural soils, we here propose a standardized and accurate step-by-step method for estimating the functional diversity of a soil bacterial community by Biolog® EcoPlates™ assay. A case study of soils sampled in a Mediterranean olive orchard managed accordingly to sustainable/conservation practices was reported for justifying the standardized method here used. The results of this methodological paper could be important for correctly evaluating and comparing the microbiological fertility of soils managed by sustainable/conservation or conventional/non-conservation systems.

Keywords: Biolog®; community-level physiological profiling (CLPP); functional diversity indices; metabolic bacterial diversity; olive; soil fertility; soil quality

---

1. Introduction

Microorganisms are present in all ecosystems and due to their rapid responses to physical and chemical changes, they can be used as bioindicators of environmental quality. The Community-Level Physiological Profiling (CLPP) is a rapid and relatively inexpensive technique to relate microbial functional diversity over space and time to changes in the environment [1–4].

Biolog® EcoPlates™ (Biolog Inc., Hayward, CA, USA) were developed to analyse the functional diversity of bacterial communities by means of measuring their ability to oxidize carbon substrates. An EcoPlate is a 96-well microplate that contains 31 common carbon sources from altogether five compound groups—that is, carbohydrates, carboxylic and ketonic acids, amines and amides, amino acids and polymers—plus a blank well as a control, all these replicated thrice to control variation in inoculum densities. Each EcoPlate is filled with a dilution of one soil suspension, thus representing one soil sample. The utilization rates of carbon compounds in the wells are quantified spectrophotometrically by following the reduction of water-soluble colourless triphenyl tetrazolium chloride to purple triphenyl formazan. For the measurements of optical density (OD), two filters are used: (a) 590 nm (absorbance peak of tetrazolium) to evaluate colour development plus turbidity values and (b) 750 nm to measure turbidity values only. The turbidity of dilutions is due to clay and humic particles in soil colloidal suspension.

Every bacterial community has a characteristic reaction pattern with different OD values for different carbon compounds, called a 'metabolic fingerprint' [2,5]. While the inoculated bacterial density significantly affects the rate of colour development in the wells, on the other side the choice of a wrong inoculum can compromise the results of this techniques [5,6]. For this reason, it should be necessary to accurately choose the inoculum densities for different soil samples using a plate count culture-based method. Moreover, a correct definition of blank is essential for calculating the related diversity indices. Biolog® EcoPlates™ have been successfully adopted in our laboratories for studying bacterial soil communities from different soil environments, polluted soils and soils subjected to various agronomic treatments [6–9]. Unfortunately, the studies regarding soil as a matrix for bacterial communities often include a Biolog® analysis but the reagents and apparatus for soil preparation, dilution and incubation time used, standard deviations and number of replicates and precise formulas for calculations are often not reported, so it becomes difficult for other researchers to understand how to correctly use this method.

On this basis and taking into account our previous research on soils, we here propose for the first time a standardized and accurate step-by-step method for estimating the functional diversity of a soil bacterial community by Biolog® EcoPlates™ assay. We tried to adopt a clear terminology and parameters, explaining them in detail, in order to facilitate the calculation of the most relevant and used Biolog®-related indices of microbial functional diversity. Even with its limitations [10], the Biolog® EcoPlates™ method remains a quick and relatively simple and inexpensive technique for comparing microbial communities. The standardization of this technique could be essential for indicating differences in CLPPs between samples within a single experiment, for comparing different experiments, soil types and management systems or simply soils sampled at different times. The aim of this report is not describing the limitations of the method itself but proposing a detailed methodological procedure, easy to be followed and reproduced and based on previous data and experiments. A case study of soils of a Mediterranean olive orchard was reported for comparing the methods and justify the methodological protocol here proposed.

2. Materials and Methods

2.1. Methodology

2.1.1. Culture-Based Plate Count Method

Plate count must be conducted under sterile conditions in a laminar-flow hood using single-use sterile plastic material and autoclaved solutions and glassware. The following materials are used: sterile flasks and tubes, sterile spatula, sterile pipettes, P1000 and P100 micropipettes with sterile tips, Petri dishes (size 90 mm, polystyrene, γ-irradiated), sterile hockey stick (disposable cell spreaders). The apparatus includes laminar flow hood, autoclave, incubator, water bath, magnetic stirrer, ultrasonic bath. The reagents are: Tryptic Soy Agar (TSA), cycloheximide (to inhibit fungal growth), 25% sterile

Ringer solution (NaCl 2.25 g L^{-1}, KCl 0.105 g L^{-1}, CaCl$_2$ 0.045 g L^{-1} and NaHCO$_3$ 0.05 g L^{-1}), 1.8% (*w/v*) sterile sodium pyrophosphate (Na$_4$P$_2$O$_7$ • 10 H$_2$O) solution.

For making and plating, 3 g of TSA powder were added to 1 L of distilled water in a 2-L glass bottle. This solution was sterilized (autoclaved) at 121 °C for 20 min, cooled at 50 °C and then 100 µg cycloheximide mL^{-1} were added mixing thoroughly. Finally, 20 mL of the solution were poured in each Petri dish.

For plate counting, in order to obtain a 0.18% (*w/v*) Na$_4$P$_2$O$_7$ • 10 H$_2$O final concentration, 4.5 mL of 1.8% sodium pyrophosphate and 40.5 mL of 25% Ringer solution were added to 5 g (dry weight equivalent) of fresh soil. The suspension was sonicated for 2 min and soil particles were allowed to settle at 4 °C for 15 min. Then, ten-fold serial dilutions of the supernatant up to 10^{-7} in sterile Ringer solution were done, spreading a 100 µL-aliquot of each dilution onto a TSA plate (3-5 replicates for each dilution) and incubating at 28 °C for 72 h. For each sample the suited dilution to enumerate colony forming units (CFUs) for g of dried weight soil was chosen. Then, for microplate incubations, the dilution leading to ~10^4 CFUs mL^{-1} solution was selected for microplate incubation.

2.1.2. Microplate Incubation

The following materials are used: multichannel pipet and sterile tips, sterile plastic multichannel reservoir, Biolog® EcoPlates™ (Biolog Inc., Hayward, CA, USA). The apparatus includes laminar flow hood, incubator, agitator and Biolog® Microplate Reader™ equipped with 750-nm and 590-nm filters.

For preparing the microplate, 10 mL of the dilution that was chosen in plate counting was poured into a sterile reservoir of an 8-channel pipet (be careful there are no bubbles in the dilution) and 120 µL of the dilution were inoculated into each well of a microplate. Not the same dilutions of the counting assay were used but new fresh dilutions were prepared. Then, the microplate was placed in its bag to avoid desiccation and incubated at 25 °C in dark, continuously shaking at 50 rpm on an agitator with tilting platform to obtain a uniform distribution of triphenyl formazan. Finally, spectrophotometric readings at both 590 nm (OD$_{590}$) and 750 nm (OD$_{750}$) were taken at time 0 and at 12-h increments ($\pm 12_{Xh}$) up to a 144-h incubation.

In order to select the optimal incubation time for microplate analyses (as explained later in the Results and Discussion Section), it is recommended to follow this pattern for each incubation time (as an example, measuring times at 0 h and X h are given here):

- calculate a colour value for each substrate well i and the blank (water) well b for each incubation time by subtracting the OD$_{750}$ value from the OD$_{590}$ value:

$$0\text{ h:}\quad i_{0h} = \text{OD}_{590} - \text{OD}_{750} \quad \text{and} \quad b_{0h} = \text{OD}_{590} - \text{OD}_{750}$$

$$X\text{ h:}\quad i_{Xh} = \text{OD}_{590} - \text{OD}_{750} \quad \text{and} \quad b_{Xh} = \text{OD}_{590} - \text{OD}_{750}$$

- subtract the blank well OD reading from the OD value of each substrate well to obtain a blank-corrected value (i_{bc}) for each well:

$$0\text{ h:}\quad i_{bc0h} = i_{0h} - b_{0h}$$

$$X\text{ h:}\quad i_{bcXh} = i_{Xh} - b_{Xh}$$

- subtract the blank-corrected OD reading at time 0 from subsequent blank-corrected readings at $\pm 12_{Xh}$ to obtain colour development values (c_i) for each well for each incubation time: for example, $c_{iXh} = i_{bcXh} - i_{bc0h}$ and set negative values to 0
- calculate the average well colour development (*AWCD*) for all incubation times separately using the equation:

$$AWCD = \sum \frac{c_i}{93}$$

2.1.3. Utilizing the AWCD and c_i Values

The *AWCD* calculated above is an estimate of the total capacity of a bacterial community to use different carbon compounds. Using the c_i values of the chosen incubation time, it is possible to further calculate indices of bacterial functional diversity [2,11], such as:

(a) Richness (S), which is the number of utilized carbon substrates, using an OD value of 0.250 as threshold for positive response

(b) Shannon's diversity index (H'), which is related to the number of carbon substrates the bacterial community is able to degrade

$$H' = -\sum p_i \, (\ln p_i)$$

where p_i is c_i divided by the sum of all the c_i values.

(c) and Shannon's evenness index (E), which particularly focuses on the evenness of ci values across all utilized substrates

$$E = \frac{H'}{\ln S}$$

For a more detailed analysis, the carbon substrates can eventually be divided into eight classes of compounds (polysaccharides and complex compounds, cellulose, hemicellulose, chitin, phosphorylated compounds, organic acids, amino acids and biogenic amines) and the *AWCD* and diversity indices calculated for each group separately (see examples in References [4,5] and in Figure 1).

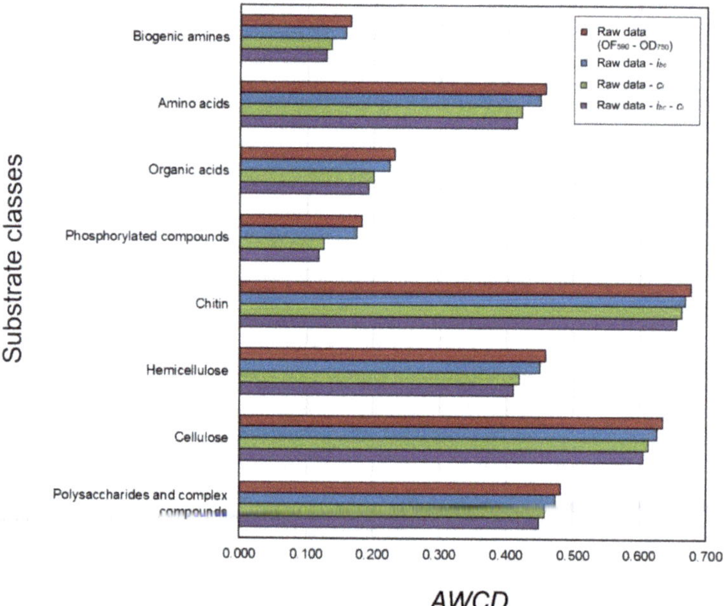

Figure 1. Mean values ($n = 5$) of average well colour development (*AWCD*) measured after a 96-h incubation in dilutions ($\sim$$10^4$ CFUs mL^{-1}) of soils from the experimental olive orchard. The values were calculated subtracting from the raw data (OD$_{590}$ – OD$_{750}$) the blank well optical density (OD) reading from the OD value of each substrate well (i_{bc}) for each well and successively subtracting the blank-corrected OD reading at time 0 from subsequent blank-corrected readings at $\pm12_{Xh}$ (c_i) (in purple). The same methodology was carried out without subtracting the blank well OD reading (in green), without subtracting the blank-corrected OD reading at time 0 (in blue) or without subtracting both (raw data in red).

2.2. Case Study and Methods Comparison

Experimental Field and Soil Sampling

The trial was done in a 1-ha olive orchard (*Olea europaea* L., cv. 'Maiatica'; 70-year-old plants with a distance of 8 × 8 m; NE orientation) located in Ferrandina (Southern Italy, Basilicata region; N 40°29′; E 16°28′). The area has a semi-arid climate, with an annual rainfall of 560 mm (mean 1995–2018), concentrated mostly in the winter and a mean annual temperature of 16.3 °C (mean 1995–2018). The soil is a sandy loam, a Haplic Calcisol, according to the World Reference Base for Soil Resources, with a mean bulk density of 1.30 g cm^{-3} and sediment as parental material.

The orchard was managed accordingly to sustainable/conservation agricultural practices, namely: (a) minimum tillage and cover crop application (30 kg ha^{-1} of *Trifolium subterraneum* seeds and spontaneous grass); (b) guided fertilization based on plant nutrient demand evaluated by leaf mineral analyses and on soil measured nitrogen levels; (c) compost amendment (15 t ha^{-1} fresh weight); (d) incorporation of cover crop and pruning residues into the soil (light harrowing at a depth of 10 cm carried out in Autumn); (e) guided drip irrigation based on crop evapotranspiration (3 drip emitters per plant along the tree lines with a capacity of 4 L h^{-1} each); (f) pruning aimed to vegetative-productive equilibrium of plants (winter pruning based on the selection of shoots with a high number of floral buds and on a better light interception in the canopy).

In October 2018, soil sampling was done in the undisturbed inter-row area. Soil sub-samples were picked in 10 points over a 1-m radius area around each olive tree from the topsoil layer (0–10 cm) for bacterial communities' analysis by Biolog® assay. The 10 sub-samples were pooled on site to constitute a composite soil sample of about 1 kg. Five composite samples ($n = 5$), each composed of 10 different sub-samples, were prepared, in order minimize spatial variability. After removing visible crop residues, roots and pebbles with sterile tweezers and slightly mixing and homogenizing with a sterile spatula, the soil composite samples were immediately stored in sterilized plastic bags at 4 °C and subsequently analysed within 5 days using the procedure described in the previous paragraph.

3. Results and Discussion

3.1. Influence of Bacterial Inoculum Densities

For the best discrimination of bacterial communities, it is of fundamental importance to choose the shortest incubation time at which *AWCD* reaches a peak value before the following constant phase. Figure 2A shows the *AWCD* values measured every 12 h during a 144-h incubation from different soil dilutions. In our case the dilutions examined were 10^{-5}, corresponding to ~10^4 colony forming units (CFUs) mL^{-1} measured by plate counting, 10^{-4} and 10^{-6}. It appears that the dilution with 10^{-6} dilution contained a relatively low number of bacteria (~10^3 CFUs mL^{-1}) causing an increasing *AWCD* trend over time, without reaching a constant phase before 144 h and high variability among replicates, as showed by the high values of standard deviation (Figure 2A). On the other side, the 10^{-4} dilution likely had too many bacteria (10^5 CFUs mL^{-1}) and *AWCD* reached a peak after only 48 h, followed by a slight decline likely due to substrates or triphenyl formazan degradation, without a clear constant phase (Figure 2A). Instead, the dilution of 10^{-5} (~10^4 CFUs mL^{-1}) reached a constant phase at 96 h and so it was chosen as the time for measuring *AWCD* values (Figure 2A). Similar trends were observed both for H' (Figure 2B) and E values (Figure 2C), that had a constant phase starting from 96 h in the dilution of 10^{-5} (~10^4 CFUs mL^{-1}), with low values of standard deviation from 96 to 120 h of incubation (Figure 2B,C). The common pattern of the graphs of Figure 2 likely depends on the fact that the calculation of *AWCD*, H' and E are all based on the values of c_i, as previously explained.

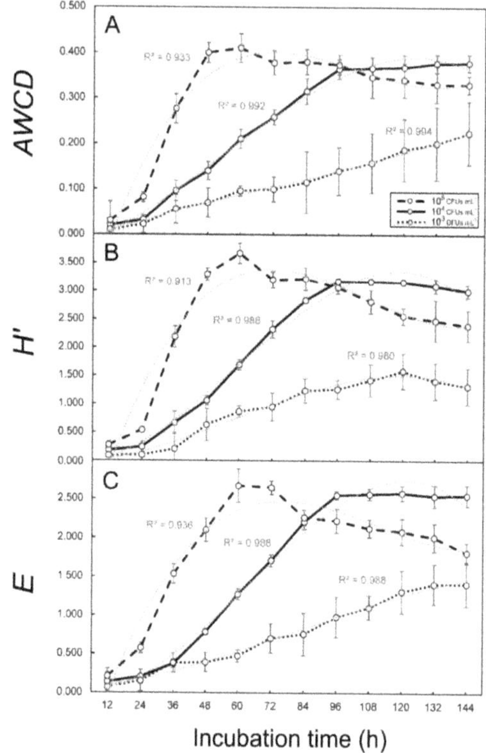

Figure 2. Mean values (n = 5) ± SD of (**A**) *AWCD*, (**B**) *H′* and (**C**) *E* measured every 12 h during a 144-h incubation in different dilutions (~10^5 CFUs mL^{-1}, dashed line; ~10^4 CFUs mL^{-1}, continuous line; ~10^3 CFUs mL^{-1}, dotted line) of soils from the experimental olive orchard. Na$_4$P$_2$O$_7$ • 10 H$_2$O concentration was 1.8% (w/v). The R^2 values and the respective trend lines (dotted lines in grey) were calculated on the basis of a polynomial function of degree three.

Comparing our results with those of other studies [10–16] using our protocol, we managed to obtain more stable growth curves for *AWCD*, particularly in the constant phase and relatively low values of SD (Figure 2A). Another important remark is that the inoculum could be relativized simply resuspending soil mass, in order to not to artificially homogenize differences that are intended to be detected among the different soil samples of the same experiment [17–20]. In order to avoid this possible artefact, in our case, the same dilution (leading to ~10^4 CFUs mL^{-1}) was chosen for all the samples analysed in the same experiment.

3.2. Correct Calculation of Blank

The *AWCD* values measured on the 10^{-5} dilutions at 96 h were calculated subtracting from the raw data (OD$_{590}$ – OD$_{750}$) the blank well OD reading from the OD value of each substrate well (i_{bc}) for each well and successively subtracting the blank-corrected OD reading at time 0 from subsequent blank-corrected readings at ±12$_{Xh}$ (c_i) for both 0-h and 96-h incubation times. The same methodology was carried out (a) without subtracting the blank well OD reading, (b) without subtracting the blank-corrected OD reading at time 0 or (c) without subtracting both. The results, depicted in Figure 1, were quite different on the basis of the type of calculation done, particularly for some classes of carbon substrates (e.g., phosphorylated compounds), showing that it is necessary to apply the sequential subtractions and that non-specific OD values must be taken into account for obtaining reliable *AWCD* data.

3.3. Influence of Soil Preparation

On the same soil samples (10^{-5} dilution, reading at 96 h, subtraction of both blank well and blank-corrected OD readings) and on the basis of previous experiments [7,9], eight concentrations of $Na_4P_2O_7 \bullet 10\,H_2O$ ranging from 0.0% to 3.0% (*w/v*) (0, 0.6, 1.0, 1.4, 1.8, 2.2, 2.6 and 3.0%) were used for finding the optimal concentration to be used. Due to its high number of electric charges once in solution, it is known that this compound has the ability to disperse charged soil colloids. This is particularly important for soil microbiological assays, as bacteria are not only present on the surface of soil aggregates but also inside the aggregates, that is, on the micropore and macropore walls and in their internal spaces [6]. Figure 3 shows that with more than 1.8% (*w/v*) $Na_4P_2O_7$ soils did not show an increase in *AWCD*, *H'* and *E*, probably because the maximum possible number of bacteria was extracted from soil. Of course, this value could vary for different soil types but we can suggest that a concentration of around 2.0% (*w/v*) is necessary for not underestimating *AWCD* values. The use of the appropriate concentration of $Na_4P_2O_7$ allows to overcome the technical problem in extracting the highest number of microorganisms from soil samples, that was one of the main remarks of this technique highlighted by Preston-Mafham et al. [10].

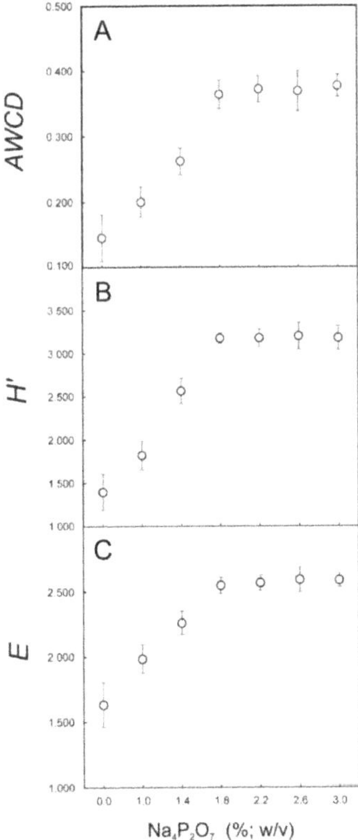

Figure 3. Mean values (*n* = 5) ± SD of (**A**) *AWCD*, (**B**) *H' and* (**C**) *E* measured a 96-h incubation in dilutions (~10^4 CFUs mL^{-1}) of soils from the experimental olive orchard. The dilutions were obtained using different $Na_4P_2O_7 \bullet 10\,H_2O$ concentrations.

4. Conclusions

A detailed and step-by-step description in the materials and methods sections of the scientific articles dealing with Biolog$^{®}$ is often lacking, making this method hardly reproducible. For this reason, we hope that the description of this method can help to solve the main issues related to the Biolog$^{®}$ EcoPlates™ assay: (1) soil preparation, dispersion and dilution (importance of using composite samples and $Na_4P_2O_7$); (2) the optimal bacterial density of the inoculum (~10^4 CFUs mL^{-1}); (3) a correct definition for blank ($c_{iXh} = i_{bcXh} - i_{bc0h}$); and (4) the use of unequivocal terminology and parameters in the calculation of diversity indices, as reported in the formulas. The previously-cited problems make this technique apparently difficult to reproduce and hard to standardize. This notwithstanding, we tried to demonstrate that, if rigorously executed, Biolog$^{®}$ EcoPlates™ assay can be a relatively simple and powerful technique, extremely useful for evaluating soil bacterial functional and metabolic diversity. The results of this methodological paper could be important for correctly evaluating and comparing the microbiological fertility of soils managed by sustainable/conservation or conventional/non-conservation systems.

Author Contributions: Conceptualization, A.S.; Methodology, P.R. and A.S.; Validation, P.R.; Data Curation, A.S.; Writing—Original Draft Preparation, A.S.; Writing—Review & Editing, P.R. and A.S.

Funding: This work was partly supported by an OECD Co-operative Research Programme grant: Biological Resource Management for Sustainable Agricultural Systems. Directorate: T AD/CRP; Contract: JA00091460.

Conflicts of Interest: The authors declare no conflict of interest.

References

1. Garland, J.L.; Mills, A.L. Classification and characterization of heterotrophic microbial communities on the basis of patterns of community-level sole-carbon-source utilization. *Appl. Environ. Microbiol.* **1991**, *57*, 2351–2359. [PubMed]
2. Zak, J.C.; Willig, M.R.; Moorhead, D.L.; Wildman, H.G. Functional diversity of microbial communities: A quantitative approach. *Soil Biol. Biochem.* **1994**, *26*, 1101–1108. [CrossRef]
3. Calbrix, R.; Laval, K.; Barray, S. Analysis of the potential functional diversity of the bacterial community in soil: A reproducible procedure using sole-carbon-source utilization profiles. *Eur. J. Soil Biol.* **2005**, *41*, 11–20. [CrossRef]
4. Gelsomino, A.; Badalucco, L.; Ambrosoli, R.; Crecchio, C.; Puglisi, E.; Meli, S.M. Changes in chemical and biological soil properties as induced by anthropogenic disturbance: A case study of an agricultural soil under recurrent flooding by wastewaters. *Soil Biol. Biochem.* **2006**, *38*, 2069–2080. [CrossRef]
5. Zhang, X.; Xu, S.; Li, C.; Zhao, L.; Feng, H.; Yue, G.; Ren, Z.; Cheng, G. The soil carbon/nitrogen ratio and moisture affect microbial community structures in alkaline permafrost-affected soils with different vegetation types on the Tibetan plateau. *Res. Microbiol.* **2014**, *165*, 128–139. [CrossRef] [PubMed]
6. Sofo, A.; Ricciuti, P.; Fausto, C.; Mininni, A.N.; Crecchio, S.; Scagliola, M.; Malerba, A.D.; Xiloyannis, C.; Dichio, B. The metabolic and genetic diversity of soil bacterial communities is affected by carbon and nitrogen dynamics: A qualitative and quantitative comparison of soils from an olive grove managed with sustainable or conventional approaches. *Appl. Soil Ecol.* **2019**, *137*, 21–28.
7. Sofo, A.; Ciarfaglia, A.; Scopa, A.; Camele, I.; Curci, M.; Crecchio, C.; Xiloyannis, C.; Palese, A.M. Soil microbial diversity and activity in a Mediterranean olive orchard using sustainable agricultural practices. *Soil Use Manag.* **2014**, *30*, 160–167. [CrossRef]
8. Sofo, A.; Palese, A.M.; Casacchia, T.; Celano, G.; Ricciuti, P.; Curci, M.; Crecchio, C.; Xiloyannis, C. Genetic, functional, and metabolic responses of soil microbiota in a sustainable olive orchard. *Soil Sci.* **2010**, *175*, 81–88. [CrossRef]
9. Fourati, R.; Scopa, A.; Ben Ahmed, C.; Ben Abdallah, F.; Terzano, R.; Gattullo, C.E.; Allegretta, I.; Galgano, F.; Caruso, M.C.; Sofo, A. Biochemical responses and oil quality parameters in olives plants and fruits exposed to airborne metal pollution. *Chemosphere* **2017**, *168*, 514–522. [CrossRef] [PubMed]
10. Preston-Mafham, J.; Boddy, L.; Randerson, P.F. Analysis of microbial community functional diversity using sole-carbon-source utilisation profiles: A critique. *FEMS Microbiol. Ecol.* **2002**, *42*, 1–14. [PubMed]

11. Xu, W.; Ge, Z.; Poudel, D.R. Application and optimization of Biolog Ecoplates in functional diversity studies of soil microbial communities. *MATEC Web Conf.* **2015**, *22*, 04015. [CrossRef]

12. Bending, G.D.; Turner, M.K.; Jones, J.E. Interactions between crop residue and soil organic matter quality and the functional diversity of soil microbial communities. *Soil Biol. Biochem.* **2002**, *34*, 1073–1082. [CrossRef]

13. Bucher, A.E.; Lanyon, L.E. Evaluating soil management with microbial community-level physiological profiles. *Appl. Soil Ecol.* **2005**, *29*, 59–71. [CrossRef]

14. Govaerts, B.; Mezzalama, M.; Sayre, K.D.; Crossa, J.; Litcher, K.; Troch, V.; Vanherck, K.; De Corte, P.; Deckers, J. Long-term consequences of tillage, residue management, and crop rotation on selected soil micro-flora groups in the subtropical highlands. *Appl. Soil Ecol.* **2008**, *38*, 197–210. [CrossRef]

15. Liu, B.; Li, Y.; Zhang, X.; Wang, J.; Gao, M. Effects of chlortetracycline on soil microbial communities: Comparisons of enzyme activities to the functional diversity via Biolog EcoPlates™. *Eur. J. Soil Biol.* **2015**, *68*, 69–76. [CrossRef]

16. Feigl, V.; Ujaczki, É.; Vaszita, E.; Molnár, M. Influence of red mud on soil microbial communities: Application and comprehensive evaluation of the Biolog EcoPlate approach as a tool in soil microbiological studies. *Sci. Total Environ.* **2017**, *595*, 903–911. [CrossRef] [PubMed]

17. Orlando, J.; Chávez, M.; Bravo, L.; Guevara, R.; Carú, M. Effect of *Colletia hystrix* (Clos), a pioneer actinorhizal plant from the Chilean matorral, on the genetic and potential metabolic diversity of the soil bacterial community. *Soil Biol. Biochem.* **2007**, *39*, 2769–2776. [CrossRef]

18. Orlando, J.; Carú, M.; Pommerenke, B.; Braker, G. Diversity and activity of denitrifiers of Chilean arid soil ecosystems. *Front. Microbiol.* **2012**, *3*, 101. [CrossRef] [PubMed]

19. Almasia, R.; Carú, M.; Handford, M.; Orlando, J. Environmental conditions shape soil bacterial community structure in a fragmented landscape. *Soil Biol. Biochem.* **2016**, *103*, 39–45. [CrossRef]

20. Leiva, D.; Clavero-León, C.; Carú, M.; Orlando, J. Intrinsic factors of *Peltigera* lichens influence the structure of the associated soil bacterial microbiota. *FEMS Microbiol.* **2016**, *92*, fiw178. [CrossRef] [PubMed]

Article

Influence of Increasing Tungsten Concentrations and Soil Characteristics on Plant Uptake: Greenhouse Experiments with *Zea mays*

Gianniantonio Petruzzelli * and Francesca Pedron

Institute of Research on Terrestrial Ecosystems (IRET), National Council of Research, 56124 Pisa, Italy; francesca.pedron@cnr.it
* Correspondence: gianniantonio.petruzzelli@cnr.it

Received: 4 September 2019; Accepted: 20 September 2019; Published: 24 September 2019

Featured Application: This work can contribute to the knowledge on the behavior of tungsten in the soil in relation to its potential transfer into the food chain through plant uptake. The results are aimed to bring the attention of the legislator to this element, which is completely neglected in environmental legislation.

Abstract: Tungsten is largely used in high-tech and military industries. Soils are increasingly enriched in this element, and its transfer in the food chain is an issue of great interest. This study evaluated the influence of soil characteristics on tungsten uptake by *Zea mays* grown on three soils, spiked with increasing tungsten concentrations. The soils, classified as Histosol, Vertisol, and Fluvisol, are characteristic of the Mediterranean area. The uptake of the element by *Zea mays* was strictly dependent on the soil characteristics. As the pH of soils increases, tungsten concentrations in the roots and shoots of the plants increased. Also, humic substances showed a great influence on tungsten uptake, which decreased with increasing organic matter of soils. Tungsten uptake by *Zea mays* can be described by a Freundlich-like equation. This soil-to-plant transfer model may be useful in promoting environmental regulations on the hazards of this element in the environment.

Keywords: Tungsten; corn uptake; soil characteristics; Freundlich model

1. Introduction

Tungsten (W) is a metal characterized by a very low abundance in the Earth's crust, and thus occurs in soils in small concentrations of less than 3 mg kg^{-1} [1]. Tungsten occurs naturally in several minerals such as wolframite (Fe,Mn)WO$_4$, hübnerite MnWO$_4$, ferberite FeWO$_4$, and scheelite (CaWO$_4$). It is used in a variety of industrial activities for the production of high-tech components [2] due to its high melting point and resistance to corrosion.

Tungsten has also been used in several military applications as a substitute for the dangerous depleted uranium [1,3]. The most important use of tungsten is in the production of tungsten/nylon bullets as alternatives to lead-based munitions [4]. Soils are an important sink and source of tungsten in the environment as a result of the transport of dust deriving from erosion processes. These last have been recognized as one of the most important contributions to the dispersion of tungsten into the environment [1].

In soil, tungstate is the most stable form of the element. This oxyanion is chemically similar to molybdate, however, while molybdenum is an essential element [5], tungsten is not an essential micronutrient, and it has been historically regarded as inert in the soil environment and classified as nontoxic for human health [6].

However, several ecotoxicological studies have discovered that under certain environmental conditions, tungsten compounds can be solubilized and enter the biogeochemical cycles [7,8]. Occupational studies on workers' exposure in industry where tungsten is used have revealed several adverse health effects [9]. The presence of tungsten in the atmosphere has been particularly studied in Nevada, due to the possible link between tungsten in the atmosphere and cases of childhood leukemia [10,11].

Knowledge regarding the effects of tungsten in the soil, on its transfer in the food chain, and thus on human health is still incomplete [1]. In some cases, tungsten has been found to be nontoxic and substantially inert in the soil [12], whereas in other studies, very negative hypotheses have been put forward on the possible action of tungsten in the formation of neoplasms in animal cells [7]. The growing interest in this element is further highlighted by several studies on the impact of W on soil microorganisms. Several works have shown that under certain conditions, the presence of tungsten increases the microbial biomass [13]. Conversely, in other cases, negative impacts on microbial activity have been verified [14].

The increase in the industrial and military use of tungsten has become of interest due to its potential entry into the food chain following uptake by agricultural crops. Moreover, in agricultural soils tungsten is also added to fertilizers [15], and may be present in sewage sludge due to its presence in many household products [16]. Many plants such as oats, radishes, and lettuce are able to grow in soil with high tungsten concentrations [17] and can accumulate significant quantities of the metal also due to its similarity with molybdenum [14]. At high concentrations, tungsten has been reported to decrease root elongation in *Pisum sativum* and *Gossypium hirsutum* [18–20]. High concentrations of tungsten in plants have been reported for species growing on the soils of abandoned mine [21–23]. Near tungsten mines, its transfer from soil to rice has been reported as a possible health risk via the food chain [24]. In general, the tungsten content in roots is higher than that in the aerial parts of plants [25,26].

Studies on tungsten are very lacking in Italy where the potential problems arising from this element are completely neglected. This element is not taken into consideration in any environmental legislation, nor in the regulations that define the quality of fertilizers to be used in agriculture. It therefore seems no longer unpostponable to study the behavior of this element in Italian soils with a view to protecting the environment and human health.

The aim of this study was to investigate the influence of different soil characteristics pH, organic matter (O.M.), and texture on tungsten uptake by *Zea mays*, and to provide a simple model to evaluate the uptake of increasing tungsten concentration in soil. *Zea mays* was selected due to its relevant production in the Mediterranean area, which highlights the growing interest in this crop. In Italy, the average production for over a decade has been ~9.5 t ha^{-1}, which is higher than the mean European production. The experiment was carried out at a greenhouse scale by growing the plants on three agricultural soils characteristic of Mediterranean areas spiked with increasing concentrations of sodium tungstate. The concentrations were chosen to be close to natural soil values conditions and to take into account potential anthropogenic contamination, which can reach 2000 mg/kg in particular areas [24]. The chosen range of concentrations can be considered environmentally significant due to the increase in electronic waste, which is a feature of all industrialized countries [14]. Obtained results show that soil properties regulate tungsten plants uptake, which increases with increasing pH and decreasing organic matter content.

2. Materials and Methods

2.1. Soils

Samples were collected from three Italian soils, classified as Histosol (soil A), Vertisol (soil B), and Fluvisol (soil C), according to FAO classification. The soil samples were air-dried, sieved at 2 mm, and analyzed for pH [27], organic matter [28], and texture [29]. The studied soils characteristics are reported in Table 1.

Table 1. Characteristics of the soils A, B, and C.

Soil	pH	OM (%)	Clay (%)	Silt (%)	Sand (%)	Total W (mg kg^{-1})
A	4.50	5.81	7.10	24.8	67.6	0.38
B	5.80	2.10	12.5	28.8	59.3	0.44
C	7.40	1.43	7.65	14.0	78.3	0.29

2.2. Experimental Soils

Increasing amounts of $Na_2WO_4 \cdot H_2O$ solutions were added to each soil to obtain final concentrations of 10, 30, 40, 50, and 100 mg kg^{-1} of tungsten.

To promote the interactions between soil and the added tungsten salt and to simulate the ageing effects, the soils were incubated at 50% maximum water holding capacity for 12 months at room temperature.

During this period, the spiked soils were maintained in loosely covered polypropylene containers. To obtain the greatest homogeneity, once a week soils were gently mixed by a shovel, and watered with deionized water once a month as necessary [30].

At the end of this period, before starting the microcosm experiments the pH of each spiked soils was determined [27] to check if there were any changes in the values of this parameter. No changes were detected.

To determine tungsten in the soil solution, deionized water was added to each spiked soil samples (1:0.5 soil/water), and after 24h of incubation, tungsten in soil solution was evaluated by centrifugation at 21,000 g for 15 min [30]. Three replicates for each concentration were carried out.

2.3. Microcosm Experiment

The experiment was carried out at a greenhouse scale with a 12h photoperiod (320 µmol m^{-2} s^{-1} photosynthetic active radiation). Microcosms were filled with 400 g of soil. Six *Zea mays* seeds per pot were sown. Three replicates for each concentration were carried out with controls (CT) running simultaneously. During the growing period, plants were watered daily. The whole experiment lasted 60 days. At plant harvest, the aerial parts were separated from the roots and pooled to a composite sample representative of each pot. All samples were washed with deionized water. The roots were further washed in an ultrasound bath (Branson Sonifier 250 ultrasonic processor, Branson Ultrasonics Corporation, USA) for 10 min, to eliminate any soil particles remaining on root surfaces. The dry biomass of vegetal samples was then gravimetrically determined after the samples had been dried in a ventilated oven at 60°C until a constant weight was achieved.

2.4. Tungsten Analysis in Plants and Soil

The dry plant samples were ground, homogenized, and digested with an acid mixture ($HNO_3 + H_2O_2$) for tungsten determination, according to the procedure described by Poykio et al. [31] and Oburger et al. [30].

Concentrations of tungsten in spiked soils and soil solution were determined by inductively coupled plasma optical emission spectroscopy (ICPOES Varian AX Liberty, Milano, Italy). Due to the specific tungsten properties in soils that can lead to the formation of insoluble species, the procedure based on the addition of phosphoric acid to nitric acid developed by Dermatas et al. (2004) [32] and reported by Bednar et al. [33] was used. All chemicals used were of reagent grade.

Quality assurance and quality control were carried out by testing a standard solution every 10 samples. Because no reference plant material for tungsten is available, for quality control the procedure developed and described in detail by Oburger et al. [30] was strictly followed using Oriental

Basma Tobacco Leaves, INCT-OBTL-5, as reference material. As a certified reference soil material, NIST SRM 2710 was used to control the quality of the analytical system. The detection limit for tungsten was 0.05 mg L^{-1}. The recovery of spiked samples ranged from 95 to 101%, with an RSD of 1.84 of the mean.

2.5. Statistical Analysis

Statistical analysis was performed using STATISTICA version 6.0 (Statsoft, Inc., Tulsa, OK, USA). Treatment effects were analyzed using one-way analysis of variance (ANOVA). Differences among means were compared and a post hoc analysis of variance was performed using the Tukey Honestly Significant Difference test ($p < 0.05$).

3. Results and Discussion

3.1. Biomass Production

The biomass of plants grown in the three soils with different tungsten concentrations is reported in Table 2. Shoot and root productions are considered separately.

Table 2. Biomass production (mg dry weight) of shoot and root grown in the soils with increasing tungsten concentrations. CT is the original soil. Values are the mean ± standard deviation.

W in soil (mg kg^{-1})	Shoot			Root		
	Soil A	Soil B	Soil C	Soil A	Soil B	Soil C
CT	283 ± 13.2a	279 ± 10.4a	285 ± 13.8a	118 ± 7.2a	116 ± 12.4a	126 ± 9.0a
10	281 ± 14.6a	275 ± 14.7a	295 ± 19.3a	115 ± 9.1a	114 ± 10.1a	124 ± 8.3a
30	276 ± 15.1a	257 ± 14.1a	278 ± 15.4a	123 ± 4.3a	117 ± 7.1a	118 ± 7.6a
40	263 ± 11.2a	263 ± 12.9a	251 ± 16.1a	129 ± 8.6a	121 ± 8.6a	122 ± 8.1a
50	284 ± 18.6a	259 ± 10.1a	245 ± 18.5a	110 ± 9.6a	125 ± 5.9a	119 ± 7.4a
100	251 ± 16.3a	269 ± 16.5a	269 ± 13.6a	119 ± 5.1a	116 ± 4.9a	115 ± 5.9a

Note: At each tungsten soil concentration, values with different letter in the same rows separately for shoots and roots are statistically different at $p < 0.05$.

Despite the different characteristics, such as pH and organic matter content, no significant differences in biomass were observed among the three soils. At the tungsten concentrations used in this experiment, no differences in biomass production were found between the control and tungsten enriched soils. The results are in agreement with previous findings for another species, *Triticum aestivum*, which reported that biomass was not influenced by low tungsten concentrations in soil [34,35]. A reduction in biomass production was discovered in soils only with very high tungsten concentrations of up to 5000 mg kg^{-1} for *Glycine max* [30] and 2600 mg kg^{-1} for *Helianthus annuus* [26]. Under the experimental conditions adopted, tungsten did not show any visual toxicity effects on plants.

3.2. Tungsten Uptake by Plants

The concentration of tungsten in plants grown in the control soils (CT) was always under the limit of detection, thus the data are not reported. Considering the plants grown in the spiked soils, the uptake of tungsten was lowest in soil A and highest in soil C, with intermediate values in soil B. Data are reported in Table 3.

Considering the mean of the five concentrations (from 10 to 100 mg kg^{-1}), the tungsten concentration in the root portions increased with respect to soil A by an average factor of 2 in soil B and of 9 in soil C, respectively.

A similar trend was found in the aerial part of the plants with an increase with respect to plants grown in soil A by an average factor of 3 and 15 in plants grown in soils B and C, respectively.

Table 3. Tungsten concentration in shoots and roots of *Zea mays* grown in the different spiked soils. Values are the mean ± standard deviation.

W in Soil (mg kg^{-1})	Shoot			Root		
	Soil A	Soil B	Soil C	Soil A	Soil B	Soil C
10	0.12 ± 0.02a	0.45 ± 0.04b	2.69 ± 0.02c	1.30 ± 0.07a	2.38 ± 0.6b	5.10 ± 0.4c
30	0.34 ± 0.08a	1.26 ± 0.3b	5.42 ± 0.4c	2.04 ± 0.3a	6.50 ± 0.9b	19.8 ± 1.1c
40	0.98 ± 0.07a	3.05 ± 0.5b	11.7 ± 0.8c	4.51 ± 0.4a	9.80 ± 2.3b	45.3 ± 3.8c
50	1.26 ± 0.08a	3.98 ± 0.3b	13.5 ± 0.6c	5.15 ± 0.8a	13.6 ± 2.1b	52.6 ± 3.0c
100	1.80 ± 0.04a	5.43 ± 0.6b	21.0 ± 1.3 c	7.10 ± 0.9a	18.6 ± 4.2b	95.5 ± 6.1c

Note: At each tungsten soil concentration, values with different letter in the same rows separately for shoots and roots are statistically different at $p < 0.05$.

The effects of tungsten on the physiological responses of plants were not studied, unless exclusively concerning the production of biomass. There are many studies that address these aspects to which refer for further deepening [30,36]. Finally, we considered the dose–uptake relationship and the internal distribution of tungsten between the roots and shoots, which may have occurred within the time frame of this experiment. Table 3 clearly shows that the amount of tungsten in the plants differed significantly in the three soils at each tungsten concentration. Regardless of the type of soil, tungsten mainly accumulates in the roots, probably as a protective action, as tungsten does not play any essential role in the plant [37]. In fact, the translocation factor (TF), defined as the ratio between the tungsten concentration in the shoots and in roots, is always lower than 1 (Figure 1).

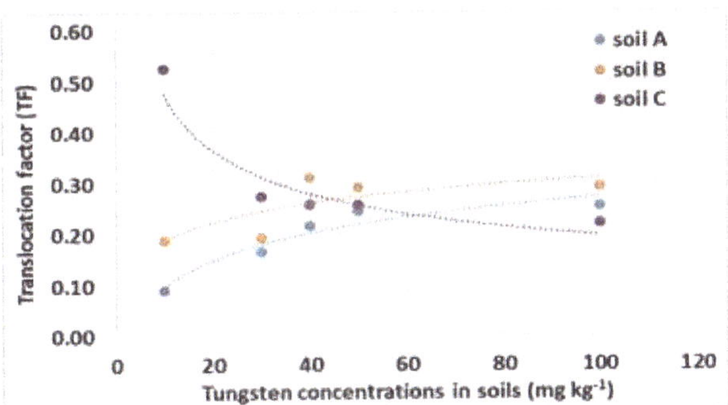

Figure 1. Trend of translocation factor at increasing tungsten concentrations in soils.

Taken as the average at the different concentrations of tungsten in the soil, the TF values ranged from 0.19 (soil A) to 0.30 (soil C), with an intermediate value (0.25) in soil B. A different behavior of TF was observed between the three soils. It is worth noting that in soil C, as the concentration of tungsten increased, the TF decreased. Since, TF describes the passage of tungsten from the roots to the aerial part of the plants, the results obtained suggest that in the presence of high concentrations of tungsten in soils, the roots play a protective role acting as a filter and accumulating organ.

These results are in agreement with those reported for *Raphanus sativus* L. (radish) grown in an alkaline substrate [38]. Conversely, in soil A, as the concentration of tungsten increases, the TF increases. Soil B showed an intermediate trend; in fact, the TF increases up to a concentration of 40 mg kg^{-1} of tungsten added to the soil, and then decreases. The soil plant barrier concept [39] does not appear to be entirely applicable to tungsten. In fact, tungsten is not strongly held by the soil surfaces and can be absorbed by the plant roots.

3.3. Influence of Soil Characteristics on Plant Uptake

There are few studies on tungsten uptake by plants and the reported tungsten concentrations in plants are highly variable due to the differences between plant species and experimental conditions [26,30,34]. Although in hydroponic experiments, the tungsten uptake can be directly linked to the concentration of metal in solution and to specific tolerance mechanisms, the interpretation of the phytoavailability in soil is more complex since it is essentially determined by tungsten species in the soil liquid phase. It is thus fundamental to evaluate the effects of soil properties on tungsten solubility and plant uptake, in particular of soil pH, which is considered the most important parameter that determines the amount and the forms of tungsten in soil solutions [9].

Considering the different soil pH, it appears that by increasing pH, plant uptake significantly increased (Figure 2).

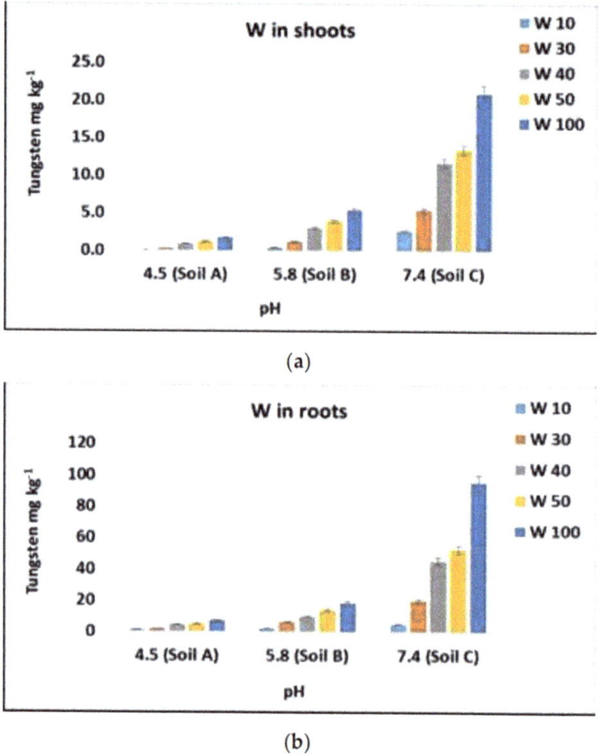

Figure 2. Mean values of tungsten concentrations in plants in relation to pH of the different soils. Bars indicate standard deviation. (**a**) shoots; (**b**) roots.

At the highest concentration (100 mg kg^{-1}) of tungsten in soil, in plants grown in soil C with a pH value of 7.40, the amount of tungsten in the root portion (95.5 mg kg^{-1}) was approximately thirteen times greater than that (7.1 mg kg^{-1}) in the roots of plants grown in soil A. The same trend was found by comparing the concentration of tungsten in the roots of plants grown in soil B (18.6 mg kg^{-1}) and soil A (7.1 mg kg^{-1}). In this case, the concentration increased by more than twice as the soil pH increased from 4.50 to 5.80. Finally, the concentration of tungsten in the roots of plants grown in soil C increased by 5.1 times compared to that found in the roots of plants grown in soil B (from 18.6 to 95.5 mg kg^{-1}).

The aboveground parts of the plant behaved similarly. As the pH increased, from soil A (4.5) to soil C (7.4), the concentration of tungsten in the aerial part of the plants increased at each concentration of tungsten in the soil. This increase was evident above all at the highest tungsten concentration (100 mg kg^{-1}), from 1.8 to 21 mg kg^{-1} of tungsten in the aerial part of the plants grown respectively in soils A and C, with an increase of about 12 times. In soil B, the concentration of tungsten in the aerial part of the plants was ~5.43 mg kg^{-1}, with an increase with respect to soil A of approximately three times. In the aerial part of the plants grown in soil C, the amount of tungsten was about four times higher than that in plants from soil B.

Increasing pH promotes the deprotonation of soil surfaces, thus reducing the retention of tungstate ions in the solid phase of soil [40]. The consequent increase in tungsten in the soil solution led to an increase in bioavailability and therefore a higher uptake by the plants. The greatest uptake occurred in the Fluvisol (soil C), which is characterized by alkaline pH values and a low content of organic matter, conditions that are very common in soils of the Mediterranean area.

The organic matter is another parameter that seems to play a key role in the absorption of tungsten by plants. Figure 3 reports the uptake trend in relation to the different values of organic matter in the soils.

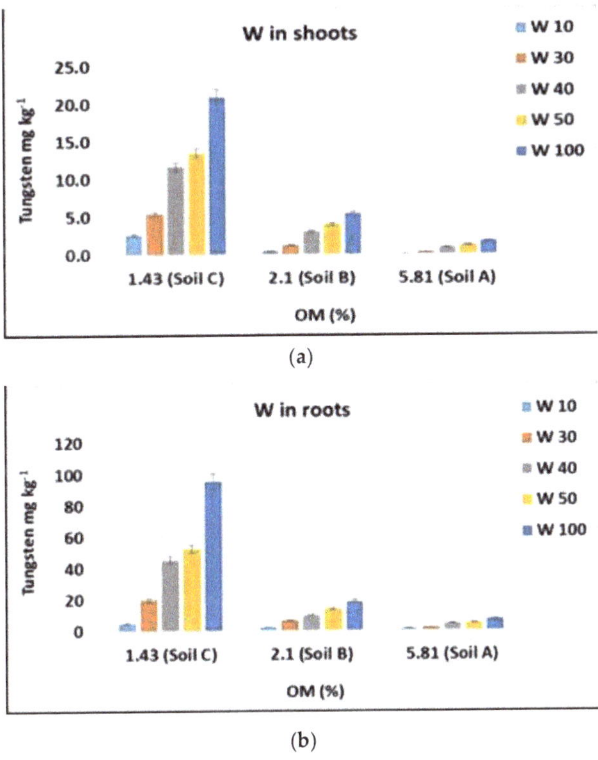

Figure 3. Mean values of tungsten concentrations in plants in relation to organic matter (OM%) of the different soils. Bars indicate standard deviation. (**a**) shoots; (**b**) roots.

The data reported in Figure 3 show that the different organic matter content in the three soils influences the concentration of tungsten in the aerial part and in the roots of the plants. As the amount of organic matter increases, the concentration of tungsten in plant tissues decreases. The action of organic matter can be ascribed to the adsorption reactions of tungstate on humic materials [41], with the formation of stable complexes [42] that reduce the tungsten mobility and bioavailability in soil. This trend is particularly notable in soil A due to the specific characteristics of this Histosol: low pH with a high organic matter content.

Humic compounds can greatly influence the distribution of metallic elements, such as tungsten, between the solid and liquid phase of the soil. This is also in relation to the environmental variations in the soil, such as dynamic redox conditions [43], that can modify the interactions between humic compounds and the surfaces of oxides and hydroxides of iron on which the metal can be adsorbed. Similar to what was found in the case of arsenic and antimony in the soil [44–46], organic materials with a high binding capacity, strongly influence the adsorption of anions on variable charge minerals in soils.

The content of tungsten in plants does not seem to be linked to the texture of the soils. For example, Figure 4 reports the uptake trend in relation to the clay content of the three studied soils. The highest clay content in the Vertisol (Soil B) does not influence plant uptake. Similar trends were also obtained considering the plant tungsten concentration in relation to the silt and sand fraction of the three soils (data not reported).

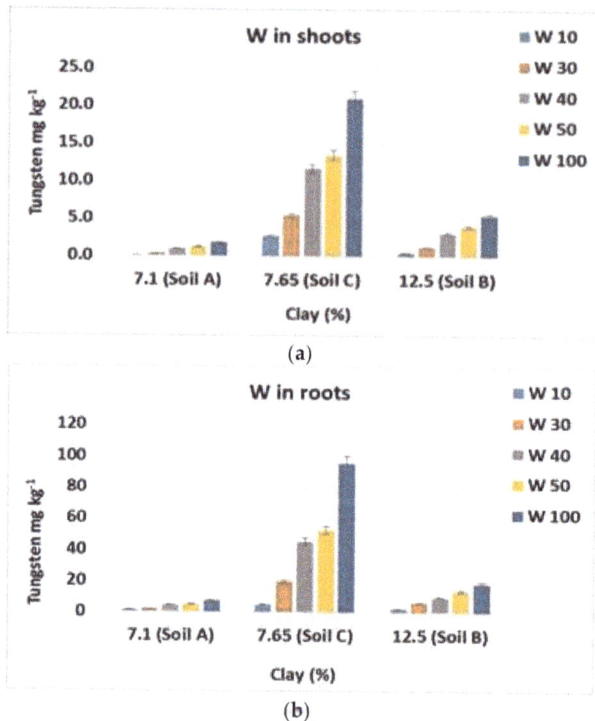

Figure 4. Mean values of tungsten concentrations in plants in relation to clay content of the different soils. Bars indicate standard deviation. (**a**) shoots; (**b**) roots.

Another interesting aspect to tackle is the trend in plant uptake of tungsten from the soil. In the study of plant uptake from soil, there is often an assumption of linearity: the concentration of a substance in plants increases with an increasing of its soil concentrations. However, deviation from linearity has been observed in several contaminant uptake studies [47–49]. Plants generally take up elements more efficiently at low than at high soil concentrations, thus deviating from linearity with a decrease in uptake with increasing soil concentrations. The use of nonlinear functions to take into account this behavior has been proposed by several authors [48,50,51].

To evaluate tungsten uptake, it may be useful to estimate the increase in concentration in plants with increasing both soil concentrations and soil solution concentrations. The trend of tungsten uptake for shoots and roots was analogous in both cases, and it is reported in Figure 5, as an example, in relation to soil solution concentration.

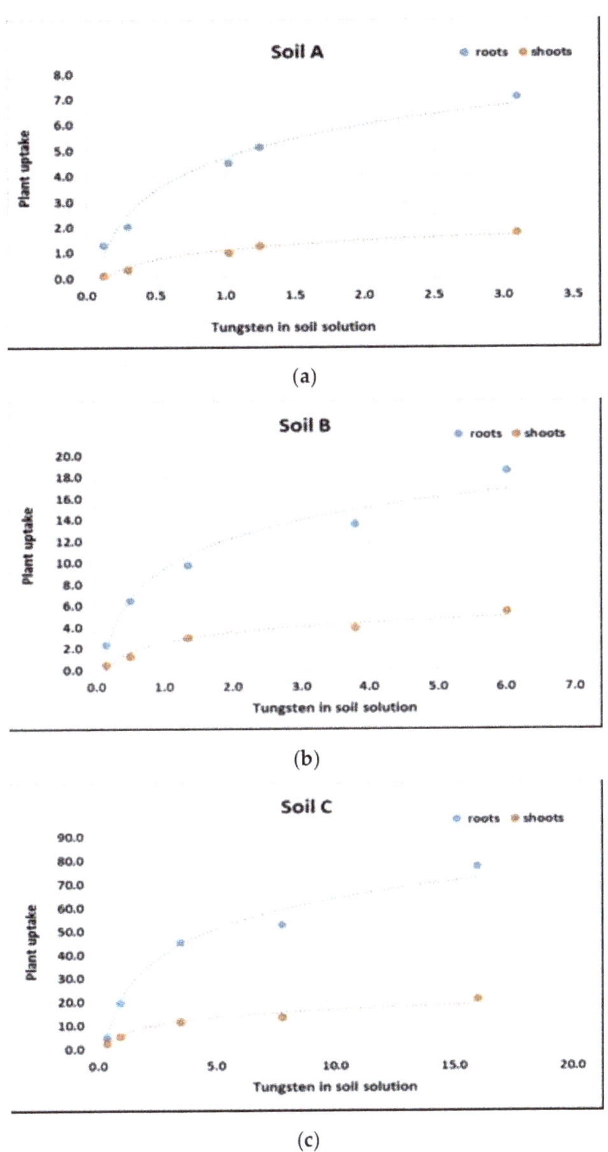

Figure 5. Trend of tungsten uptake for shoots and roots at increasing W concentrations in soil solution. Data are expressed as mg kg^{-1}. (a) Soil A; (b) Soil B; (c) Soil C.

Also, in this experiment, the increase in tungsten in the plants was not linearly correlated with an increase in tungsten concentration in soil. The uptake of tungsten by the plants, in relation to the concentration of the element in the soil, and in the soil solution although not far from linearity, is described more accurately by a trend that can be modeled using a Freundlich-like equation. The absorption is higher when the concentration in the soil is lower, whereas it is reduced when the concentration of an element in the soil increases.

In recent years, models of plant uptake have been proposed for organic [52,53] and inorganic [54,55] contaminants at contaminated sites to define the risk to human health.

In agricultural soils, where the concentration of potentially toxic elements is much lower, simpler models can be successfully used to provide useful information on the soil to plant transfer process. One of the models is based on the use of a Freundlich-like equation to evaluate plant uptake in the presence of increasing quantities of an element. This approach has been successfully used also in the case of contaminated soils [56].

The Freundlich-like equation (1):

$$q = K\,C^{\frac{1}{n}} \tag{1}$$

where q is the tungsten concentration in plants (mg kg^{-1}) and C is the tungsten concentration in the soil or soil solution (mg kg^{-1}).

It is important to underline that no thermodynamic value can be attributed to the parameters of the Freundlich-like equation, which can be used exclusively as an operational tool to relate plant uptake and tungsten soil concentration. In broad terms, similarly to the Freundlich equation, K can be considered as the uptake capacity (a larger K indicates a greater uptake).

The Freundlich parameters (Table 4) can be estimated from the linearized form of the equation (2):

$$\log q = \log K_f + \frac{1}{n}\log C \tag{2}$$

Table 4. Relationship between tungsten concentration in shoots and roots and tungsten concentration in soil. Parameters of the Freundlich-like model.

Soil	Shoots			Roots		
	Log K	1/n	R^2	Log K	1/n	R^2
A	−2.17	1.26	0.922	−0.718	0.791	0.896
B	−1.48	1.15	0.932	−0.5357	0.933	0.937
C	−0.522	0.936	0.942	−0.4886	1.24	0.944

By operationally using the Freundlich model parameters, we can infer that the uptake capacity increased when the value of K was increased. As expected, the results from this model confirm that the uptake always increased with increasing soil pH due to the alkaline pH of soil influencing the desorption of tungsten from soil surfaces, and its release in soil solution. The coefficient 1/n has been interpreted as an index of a plant's ability to control metal accumulation [47]. A value of less than 1 of the coefficient 1/n is considered an index of active tungsten uptake, whereas higher values of this index suggest the reduced ability of the plant to control the tungsten uptake. The obtained data show a different behavior of this index among soils and between the roots and shoot uptake of the plants.

The correlation between the concentrations calculated by the model and measured in the shoots was high, with R^2 values ranging from 0.922 (soil A) to 0.942 (soil C). Similar results were obtained for the root portion with R^2 values ranging from 0.896 (soil A) to 0.944 (soil C).

The same Freundlich-like model was also applied by correlating the amount absorbed by the plants with tungsten concentration in soil solution. This amount is determined by the retention/release processes, which influence the distribution of tungsten between the solid and liquid phases, and thus regulate the amount of tungsten in the soil solution from which the plants absorb the element. The results are reported in Table 5 for shoots and roots, respectively.

Also in this case, a Freundlich-like equation can be used to describe the pattern of plant uptake from soil solution with correlation coefficients higher than those of uptake vs. total content (see Table 4) with R^2 values ranging from 0.961 (soil B) to 0.973 (soil C); whereas, for the root portion, R^2 values range from 0.959 (soil B) to 0.988 (soil A).

Table 5. Relationship between tungsten concentration in shoots and roots and tungsten concentration in soil solution. Parameters of the Freundlich-like model.

Soil	Shoots			Roots		
	Log K	1/n	R^2	Log K	1/n	R^2
A	−0.066	0.863	0.970	0.619	0.553	0.988
B	0.260	0.670	0.961	0.874	0.528	0.959
C	0.6992	0.534	0.973	1.14	0.685	0.968

The K and 1/n coefficients changed, but the trend was similar to the uptake vs. total concentration, with the highest K value in soil C, and the lowest in soil A. We have to consider that soil pH exerts a great influence on W solubility due to increased sorption under acidic conditions [57], and this parameter appears to mainly determine plant uptake.

Although speciation of tungsten in soil is of great importance [9,58], especially for the polymerization phenomena [59,60], we decided to evaluate the trend of the tungsten uptake using a model as simple as possible. This approach of course can only be considered as a preliminary step in developing mechanistic models. More experiments need to be performed by varying plant species and using soils with different characteristics to improve the modeling of tungsten uptake by plants, or to develop more complex models. Clearly, the plant uptake model and the model parameters applied in this study are not necessarily valid for different plant species and soils. For example, in the case of soils subject to seasonal variations in the oxidation–reduction conditions, it will be essential to take into consideration also the redox potential [61], which may influence the mobility of the element and therefore its transferability to plants. However, this approach is very simple and can be easily adjusted in different environmental contexts where the tungsten uptake by plants needs to be monitored.

4. Conclusions

There is still a limited understanding of the behavior of tungsten in the environment [62], both because it is a not essential element and because until recently it was considered completely inert. However, tungsten has placed among the possible emerging pollutants by EPA [63]. The present work, to our knowledge, is the only one dealing with the problem of tungsten in Italian soils. The study of this element in different types of Mediterranean soils and the potential passage in the food chain, following the transfer from the soil to the plant, is very important to focus attention on an environmental problem that is currently very undervalued. Our results highlight the importance of soil properties in determining the transfer of tungsten from the soil to plants. The obtained data show that with decreasing pH and increasing organic matter content the plant uptake decreases, due to lower tungsten concentrations in soil solution in bioavailable forms.

In this work, we have taken into consideration those characteristics of the soil that are generally considered in environmental regulations. However, knowledge of the behavior of this element in the soil requires further investigation, so as to better understand the processes that regulate its bioavailability. Among these, a role of primary importance can be played by the redox potential because the oxidizing and reducing conditions determine the behavior of trace elements in soil. Redox variations in soil and the influence of these variations on the mineralogy [61,64] is an essential aspect for future research on the geochemical behavior of tungsten, which is strictly linked to plant bioavailability.

In the last few years, several studies on tungsten in plants have revealed several toxic effects related to the Mo-enzyme inhibitor and gene expression [30,36]. However, in our experiment, the addition of increasing amounts of tungsten to the soil did not cause a biomass reduction or toxicity effects in plants. It is possible that considerable amounts of tungsten accumulate in plants, especially in alkaline conditions. Under the worst conditions, plants may be insensitive to tungsten at levels that are already toxic to animals. This phenomenon has also been detected for molybdenum in the case of forage for ruminants [65].

The low phytotoxicity of tungsten highlights the need to find appropriate models aimed at predicting the passage of tungsten from the soil to plants. We applied a Freundlich-like model for *Zea mays* uptake at increasing tungsten concentrations in three different Italian soils. The aim of this approach is to respond as simply as possible to the need to assess the potential effects derived from the transfer of W into the food chain, and also to enable the regulatory authorities to better define the hazards of this element in the environment. The transfer mechanism of tungsten from the soil to roots and shoots is of great interest and should be further investigated to understand how the environmental source of this element may impact on human health, which is an urgent issue that must be addressed in the near future.

Author Contributions: Conceptualization, G.P. and F.P.; Writing – original draft, G.P. and F.P.

Funding: This research received no external funding. This study was supported by the Italian National Research Council CNR.

Acknowledgments: The authors thank Irene Rosellini for technical support.

Conflicts of Interest: The authors declare no conflict of interest.

References

1. Koutsospyros, A.D.; Braida, W.J.; Christodoulatos, C.; Dermatas, D.; Strigul, N.S. A review of tungsten: From environmental obscurity to scrutiny. *J. Hazard. Mater.* **2006**, *136*, 1–19. [CrossRef] [PubMed]

2. Koutsospyros, A.D.; Strigul, N.S.; Braida, W.J.; Christodoulatos, C. Tungsten: Environmental pollution and health effects. In *Encyclopedia of Environmental Health*; Nriagu, J.O., Ed.; Elsevier: Burlington, UK, 2011; pp. 418–426.

3. Clausen, J.L.; Korte, N. Environmental fate of tungsten from military use. *Sci. Tot. Environ.* **2009**, *407*, 2887–2893. [CrossRef] [PubMed]

4. USEPA. *Technical Fact Sheet-Tungsten*; Environmental Protection Agency: Washington, DC, USA, 2017.

5. Kaiser, B.N.; Gridley, K.L.; Ngaire Brady, J.; Phillips, T.; Tyerman, S.D. The role of molybdenum in agricultural plant production. *Ann. Bot.* **2005**, *96*, 745–754. [CrossRef] [PubMed]

6. Van der Voet, G.B.; Todorov, T.I.; Centeno, J.A.; Jonas, W.; Ives, J.; Mullick, F.G. Metals and Health: A Clinical Toxicological Perspective on Tungsten and Review of the Literature. *Mil. Med.* **2007**, *172*, 1002–1005. [CrossRef] [PubMed]

7. Kelly, A.D.R.; Lemaire, M.; Young, Y.K.; Eustache, J.H.; Guilbert, C.; Molina, M.F.; Mann, K.K. In vivo tungsten exposure alters B-cell development and increases DNA damage in murine bone marrow. *Toxicol. Sci.* **2013**, *131*, 434–446. [CrossRef] [PubMed]

8. Laulicht, F.; Brocato, J.; Cartularo, L.; Vaughan, J.; Wu, F.; Kluz, T.; Sun, H.; Oksuz, B.A.; Shen, S.; Peana, M.; et al. Tungsten-induced carcinogenesis in human bronchial epithelial cells. *Toxicol. Appl. Pharm.* **2015**, *288*, 33–39. [CrossRef] [PubMed]

9. Strigul, N. Does speciation matter for tungsten ecotoxicology? *Ecotoxicol. Environ. Saf.* **2010**, *73*, 1099–1113. [CrossRef] [PubMed]

10. Sheppard, P.R.; Ridenour, G.; Speakman, R.J.; Witten, M.L. Elevated tungsten and cobalt in airborne particulates in Fallon, Nevada: Possible implications for the childhood leukemia cluster. *Appl. Geochem.* **2006**, *21*, 152–165. [CrossRef]

11. Sheppard, P.R.; Speakman, R.J.; Ridenour, G.; Witten, M.L. Using lichen chemistry to assess airborne tungsten and cobalt in Fallon, Nevada. *Environ. Monit. Assess.* **2007**, *130*, 511–518. [CrossRef]

12. Thomas, V.G.; Roberts, M.J.; Harrison, P.T. Assessment of the environmental toxicity and carcinogenicity of tungsten-based shot. *Ecotoxicol. Environ. Saf.* **2009**, *72*, 1031–1037. [CrossRef] [PubMed]

13. Ringelberg, D.B.; Reynolds, C.M.; Winfield, L.E.; Inouye, L.S.; Johnson, D.R.; Bednar, A.J. Tungsten effects on microbial community structure and activity in a soil. *J. Environ. Qual.* **2009**, *38*, 103–110. [CrossRef] [PubMed]

14. Strigul, N.; Koutsospyros, A.D.; Arienti, P.; Christodoulatos, C.; Dermatas, D.; Braida, W.J. Effects of tungsten on environmental systems. *Chemosphere* **2005**, *61*, 248–258. [CrossRef] [PubMed]

15. Senesi, N.; Padovano, G.; Brunetti, G. Scandium, titanium, tungsten and zirconium content in commercial inorganic fertilizers and their contribution to soil. *Environ. Technol. Lett.* **1988**, *9*, 1011–1020. [CrossRef]

16. Wik, A.; Lycken, J.; Dave, G. Sediment quality assessment of road runoff detention systems in Sweden and the potential contribution of tire wear. *Water Air Soil Pollut.* **2008**, *194*, 301–314. [CrossRef]

17. Bamford, J.E.; Butler, A.D.; Heim, K.E.; Pittinger, C.A.; Lemus, R.; Staveley, J.P.; Lee, K.B.; Venezia, C.; Pardus, M.J. Toxicity of sodium tungstate to earthworm, oat, radish and lettuce. *Environ. Toxicol. Chem.* **2011**, *30*, 2312–2318. [CrossRef] [PubMed]

18. Adamakis, I.D.S.; Eleftheriou, E.P.; Rost, T.L. Effects of sodium tungstate on the ultrastructure and growth of pea (*Pisum sativum*) and cotton (*Gossypium hirsutum*) seedlings. *Environ. Exp. Bot.* **2008**, *63*, 416–425. [CrossRef]

19. Adamakis, I.D.S.; Panteris, E.; Eleftheriou, E.P. Tungsten affects the cortical microtubules of Pisum sativum root cells: Experiments on tungsten-molybdenum antagonism. *Plant Biol.* **2010**, *12*, 114–124. [CrossRef]

20. Adamakis, I.D.S.; Panteris, E.; Eleftheriou, E.P. The fatal effect of tungsten on Pisum sativum L. root cells: Indications for endoplasmic reticulum stress-induced programmed cell death. *Planta* **2011**, *234*, 21–34. [CrossRef]

21. Pratas, J.; Prasad, M.N.V.; Freitas, H.; Conde, L. Plants growing in abandoned mines of Portugal are useful for exploration of arsenic, antimony, tungsten and mine reclamation. *J. Geochem. Explor.* **2005**, *85*, 99–107. [CrossRef]

22. Wilson, B.; Pyatt, F.B. Bioavailability of tungsten in the vicinity of an abandoned mine in the English Lake District and some potential health implications. *Sci. Total Environ.* **2006**, *370*, 401–408. [CrossRef]

23. Wilson, B.; Pyatt, F.B. Bioavailability of tungsten and associated metals in calcareous soils in the vicinity of an ancient metalliferous mine in the Corbieres area, Southwestern France. *J. Toxicol. Environ. Heal. A* **2009**, *72*, 807–816. [CrossRef] [PubMed]

24. Lin, C.; Li, R.; Cheng, H.; Wang, J.; Shao, X. Tungsten Distribution in Soil and Rice in the Vicinity of the World's Largest and Longest-Operating Tungsten Mine in China. *PLoS ONE* **2014**, *9*, e91981. [CrossRef] [PubMed]

25. Jiang, F.; Heilmeier, H.; Hartung, W. Abscisic acid relations of plants grown on tungsten enriched substrates. *Plant Soil* **2007**, *301*, 37–49. [CrossRef]

26. Johnson, D.R.; Inouye, L.S.; Bednar, A.J.; Clarke, J.U.; Winfield, L.E.; Boyd, R.E.; Ang, C.Y. Tungsten bioavailability and toxicity in sunflowers (*Helianthus annuus*). *Land Contam. Reclamat.* **2009**, *17*, 141–151. [CrossRef]

27. Thomas, G.W. Soil pH and soil acidity. In *Methods of Soil Analysis. Chemical Methods*; Sparks, D.L., Ed.; Soil Science Society of America Book Series; Soil Science Society of America Inc.: Madison, WI, USA, 1996; pp. 475–490.

28. Nelson, D.W.; Sommers, L.E. Total carbon, organic carbon and organic matter. In *Methods of Soil Analysis. Chemical Methods*; Sparks, D.L., Ed.; Soil Science Society of America book series; Soil Science Society of America Inc.: Madison, WI, USA, 1996; pp. 961–1010.

29. Gee, G.W.; Bauder, J.W. Particle-size analysis. In *Methods of Soil Analysis. Physical and Mineralogical Methods*; Klute, A., Ed.; Soil Science Society of America Book Series; Soil Science Society of America Inc.: Madison, WI, USA, 1986; pp. 383–411.

30. Oburger, E.; Vergara Cid, C.; Preiner, J.; Hu, J.; Hann, S.; Wanek, W.; Richter, A. pH-Dependent Bioavailability, Speciation, and Phytotoxicity of Tungsten (W) in Soil Affect Growth and Molybdoenzyme Activity of Nodulated Soybeans. *Environ. Sci. Technol.* **2018**, *52*, 6146–6156. [CrossRef] [PubMed]

31. Poykio, R.; Torvela, H.; Peramaki, P.; Kuokkanen, T.; Ronkkomaki, H. Comparison of dissolution methods for multielement analysis of some plant materials used as bioindicator of sulphur and heavy metal deposition determined by ICP-AES and ICPMS. *Analusis* **2000**, *28*, 850–854. [CrossRef]

32. Dermatas, D.; Braida, W.; Christodoulatos, C.; Strigul, N.; Panikov, N.; Los, M.; Larson, S. Solubility, sorption, and soil respiration effects of tungsten and tungsten alloys. *Environ. Forensics* **2004**, *5*, 5–13. [CrossRef]

33. Bednar, A.J.; Jones, W.T.; Chappell, M.A.; Johnson, D.R.; Ringelberg, D.B. A modified acid digestion procedure for extraction of tungsten from soil. *Talanta* **2010**, *80*, 1257–1263. [CrossRef] [PubMed]

34. Kumar, A.; Aery, N.C. Effect of tungsten on growth, biochemical constituents, molybdenum and tungsten contents in wheat. *Plant Soil Environ.* **2011**, *57*, 519–525. [CrossRef]

35. Kumar, A.; Aery, N.C. Effect of tungsten on growth, dry-matter production, and biochemical constituents of cowpea. *Commun. Soil Sci. Plan.* **2012**, *43*, 1098–1107. [CrossRef]

36. Adamakis, I.D.S.; Panteris, E.; Eleftheriou, E.P. Tungsten toxicity in plants. *Plants* **2012**, *1*, 82–99. [CrossRef] [PubMed]

37. Clemente, R.; Nicholas, W.L. Tungsten. In *Heavy Metals in Soils: Trace Metals and Metalloids in Soils and Their Bioavailability*; Alloway, B.J., Ed.; Springer Science and Business Media: Dordrecht, The Netherlands, 2013; pp. 559–564.

38. Semhi, K.; Boutin, R.; Nallusamy, S.; Al Busaidi, W.; Al Hamdi, A.; Al Dhafri, K.; Al Busaidi, A. Impact of a Variable Tungsten Pollution on the Elemental Uptake of Two Plant Species. *Water Air Soil Pollut.* **2018**, *229*, 294. [CrossRef]

39. Reeves, P.G.; Chaney, R.L. Bioavailability as an issue in risk assessment and management of food cadmium: A review. *Sci. Total Environ.* **2008**, *398*, 13–19. [CrossRef] [PubMed]

40. Gustafsson, J.P. Modelling molybdate and tungstate adsorption to ferrihydrite. *Chem. Geol.* **2003**, *200*, 105–115. [CrossRef]

41. Bednar, A.J.; Boyd, R.E.; Jones, W.T.; McGrath, C.J.; Johnson, D.R.; Chappell, M.A.; Ringelberg, D.B. Investigations of tungsten mobility in soil using column tests. *Chemosphere* **2009**, *75*, 1049–1056. [CrossRef] [PubMed]

42. Sen Tuna, G.; Braida, W. Evaluation of the adsorption of mono- and poly- tungstates onto different types of clay minerals and Pahokee peat. *Soil Sediment Contamin.* **2014**, *23*, 838–849. [CrossRef]

43. Thomas Arrigo, L.K.; Mikutta, C.; Byrne, J.; Kappler, A.; Kretzschmar, R. Iron(II)-catalyzed iron atom exchange and mineralogical changes in Iron-rich organic freshwater flocs: An Iron isotope tracer study. *Environ. Sci. Technol.* **2017**, *51*, 6897–6907. [CrossRef]

44. Karimian, N.; Johnston, S.G.; Burton, E.D. Antimony and arsenic behavior during Fe(II)-induced transformation of Jarosite. *Environ. Sci. Technol.* **2017**, *51*, 4259–4268. [CrossRef]

45. Karimian, N.; Johnston, S.G.; Burton, E.D. Antimony and arsenic partitioning during Fe2+-induced transformation of jarosite under acidic conditions. *Chemosphere* **2018**, *195*, 515–523. [CrossRef]

46. Karimian, N.; Burton, E.D.; Johnston, S.G.; Hockmann, K.; Choppala, G. Humic acid impacts antimony partitioning and speciation during iron(II)-induced ferrihydrite transformation. *Sci. Total Environ.* **2019**, *683*, 399–410. [CrossRef]

47. Krauss, M.; Wilcke, W.; Kobza, J.; Zech, W. Predicting heavy metals transfer from soil to plant: Potential use of Freundlich-type functions. *J. Plant Nutr. Soil Sci.* **2002**, *165*, 3–8. [CrossRef]

48. Han, F.X.; Su, Y.; Monts, D.L.; Waggoner, C.A.; Plodinec, M.J. Binding, distribution, and plant uptake of mercury in a soil from Oak Ridge, Tennessee, USA. *Sci. Total Environ.* **2006**, *368*, 753–768. [CrossRef] [PubMed]

49. Kalis, E.J.J.; Temminghoff, E.J.M.; Visser, A.; Van Riemsdijk, W.H. Metal uptake by Lolium perenne in contaminated soils using a four-step approach. *Environ. Toxicol. Chem.* **2007**, *26*, 335–345. [CrossRef] [PubMed]

50. Blanco Rodriguez, P.; Vera Tomé, F.; Lozano, J.C. About the assumption of linearity in soil-to-plant transfer factors for uranium and thorium isotopes and ^{226}Ra. *Sci. Total Environ.* **2002**, *284*, 167–175. [CrossRef]

51. Tuovinen, T.S.; Roivainen, P.; Makkonen, S.; Kolehmainen, M.; Holopainen, T.; Juutilainen, J. Soil-to-plant transfer of elements is not linear: Results for five elements relevant to radioactive waste in five boreal forest species. *Sci. Total Environ.* **2011**, *410*, 191–197. [CrossRef] [PubMed]

52. Fantke, P.; Charles, R.; de Alencastro, L.F.; Friedrich, R.; Jolliet, O. Plant uptake of pesticides and human health: Dynamic modeling of residues in wheat and ingestion intake. *Chemosphere* **2011**, *85*, 1639–1647. [CrossRef]

53. Trapp, S.; Eggen, T. Simulation of the plant uptake of organophosphates and other emerging pollutants for greenhouse experiments and field conditions. *Environ. Sci. Pollut. Res.* **2013**, *20*, 4018–4029. [CrossRef] [PubMed]

54. Molina, M.; Escudey, M.; Chang, A.C.; Chen, W.; Arancibia-Miranda, N. Trace element uptake dynamics for maize (Zea mays L.) grown under field conditions. *Plant Soil* **2013**, *370*, 471–483. [CrossRef]

55. Liang, Z.; Ding, Q.; Wei, D.; Li, J.; Chen, S.; Ma, Y. Major controlling factors and predictions for cadmium transfer from the soil into spinach plants. *Ecotoxicol. Environ. Saf.* **2013**, *93*, 180–185. [CrossRef]

56. Pedron, F.; Grifoni, M.; Barbafieri, M.; Petruzzelli, G.; Rosellini, I.; Franchi, E.; Bagatin, R.; Vocciante, M. Applicability of a Freundlich-Like Model for Plant Uptake at an Industrial Contaminated Site with a High Variable Arsenic Concentration. *Environments* **2017**, *4*, 67. [CrossRef]

57. Petruzzelli, G.; Pedron, F. Tungstate adsorption onto Italian soils with different characteristics. *Environ. Monit. Assess.* **2017**, *189*, 379–388. [CrossRef] [PubMed]

58. Bostick, B.C.; Sun, J.; Landis, J.D.; Clausen, J.L. Tungsten Speciation and Solubility in Munitions-Impacted Soils. *Environ. Sci. Technol.* **2018**, *52*, 1045–1053. [CrossRef] [PubMed]

59. Cruywagen, J.J. Protonation, oligomerization, and condensation reactions of vanadate (V), molybdate (VI). In *Advances in Inorganic Chemistry*; Sykes, A.G., Ed.; Academic Press Inc.: San Diego, CA, USA, 2000; pp. 127–182.

60. Davantès, A.; Costa, D.; Lefèvre, G. Infrared study of (poly)tungstate ions in solution and sorbed into layered double hydroxides: Vibrational calculations and in situ analysis. *J. Phys. Chem. C* **2015**, *119*, 12356–12364. [CrossRef]

61. Karimian, N.; Johnston, S.G.; Burton, E.D. Iron and sulfur cycling in acid sulfate soil wetlands under dynamic redox conditions: A review. *Chemosphere* **2018**, *197*, 803–816. [CrossRef] [PubMed]

62. Datta, S.; Vero, S.E.; Hettiarachchi, G.M.; Johannesson, K. Tungsten Contamination of Soils and Sediments: Current State of Science. *Curr. Pollut. Rep.* **2017**, *3*, 55–64. [CrossRef]

63. USEPA. *Emerging Contaminant-Tungsten. Technical Fact Sheet*; Environmental Protection Agency: Washington, DC, USA, 2008.

64. Karimian, N.; Johnston, S.G.; Burton, E.D. Acidity generation accompanying iron and sulfur transformations during drought simulation of freshwater re-flooded acid sulphate soils. *Geoderma* **2017**, *285*, 117–131. [CrossRef]

65. O'Connor, G.A.; Brobst, R.B.; Chaney, R.L.; Kincaid, R.L.; Mcdowell, L.R.; Pierzynski, G.M.; Rubin, A.; Riper, G.V. A modified risk assessment to establish molybdenum standards for land application of biosolids. *J. Environ. Qual.* **2001**, *30*, 1490–1507. [CrossRef]

MDPI

St. Alban-Anlage 66

4052 Basel

Switzerland

Tel. +41 61 683 77 34

Fax +41 61 302 89 18

www.mdpi.com

Applied Sciences Editorial Office

E-mail: applsci@mdpi.com

www.mdpi.com/journal/applsci